N. Tamaki, Y. Kuge (Eds.)

Molecular Imaging for Integrated Medical Therapy and Drug Development

N. Tamaki, Y. Kuge (Eds.)

Molecular Imaging for Integrated Medical Therapy and Drug Development

Springer

Nagara Tamaki M.D., Ph.D.
Department of Nuclear Medicine
Hokkaido University Graduate School of Medicine
Kita 15 Nishi 7, Kita-ku,
Sapporo 060-8638, Japan

Yuji Kuge Ph.D.
Department of Tracer Kinetics and Bioanalysis
Hokkaido University Graduate School of Medicine
Kita 15 Nishi 7, Kita-ku,
Sapporo 060-8638, Japan

ISBN 978-4-431-54688-7 Springer Tokyo Berlin Heidelberg New York
ISBN 978-4-431-98074-2 (eBook)

Springer is a part of Springer Science+Business Media
springer.com

Softcover re-print of the Hardcover 1st edition 2010

Typesetting: SNP Best-set Typesetter Ltd., Hong Kong

Printed on acid-free paper

Preface

Molecular imaging has been employed in many fields. It combines the disciplines of molecular biology, radiochemistry, pharmacology, instrumentation, and clinical medicine into a new imaging paradigm. In particular, molecular imaging plays an important role in drug discovery and advanced medical practice. Serial assessments of molecular and cellular function are commonly used as surrogate markers of various new treatments. In this sense, molecular imaging has developed in parallel with the progress of drug development and new medical treatments.

Hokkaido University has launched the Matching Program for Innovations in Future Drug Discovery and Medical Care. This program aims to establish a center for drug discovery and medical care through molecular imaging technologies. The central pillars are our world-leading technologies in glycoprotein pharmaceuticals and optical and positron emission tomography (PET) imaging technology. The program helps our university contribute to society by improving the quality of life for patients, by creating new industries that use novel technologies, and by fostering internationally minded innovators who are able to launch new businesses. Placing particular importance on industry-academia collaboration, we have also established the Frontier Research Center for Post-Genome Science and Technology, as well as the Research Center for Cooperative Projects. With the creation of these facilities, the establishment of a core organization for drug discovery and medical care has been greatly promoted.

We have had an international symposium together with leading world specialists regarding molecular imaging for integrated medical therapy. The symposium "Molecular Imaging for Integrated Medical Therapy" held March 13–14, 2009, at Hokkaido University, Sapporo, Japan, was a joint symposium of "The 6th Symposium for Future Drug Discovery and Medical Care" and "Symposium of the Research Center for Cooperative Projects". At the symposium, more than 150 physicians and scientists attended to share our new experiences in molecular imaging for drug development and advanced medical therapy.

We consider the timely publication of the proceedings of this symposium to be important. In these proceedings, our aim is to share our experiences with many specialists in different fields all over the world.

These proceedings will be helpful for a better understanding of molecular imaging technology, and also for an understanding of the roles played by molecular imaging in drug development and integrated medical therapy. We sincerely hope this book will foster a variety of young scientists in the fields of life science and medical practice.

Nagara Tamaki and Yuji Kuge
Hokkaido University

Contents

Contributors

Fernando Albericio
Institute for Research in Biomedicine; CIBER-BBN Networking Centre on Bioengineering, Biomaterials and Nanomedicine, Barcelona Science Park, University of Barcelona, Baldiri Reixac 10, E-08028 Barcelona, Spain

Department of Organic Chemistry, University of Barcelona, Martí Franqués 1, E-08028 Barcelona, Spain

Mercedes Álvarez
Institute for Research in Biomedicine; CIBER-BBN Networking Centre on Bioengineering, Biomaterials and Nanomedicine, Barcelona Science Park, University of Barcelona, Baldiri Reixac 10, E-08028 Barcelona, Spain

Laboratory of Organic Chemistry, Faculty of Pharmacy, University of Barcelona, Joan XXII s/n, E-08028 Barcelona, Spain

Akinori Arimura
Strategic Development Department, Pharmaceutical Development Division, Shionogi & Co., Ltd, 5-12-4 Sagisu, Fukushima-ku, Osaka 553-0002, Japan

Manel Ballester
University of Lleida, Spain

Rob S.B. Beanlands
Division of Cardiology, University of Ottawa Heart Institute, 40 Ruskin Street, Ottawa, Ontario, K1Y 4W7 Canada

Ambros J. Beer
Department of Nuclear Medicine, Technische Universität München, Klinikum rechts der Isar; Ismaningerstr. 22, 81675 Munich, Germany

Gerard Bengua
Department of Medical Physics, Hokkaido University Graduate School of Medicine, Kita 15 Nishi 7, Kita-ku, Sapporo 060-8638, Japan

Francis G. Blankenberg
Department of Pediatric Radiology, Stanford University School of Medicine, 725 Welch Road, RM 1671 MC 5654, Palo Alto, CA 94304, USA

Ignasi Carrió
Department of Nuclear Medicine, Hospital Sant Pau, Autonomous University of Barcelona, Pare Claret 167, 08025-Barcelona, Spain

Yasuhiro Chiba
Department of Neurosurgery, Hokkaido University Graduate School of Medicine, Kita 15 Nishi 7, Kita-ku, Sapporo 060-8638, Japan

Carmen Cuevas
Pharma Mar S. A., Avda de los Reyes 1, E-28770 Colmenar Viejo, Madrid, Spain

Shusaku Daikoku
Mitsubisi Kagaku Institute of Life Sciences (MITILS), 11 Minami-Ooya, Machida, Tokyo 194-8511, Japan

Robert A. deKemp
Division of Cardiology, University of Ottawa Heart Institute, 40 Ruskin Street, Ottawa, Ontario, K1Y 4W7 Canada

Albert Flotats
Department of Nuclear Medicine, Hospital Sant Pau, Autonomous University of Barcelona, Pare Claret 167, 08025-Barcelona, Spain

Andres Francesch
Pharma Mar S. A., Avda de los Reyes 1, E-28770 Colmenar Viejo, Madrid, Spain

Toshihiko Fujimoto
Division of Radiopharmaceutical Chemistry, Cyclotron and Radioisotope Center, Tohoku University, 6-3 Aoba, Aramaki, Aoba-ku, Sendai 980-8578, Japan

Hiroshi Fukuda
Institute of Development, Aging and Cancer, Tohoku University, 4-1 Seiryo-machi, Aoba-ku, Sendai 980-8575, Japan

Jun-ichi Furukawa
Graduate School of Advanced Life Science, Frontier Research Center for Post-Genomic Science and Technology, Hokkaido University, Kita 21 Nishi 11, Sapporo 001-0021, Japan

Xiao-Dong Gao
Graduate School of Advanced Life Science, Frontier Research Center for Post-Genomic Science and Technology, Hokkaido University, Kita 21, Nishi 11, Kita-ku, Sapporo 001-0021, Japan

Jian Ping Gong
Division of Biological Sciences, Graduate School of Science, Hokkaido University, Kita 10 Nishi 8, Kita-ku, Sapporo 060-0810, Japan

Masakazu Hasegawa
Department of Radiology, Hokkaido University Graduate School of Medicine, Kita 15 Nishi 7, Kita-ku, Sapporo 060-8638, Japan

Yasuko Hasegawa
Tokyo Institute of Technology, Graduate School of Bioscience and Biotechnology, 4529 Nagatsuta-cho, Midori-ku, Yokohama 226-0018, Japan

Kazutoshi Hida
Department of Neurosurgery, Hokkaido University Graduate School of Medicine, Kita 15 Nishi 7, Kita-ku, Sapporo 060-8638, Japan

Hiroshi Hinou
Graduate School of Advanced Life Science, Frontier Research Center for Post-Genomic Science and Technology, Hokkaido University, Kita 21 Nishi 11, Kita-ku, Sapporo 001-0021, Japan

Kenji Hirata
Department of Nuclear Medicine, Hokkaido University, Kita 15 Nishi 5, Kita-ku, Sapporo 060-8638, Japan

Tsuyoshi Hirota
Division of Biological Sciences, University of California San Diego, 9500 Gilman Drive, La Jolla, CA 92093, USA

Masaaki Hokari
Department of Neurosurgery, Hokkaido University Graduate School of Medicine, Kita 15 Nishi 7, Kita-ku, Sapporo 060-8638, Japan

Sato Honma
Department of Physiology; Department of Chronomedicine, Hokkaido University Graduate School of Medicine, Kita 15 Nishi 7, Kita-ku, Sapporo 060-8638, Japan

Ken-ichi Honma
Department of Physiology, Hokkaido University Graduate School of Medicine, Kita 15 Nishi 7, Kita-ku, Sapporo 060-8638, Japan

Kensaku Hosoguchi
Graduate School of Life Science, Frontier Research Center for Post-Genomic Science and Technology, Hokkaido University, Kita 21 Nishi 11, Kita-ku, Sapporo 001-0021, Japan

Tomio Inoue
Department of Radiology, Yokohama City University School of Medicine, 3-9 Fukuura, Kanazawa-ku, Yokohama 236-0004, Japan

Masayori Ishikawa
Department of Medical Physics, Hokkaido University Graduate School of Medicine, Kita 15 Nishi 7, Kita-ku, Sapporo 060-8638, Japan

Takafumi Ishitsu
Central Research Laboratory, Hitachi, Ltd., 7-2-1 Omika-cho, Hitachi, Ibaraki 319-1221, Japan

Yoshinobu Iwasaki
Department of Neurosurgery, Hokkaido University Graduate School of Medicine, Kita 15 Nishi 7, Kita-ku, Sapporo 060-8638, Japan

Ren Iwata
Center for the Advancement of Higher Education, Tohoku University, 41 Kawauchi, Aoba-ku, Sendai 980-8576, Japan

Osamu Kanie
Mitsubisi Kagaku Institute of Life Sciences (MITILS), 11 Minami-Ooya, Machida, Tokyo 194-8511, Japan

Chietsugu Katoh
Department of Health Sciences, Hokkaido University Graduate School of Medicine, Kita 15 Nishi 7, Kita-ku, Sapporo 060-8638, Japan

Norio Katoh
Department of Nuclear Medicine; Department of Radiology, Hokkaido University Graduate School of Medicine, Kita 15 Nishi 7, Kita-ku, Sapporo 060-8638, Japan

Steve A. Kay
Division of Biological Sciences, University of California San Diego, 9500 Gilman Drive, La Jolla, CA 92093, USA

Nobuto Kitamura
Department of Sports Medicine and Joint Surgery, Hokkaido University Graduate School of Medicine, Kita 15 Nishi 7, Kita-ku, Sapporo 060-8638, Japan

Ran Klein
Division of Cardiology, University of Ottawa Heart Institute, 40 Ruskin Street, Ottawa, Ontario, K1Y 4W7 Canada

Keiji Kobayashi
Central Research Laboratory, Hitachi Ltd., 7-2-1 Omika-cho, Hitachi, Ibaraki 319-1221, Japan

Mariko Kobayashi
Mitsubisi Kagaku Institute of Life Sciences (MITILS), 11 Minami-Ooya, Machida, Tokyo 194-8511, Japan

Masato Koike
Department of Cell Biology and Neuroscience, Juntendo University Graduate School of Medicine, 2-1-1 Hongo, Bunkyo-ku, Tokyo 113-8421, Japan

Shinichi Kojima
Central Research Laboratory, Hitachi, Ltd., 7-2-1 Omika-cho, Hitachi, Ibaraki 319-1221, Japan

Kohtarou Konno
Department of Neuropharmacology, Hokkaido University Graduate School of Medicine, Kita 15 Nishi 7, Kita-ku, Sapporo 060-8638, Japan

Naoki Kubo
Department of Nuclear Medicine, Hokkaido University, Kita 15 Nishi 7, Kita-ku, Sapporo 060-8638, Japan

Faculty of Health Sciences, Hokkaido University, Kita 12 Nishi 5, Kita-ku, Sapporo 060-0812, Japan

Yuji Kuge
Department of Tracer Kinetics and Bioanalysis; Department of Molecular Imaging, Hokkaido University Graduate School of Medicine, Kita 15 Nishi 7, Kita-ku, Sapporo 060-8638, Japan

Satoshi Kuroda
Department of Neurosurgery, Hokkaido University Graduate School of Medicine, Kita 15 Nishi 7, Kita-ku, Sapporo 060-8638, Japan

Quynh-Thu Le
Department of Radiation Oncology, Stanford University, 875 Blake Wilbur Dr, MC 5847, Stanford, CA 94305-5847, USA

Nancy Lee
Memorial Sloan Kettering Cancer Center, 1275 York Ave, New York, NY 10021, USA

Billy W. Loo
Department of Radiation Oncology, Stanford University, 875 Blake Wilbur Dr, MC 5847, Stanford, CA 94305-5847, USA

Takahiro Maeda
Graduate School of Advanced Life Science, Frontier Research Center for Post-Genomic Science and Technology, Hokkaido University, Kita 21, Nishi 11, Kita-ku, Sapporo 001-0021, Japan

Keiichi Magota
Department of Nuclear Medicine, Hokkaido University Graduate School of Medicine, Kita 15 Nishi 7, Kita-ku, Sapporo 060-8638, Japan

Department of Radiology, Hokkaido University Hospital, Kita 14 Nishi 5, Kita-ku, Sapporo 060-8648, Japan

Osamu Manabe
Department of Nuclear Medicine, Hokkaido University Graduate School of Medicine, Kita 15 Nishi 7, Kita-ku, Sapporo 060-8638, Japan

Katsuhiko Maruichi
Department of Neurosurgery, Hokkaido University Graduate School of Medicine, Kita 15 Nishi 7, Kita-ku, Sapporo 060-8638, Japan

Kazuki Matsuzaki
Central Research Laboratory, Hitachi, Ltd., 7-2-1 Omika-cho, Hitachi, Ibaraki 319-1221, Japan

Ryogo Minamimoto
Department of Radiology, Yokohama City University School of Medicine, 3-9 Fukuura, Kanazawa-ku, Yokohama 236-0004, Japan

Nobuaki Miura
Graduate School of Advanced Life Science, Frontier Research Center for Post-Genomic Science and Technology, Hokkaido University, Kita 21, Nishi 11, Kita-ku, Sapporo 001-0021, Japan

Taisuke Miyazaki
Department of Anatomy, Hokkaido University School of Medicine, Kita 15 Nishi 7, Kita-ku, Sapporo 060–8638, Japan

Yuichi Morimoto
Central Research Laboratory, Hitachi, Ltd., 7-2-1 Omika-cho, Hitachi, Ibaraki 319-1221, Japan

Satoru Moriyama
Graduate School of Advanced Life Science, Frontier Research Center for Post-Genomic Science and Technology, Hokkaido University, Kita 21, Nishi 11, Kita-ku, Sapporo 001-0021, Japan

Kenji Narihiro
Department of Tracer Kinetics and Bioanalysis, Hokkaido University Graduate School of Medicine, Kita 15 Nishi 7, Kita-ku, Sapporo 060-8638, Japan

Masanao Naya
Department of Cardiology, Hokkaido University Graduate School of Medicine, Kita 15 Nishi 7, Kita-ku, Sapporo 060-8638, Japan

Shin-ya Nishide
Department of Physiology, Hokkaido University Graduate School of Medicine, Kita 15 Nishi 7, Kita-ku, Sapporo 060-8638, Japan

Kenichi Nishijima
Department of Tracer Kinetics and Bioanalysis, Hokkaido University Graduate School of Medicine, Kita 15 Nishi 7, Kita-ku, Sapporo 060-8638, Japan

Shin-Ichiro Nishimura
Graduate School of Advanced Life Science, Frontier Research Center for Post-Genomic Science and Technology, Hokkaido University, Kita 21, Nishi 11, Kita-ku, Sapporo 001-0021, Japan

Nobuyuki Okamura
Department of Pharmacology, Tohoku University Graduate School of Medicine, 2-1 Seiryo-machi, Aoba-ku, Sendai 980-8575, Japan

Rikiya Onimaru
Department of Radiology, Hokkaido University Graduate School of Medicine, Kita 15 Nishi 7, Kita-ku, Sapporo 060-8638, Japan

Daisuke Ono
Department of Physiology, Hokkaido University Graduate School of Medicine, Kita 15 Nishi 7, Kita-ku, Sapporo 060-8638, Japan

Yasunari Ono
Mitsubisi Kagaku Institute of Life Sciences (MITILS), 11 Minami-Ooya, Machida, Tokyo 194-8511, Japan

Yuichirou Ueno
Central Research Laboratory, Hitachi, Ltd., 7-2-1 Omika-cho, Hitachi, Ibaraki 319-1221, Japan

Kikuo Umegaki
Central Research Laboratory, Hitachi, Ltd., 7-2-1 Omika-cho, Hitachi, Ibaraki 319-1221, Japan

Reiko Usui
Department of Nuclear Medicine, Hokkaido University, Kita 15 Nishi 7, Kita-ku, Sapporo 060-8638, Japan

Masahiko Watanabe
Department of Anatomy, Hokkaido University School of Medicine, Kita 15 Nishi 7, Kita-ku, Sapporo 060-8638, Japan

Hans-Jürgen Wester
Department of Nuclear Medicine, Technische Universität München, Klinikum rechts der Isar; Ismaningerstr. 22, 81675 Munich, Germany

Kazuhiko Yanai
Division of Cyclotron Nuclear Medicine, Cyclotron and Radioisotope Center, Tohoku University, 6-3 Aoba, Aramaki, Aoba-ku, Sendai 980-8578, Japan

Department of Pharmacology, Tohoku University Graduate School of Medicine, 2-1 Seiryo-machi, Aoba-ku, Sendai 980-8575, Japan

Shunsuke Yano
Department of Neurosurgery, Hokkaido University Graduate School of Medicine, Kita 15 Nishi 7, Kita-ku, Sapporo 060-8638, Japan

Hiroshi Yasuda
Department of Neurosurgery, Hokkaido University Graduate School of Medicine, Kita 15 Nishi 7, Kita-ku, Sapporo 060-8638, Japan

Kazunori Yasuda
Department of Sports Medicine and Joint Surgery, Hokkaido University Graduate School of Medicine, Kita 15 Nishi 7, Kita-ku, Sapporo 060-8638, Japan

Koichi Yasuda
Department of Radiology, Hokkaido University Graduate School of Medicine, Kita 15 Nishi 7, Kita-ku, Sapporo 060-8638, Japan

Keiichiro Yoshinaga
Department of Molecular Imaging, Hokkaido University Graduate School of Medicine, Kita 15 Nishi 7, Kita-ku, Sapporo 060-8638, Japan

Tomoko Yoshikawa
Department of Chronomedicine, Hokkaido University Graduate School of Medicine, Kita 15 Nishi 7, Kita-ku, Sapporo 060-8638, Japan

Mitsuhiro Yoshioka
Department of Neuropharmacology, Hokkaido University Graduate School of Medicine, Kita 15 Nishi 7, Kita-ku, Sapporo 060-8638, Japan

Songji Zhao
Department of Tracer Kinetics and Bioanalysis; Department of Nuclear Medicine, Hokkaido University Graduate School of Medicine, Kita 15 Nishi 7, Kita-ku, Sapporo 060-8638, Japan

Yan Zhao
Department of Tracer Kinetics and Bioanalysis; Department of Nuclear Medicine, Hokkaido University Graduate School of Medicine, Kita 15 Nishi 7, Kita-ku, Sapporo 060-8638, Japan

Advances in Molecular Imaging

Nagara Tamaki[1] and Yuji Kuge[2]

Introduction

Molecular imaging is a rapidly growing biomedical research and clinical discipline. It aims to visualize and measure fundamental biological processes at the molecular, subcellular, and cellular levels. Molecular imaging promises not only to deepen our understanding of already-known biological processes but also to uncover still-unknown molecular and cellular events that are at the center of the initiation and evolution of disease. Compared with traditional in vitro tissue/cell culture and/or ex vivo animal studies, molecular imaging permits the noninvasive and repetitive imaging of targeted biological processes at both cellular and subcellular levels within living organs. Therefore, this offers a means for the specific targeting of abnormal biological processes for diagnosis and treatment.

In addition, molecular imaging has another important role; that is, in drug development, because molecular imaging may facilitate the understanding of pharmacokinetics in in vivo animal studies with the use of optical or radiolabeled pharmaceutical compounds. Furthermore, treatment effects may be more precisely and quantitatively assessed at molecular and cellular levels.

In the clinical setting, molecular-targeted drug therapy has been performed on the basis of tissue characterization. Molecular imaging, particularly positron emission tomography (PET), has recently been used for cancer treatment planning and for monitoring treatment effect. Thus, molecular imaging is a promising means for facilitating new drug development and for optimizing treatment strategy in individual cases (Fig. 1).

Methodological Issues

There are a number of molecular imaging techniques, including fluorescent or bioluminescent dyes for optical imaging, target-specific paramagnetic and superparamagnetic ligands for magnetic resonance imaging (MRI) and MR spectros-

[1]Department of Nuclear Medicine, Hokkaido University Graduate School of Medicine, Kita 15 Nishi 7, Kita-ku, Sapporo 060-8638, Japan
[2]Department of Tracer Kinetics and Bioanalysis, Hokkaido University Graduate School of Medicine, Kita 15 Nishi 7, Kita-ku, Sapporo 060-8638, Japan

N. Tamaki, Y. Kuge (eds.)
Molecular Imaging for Integrated Medical Therapy and Drug Development.
DOI: 10.1007/978-4-431-98074-2_1, ©Springer 2010

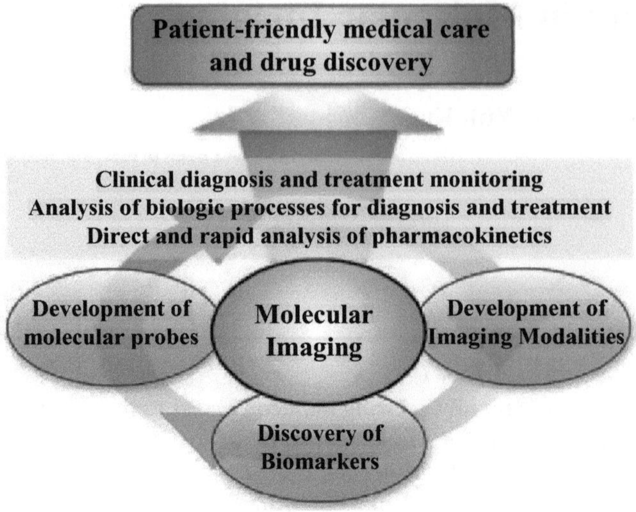

Fig. 1. Molecular imaging for patient–friendly medical care and drug discovery

copy, targeted radionuclide probes in nuclear medicine using PET/ single photon emission computed tomography (SPECT), and microbubbles for ultrasound imaging. Table 1 summarizes the characteristics of each molecular imaging technique. Optical imaging provides excellent spatial resolution and temporal resolution with high target-to-background activity. While this technique has been widely used at the molecular and cellular levels in experimental laboratory studies, it has inherent limitations for human use due to large attenuation through the body. The major advantage of MRI is high spatial resolution with no signal attenuation through the body. Because of its low sensitivity, human application with paramagnetic or superparamagnetic ligands remains limited. However, this limitation may be easily solved with the development of new and safe ligands and also imaging instruments in the near future.

Radionuclide imaging using either PET or SPECT has been most commonly used for human molecular imaging mainly due to quite high sensitivity for imaging radiolabeled ligands as compared with any other molecular imaging. Despite low spatial and temporal resolutions with high background activity and some radiation, wide clinical applications have been shown for the evaluation of various disease processes with safety. Another interesting molecular imaging is ultrasound with the use of microbubbles. This new technique holds promise for new treatment by optimizing drug delivery to the targeted lesions.

Table 1. Characteristics of various molecular imaging techniques

	Radiation exposure	Spatial resolution	Temporal resolution	Target-to-background	Attenuation	Human imaging	Treatment application
Optical imaging	No	High	High	Large	Large	Possible	No
MR imaging	No	High	Low	Medium	No	Possible	No
PET, SPECT (radionuclide imaging)	Yes	Low	Low	Low	Low	Yes	Yes
Ultrasound	No	High	High	Medium	Large	Yes	Yes

MR, magnetic resonance; PET, positron emission tomography; SPECT, single photon emission computed tomography

Applications of Molecular Imaging

There are also ever-increasing numbers of target ligands and detection devices for their visualization. Various imaging modalities may probe the very same process or, more often, different aspects or parts of the same biological process. Information gained through multimodal imaging may more comprehensively and completely delineate and characterize the biological process under investigation. For instance, suitable probes may be investigated in optical imaging and tested for in vivo imaging in small animals. By transferring a suitable optical ligand to a radioligand, radionuclide imaging may be applied in a smaller to a larger animal. On the other hand, most of the problems in radionuclide imaging may be solved with feedback to more basic optical imaging. Such translational research and feedback processes have been employed in gene expression imaging and regeneration therapy imaging, including stem-cell grafting. In addition, many of the superparamagnetic ligands used in MRI have been developed from the wide experience in radionuclide imaging. Such translational research processes should be tested more often for future clinical use and for basic molecular imaging sciences as well.

Multimodal imaging has been brought to the bedside, as is evident from the increasing use of hybrid imaging such as PET/CT, SPECT/CT, and—perhaps in the future—PET/MRI. This integrated imaging has the potential for gaining better understanding of molecular and cellular processes in specific locations, with the fusion of functional and anatomical information.

Integrated multimodal imaging in clinical practice has substantially improved the accuracy with which disease can be detected and quantified. A higher spatial resolution with use of a suitable molecular ligand permits better delineation of small lesions. In addition, precise tissue characterization is important for providing suitable treatment strategies, including molecular-targeted drug therapies and/or radiation treatment. The accurate quantification of molecular processes, such as the metabolic information provided by PET, enables the monitoring of responses to treatment and provides a means of improving disease targeting and outcome prediction.

The success of molecular imaging will likely depend on a closely integrated multidisciplinary approach by experts in many fields, including biochemistry, bioinformatics, biomathematics, cellular/molecular biology, organic chemistry, material sciences, nanotechnology, genetics, bioengineering, optics, medical imaging, medical physics, pharmacology, radiation oncology, neurology, and cardiology. Multiple collaborations by many different specialists will promote molecular imaging sciences and their wide clinical applications.

Part I
New Molecular Imaging and Image-Guided Radiotherapy

Hypoxia Imaging for Image-Guided Radiotherapy

Quynh-Thu Le[1], Billy W. Loo[1], and Nancy Lee[2]

Summary

Tumor hypoxia, or the condition of low oxygen, is a key factor for tumor progression and treatment resistance. Because hypoxic tumor cells are more resistant to ionizing radiation, hypoxia has been a focus of clinical research in radiation therapy for half a century. During this period, interest in targeting tumor hypoxia has waxed and waned as promising treatments emerged from the laboratory, only to fail in the clinics. With the development of new radiation targeting strategies, specifically intensity-modulated radiotherapy (IMRT) and image guidance technologies (image-guided RT; IGRT), there is a strong interest in imaging hypoxia for radiation targeting and dose escalation. In this review, we will discuss the role of hypoxia imaging, specifically positron emission tomography (PET)-based imaging with hypoxia-specific tracers, for directing radiation treatment in solid tumors.

Key words Hypoxia · Imaging · PET · Radiation · Dose painting

Introduction

Hypoxia is a common phenomenon in solid tumors. It arises when tumor oxygen demands exceed the oxygen supply from the vasculature. Hypoxic regions develop within solid tumors due to aberrant blood vessel formation, fluctuations in blood flow, and increasing oxygen demands from rapid tumor expansion [1]. The existence of hypoxia in human tumors was first demonstrated in 1955 by Thomlinson and Gray, who showed with histological sections that there was a constant distance (100–150 μm) across tumor tissues between blood vessels and necrosis, and that this distance was the oxygen diffusion distance based on capillary oxygen partial

[1]Department of Radiation Oncology, Stanford University, 875 Blake Wilbur Dr, MC 5847, Stanford, CA 94305-5847, USA
[2]Memorial Sloan Kettering Cancer Center, 1275 York Ave, NY, NY 10021, USA

N. Tamaki, Y. Kuge (eds.)
Molecular Imaging for Integrated Medical Therapy and Drug Development.
DOI: 10.1007/978-4-431-98074-2_2, ©Springer 2010

pressure and cellular oxygen consumption [2]. They postulated that chronically hypoxic cells existed adjacent to necrotic areas, just beyond the oxygen diffusion distance. Another form of hypoxia, known as acute hypoxia, also exists due to fluctuating flow in existing tumoral blood vessels. Injections of two different diffusible dyes minutes apart showed that temporary reduction in flow or closure of certain blood vessels could be observed in solid tumors, resulting in areas of acutely hypoxic cells [3]. It is most likely that acute and chronic hypoxia are the extremes of a continuum caused by the dynamic nature of tumor blood flow and oxygen consumption, and both can give rise to tumor cells that are prone to metastasis and resistant to conventional therapy.

Tumor hypoxia has been shown to confer universal resistance to radiotherapy (RT) damage in a wide range of cells and tissues, in studies using various endpoints [4]. The oxygen enhancement ratio, which is defined as the ratio of the radiation dose required under hypoxic conditions to the dose required under aerobic conditions to produce the same level of cell kill, is normally in the range of 2.5–3 for mammalian cells [5]. The reason for the universality of this effect is that oxygen reacts chemically with the free electron on the free radicals produced by ionizing radiation in the DNA, thereby stabilizing or fixing the damage. In the absence of oxygen, the radical damage can be repaired by hydrogen donation from sulfhydryl compounds in the cell. Thus, the lack of oxygen can severely compromise the efficacy of ionizing radiation, especially when the RT is delivered in large single fractions, without any re-oxygenation. In addition to conferring radiation resistance, tumor hypoxia can also indirectly decrease the efficacy of systemic agents through cell-cycle arrest [6] (thereby reducing tumor response to anticancer drugs), limiting drug access to the tumor due to impaired vasculature, and enhancing drug resistance via the activation of certain genes involved in drug efflux and metabolism [7–9]. At the molecular level, tumor hypoxia has been shown to play an important role in regulating cell viability, promoting cell metastatic potential, and inducing genes that are involved in invasion and metastasis [10–15]. An updated systematic review of 10 108 patients in 86 randomized trials designed to modify tumor hypoxia in patients definitively treated with primary RT suggested that hypoxia modification significantly improved the effect of RT in both locoregional control (odds ratio, 0.77; 95% confidence interval [CI], 0.71 to 0.86) and overall survival (odds ratio, 0.87; 95% CI, 0.80 to 0.95) [16]. Taken together, these data indicate that hypoxia is an attractive microenvironmental target in solid cancers.

The new developments in radiation targeting (specifically the ability of intensity-modulated radiotherapy [IMRT] for dose painting and that of image-guided RT [IGRT] for visualization of the treated areas during radiation delivery) provide a tantalizing possibility of delivering higher doses to hypoxic regions if these can be visualized and delineated at the time of RT planning. This provides a rationale for the surging interest in the volumetric imaging of hypoxia. Although there are several techniques for imaging hypoxia (please see detailed reviews by Krohn et al. [17] and Le et al. [18]), the most clinically mature approach is positron emission tomography (PET), using injectable labeled hypoxia-specific tracers, which serve as indirect molecular reporters of low oxygen tension. Hypoxia PET tracers that have been studied in humans include ^{18}F-misonidazole (FMISO), ^{18}F-nitroimidazole [2-(2-nitro-1H-imidazol-1-yl)-N-(2,2,3,3,3-pentaflouropropyl) acetamide (^{18}F-EF5), ^{18}F-fluoroazomycin-arabinofuranoside (^{18}F-FAZA), ^{18}F-fluoroetanidazole (^{18}FETA), ^{18}F-fluoroerythronitroimidazole (^{18}FETNIM), and copper (II) diacetyl-bis(N4)-methylthiosemicarbazone (^{64}Cu-ATSM). These agents have been discussed in detail previously [17].

Although all of the above agents have been studied clinically in patients, the data on their utility in guiding radiation treatment are quite meager. The premise that hypoxia imaging can be used to guide radiation dose escalation rests on two important assumptions: (1) PET-based hypoxia imaging does indeed detect hypoxia and can predict subregions of local failure after conventional radiation treatment and (2) such regions of hypoxia remain relatively stable before and during the course of RT over several weeks. In the remainder of this chapter, we will discuss the data that either support or contradict these assumptions.

Does PET Hypoxia Imaging Detect Hypoxia and Regions of Radiation Resistance in Human Cancers?

There are ample data from preclinical studies to suggest that PET-based hypoxia imaging does measure hypoxia in solid tumors, and these data have been reviewed in detail previously [17, 19]. With regard to actual human tumors, the data are less clear. In an initial study of 16 patients with cervical nodal metastases from head and neck carcinoma (HNC), comparison of the FMISO tumor-to-muscle uptake ratio (TMR) at 2 h post-injection and tumor PO_2, as measured by the Eppendorf PO_2 microelectrode, showed high correlation [20]. However, in an updated report of 38 HNC patients, the same group found that the correlation between the two parameters had dropped to either slight or moderate, due to the failure of hypoxia imaging to detect hypoxia in small nodes and the overestimation of hypoxia in necrotic nodes by the electrode [21]. Another study in 13 patients with soft-tissue sarcomas (STS) did not show a significant correlation between FMISO imaging and tumor PO_2 [22]. The authors, therefore, concluded that FMISO PET imaging would not be feasible for assessing hypoxia in STS.

With regard to the correlation between hypoxia PET imaging tracers and other approaches for hypoxia assessment, such as the expression of intrinsic tissue hypoxia markers (i.e., tumor expression of hypoxia-regulated proteins or genes), the published data for human tumors are even more meager.. Grigsby et al. [23] compared ^{64}Cu-ATSM-PET to several hypoxia-related biomarkers, including vascular endothelial growth factor (VEGF), cyclooxygenase 2 (COX2), epithelial growth factor receptor (EGFR), carbonic anhydrase IX (CAIX), and the apoptotic index in 15 patients with cervical cancer. They found that the only significant correlation was with tumoral CAIX and the apoptosis level, at $P < 0.05$. Hu et al. [24] reported that the maximum intratumoral standardized uptake value (SUVmax) of pretreatment ^{18}FETNIM in 19 patients with non-small cell lung cancer (NSCLC) correlated significantly with the intratumoral expression of hypoxia inducible factor-1α (HIF-1α), glucose transporter-1 (Glut-1), and VEGF. However, ^{18}FETNIM SUVmax also correlated with tumor volume. In contrast, Cherk et al. [25] found no significant correlation between FMISO uptake and tumoral Glut-1 expression in another group of patients with NSCLC.

Regardless of the correlation between PET-based hypoxia imaging and other approaches for hypoxia detection, the most important question is whether hypoxia imaging can detect regions of radiation resistance in human cancers for radiation targeting or treatment modification. Because it is impossible, in most situations, to determine the exact focus of recurrence within the tumor at the time of relapse, most studies use local relapse, locoregional relapse, or progression-free survival as surrogate endpoints. A recent critical review of FMISO-PET provides a useful clinical summary of the prognostic role of this tracer in more than 300 patients imaged throughout the world [26]. The largest published study to date, correlating treatment outcomes with hypoxia imaging, is that of Rajendran et al. [27], who performed pretreatment FMISO-PET in 73 HNC patients treated in a nonuniform manner. The investigators found that the FMISO tumor-to-blood ratio (T/BR) was an independent prognostic factor for survival; however, no locoregional relapse data were reported, and there was no control for treatment approach. Another study of 12 HNC patients found that pretreatment FMISO uptake was a prognostic indicator of RT response [28]. The most convincing data came from the Rischin and Hicks group ([Hicks et al. [29] and Rischin et al. [30]), who showed, in a phase II randomized study, that failure to achieve a complete metabolic response on serial FMISO imaging studies before and during hypoxia-targeted therapy (chemoradiotherapy plus tirapazamine [TPZ], a hypoxic cell cytotoxin) was a predictor for early relapse and that pretreatment FMISO signal intensity could be used to identify patients who would benefit from TPZ. In patients with malignant gliomas, preoperative FMISO scans showed uptake in all high-grade tumors and was prognostic for treatment outcomes [31]. In a group of 14 patients with NSCLC, a high TMR and tumor/mediastinum ratio on pretreatment FMISO was associated with a higher risk of relapse [32]. However, more recently, the group at Memorial Sloan Kettering Cancer Center imaged 20 HNC patients (of whom 90% had oropharyngeal primaries) with serial FMISO scans, pre- and mid-radiation treatment [33]. All patients were treated with fractionated RT to 70 Gy and concurrent high-dose cisplatin. Only 2 patients had persistent FMISO uptake on the mid-treatment scans. At a median follow up of 3 years for surviving patients, there was no local failure and neither of the 2 patients with persistent FMISO uptake at mid-treatment had relapsed. These excellent outcomes are much better than previously noted for tobacco-related HNC and are more compatible with those reported for human papilloma virus (HPV)-related oropharyngeal carcinomas, raising the question about the prognostic role of hypoxia in these more recently emerging HPV-induced tumors.

Outside of FMISO, prognostic data for other hypoxia PET-tracers are limited. Cu-ATSM imaging has been shown to predict response in 14 patients with NSCLC treated with either RT alone or RT and chemotherapy [34]; Cu-ATSM imaging was an independent prognostic factor for clinical outcomes as measured by overall survival and relapse-free survival in 15 cervical cancer patients treated with combined chemoradiation [23], and it correlated with survival and progression-free survival in 17 colorectal cancer patients treated with neoadjuvant chemoradiotherapy followed by surgical resection [35]. Although labeled ^{18}F-labeled EF5 has just recently been employed as an imaging agent, with minimal outcome data [36], unlabeled or "cold" EF5 has been shown to be a prognostic factor in several solid tumors, using an immunohistochemistry (IHC) approach. In a small study of 16 patients with soft tissue sarcoma, severe hypoxia, defined as EF5 binding of 20% or more in the primary tumors, correlated with increased risk of distant metastasis [37]. In 18 patients with supratentorial gliomas, increasing EF5 binding was associated with higher tumor grade and shorter time to recurrence [38]. Similarly, in 22 HNC patients, those whose tumors had an EF5 level corresponding to severe hypoxia (<0.1% O_2) had shorter event-free survival than those with less binding [39].

In summary, published data to date suggest that PET-based hypoxia imaging most likely reflects hypoxia in solid tumors and, more importantly, can help to identify patients who may benefit from future hypoxia targeting strategies.

Can PET-Based Hypoxia Imaging Study Be Used to Guide IMRT Dose Painting for Dose Escalation?

From the radiation-targeting standpoint, PET imaging with hypoxia tracers can theoretically be combined with IMRT for dose escalation to improve local control and possibly to prevent distant metastasis. The ability of IMRT to dose paint provides an enticing possibility of delivering higher doses to hypoxic regions within the tumor as visualized by PET tracers, without increasing normal tissue toxicity. Multiple feasibility studies on RT planning have confirmed that dose painting can be applied to target hypoxic regions in the tumor using FMISO or ^{18}F-FAZA PET/CT-guided IMRT [40–42]. An example of such a feasibility study is shown in Fig. 1a, where dose escalation up to 84 Gy to hypoxic volumes in the tumor was achievable using FMISO PET/CT-guided IMRT without increasing the dose to the neighboring normal tissues. However, such an approach requires the spatial distribution of tumor hypoxia to remain stable before and during the course of RT. Animal studies have suggested that this is not the case. Dewhirst et al. [43] monitored tumor PO_2 at one position using a continuously sampling polarographic electrode in R3230Ac tumors of Fisher 344 rats and found that temporal fluctuation in PO_2 was observed in all 13 experiments. An example of such temporal fluctuation is shown in Fig. 2. Results from these experiments suggested that hypoxia and re-oxygenation occurred at high frequency in these tumors (up to 4–7 times per hour on the average) and that the extent of PO_2 fluctuation exceeded the threshold set for hypoxia in the experiments ($PO_2 < 10$ mm Hg) in more than 50% of the tumors.

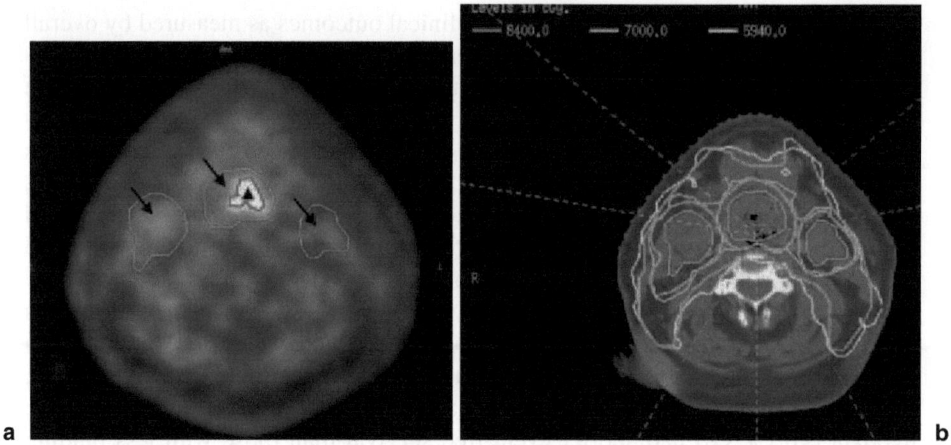

Fig. 1. **a** Axial image from an ^{18}F-misonidazole (FMISO) scan of a supraglottic carcinoma showing a small hypoxic region in the middle of the tumor (*arrowhead*). The *arrows* point to areas of the tumor and involved nodes that were avid on fluorine-18-2-fluoro-2 deoxyglucose (FDG) positron emission tomography (PET). The *red lines* delineate the gross tumor volume based on FDG-PET and computed tomography (CT), *the turquoise lines* denote the planning target volume prescribed to 70 Gy, and the *dark blue line* delineates the hypoxic region inside the tumor. **b** An example of an FMISO PET-guided intensity-modulated radiotherapy (IMRT) plan for the same patient. Note the 84-Gy isodose line (*in magenta*) covers the hypoxic region. The *orange line* represents the 70-Gy isodose line

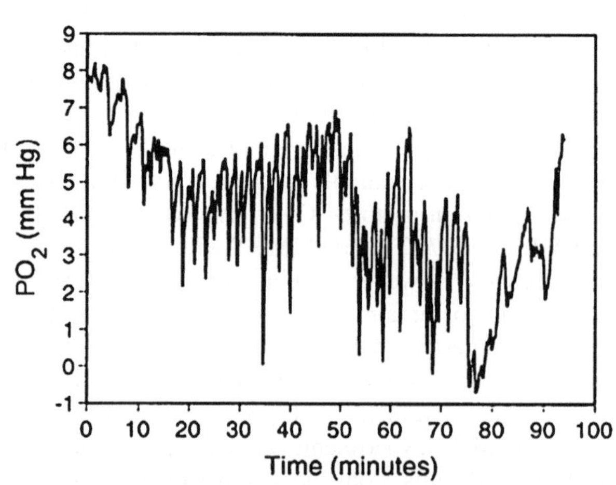

Fig. 2. An example of temporal fluctuation in tumor PO_2 as measured by a PO_2 electrode, placed in the same position over the recorded time period within a rat tumor. *A*, adapted with permission from reference 43

In theory, hypoxia-specific PET radiotracers, specifically those with the 2-nitroimidazole backbone, should accumulate only in areas of chronic hypoxia, at PO_2 levels below 10 mmHg, and are more sensitive at severe hypoxic conditions than the microelectrode [44, 45]. By extrapolation, they should not reflect the fluctuation of acute hypoxia. However, a recent study by the group at Memorial Sloan Kettering Cancer Center suggested that this is not the case. In this landmark study, 20 patients with HNC were imaged with two serial FMISO scans pretreatment, spaced 3 days apart, without any therapeutic intervention between the two scans [46]. Patients were imaged using the same immobilization system to allow for voxel-by-voxel image registration to determine the temporal variation in the FMISO distribution throughout the tumor volume (previously defined by a treatment planning PET-CT acquired with the same immobilization system). The results of this study are summarized in Table 1. Of the 13 analyzable patients, only 6 patients exhibited a correlation coefficient (R) greater than or equal to 0.5 between the two FMISO scans when only voxels with a TBR of 1.2 or more were considered. The remainder of the patients exhibited variable degrees of mismatches in the location of the most intense tumoral FMISO uptake between the two consecutive scans. The same group proceeded to evaluate the theoretical impact of such spatial changes in imaged hypoxia on RT dose escalation using the IMRT dose-painting approach [47]. They found that the spatial changes in hypoxic regions resulted in lower RT dose coverage and defeated the dose escalation intent, especially in those with significant variation between the two scans. However, because the hypoxic volumes are found within the gross-target volume (GTV), the dose to the hypoxic region is always at or above the prescribed GTV dose of 70 Gy.

Table 1. Correlation of FMISO uptake between two consecutive pretreatment FMISO scans in 14 head and neck cancer patients

Patient no.	SUVmax FMISO scan 1	SUVmax FMISO scan 2	T/B ratio FMISO scan 1	T/B ratio FMISO scan 2	R value T/B ≥ 1.2
1	2.1	2.1	16.6	19.8	NA
2	3.4	2.5	60.3	99.7	0.5
3	1.9	2.0	74.4	83.7	0.3
4	3.0	2.8	66.0	79.2	0.7
5	2.9	2.8	43.3	44.9	0.7
6	3.2	2.3	36.0	26.8	0.0
7	2.8	2.6	69.7	89.5	0.1
8	4.5	4.4	93.8	97.8	0.5
9	3.7	3.1	98.4	99.5	0.7
10	3.4	2.9	87.7	82.8	0.1
11	3.3	3.1	100	85.1	0.4
12	2.5	2.5	44.4	90.4	0.0
13	2.4	2.4	45.0	16.3	0.0
14	2.5	2.9	89.0	74.7	0.5

Adapted and modified from Table 2 of reference [46]

FMISO, ^{18}F-misonidazole; SUVmax, maximum intratumoral standardized uptake value; T/B ratio, tumor-to-blood ratio; NA, not analyzable

Several explanations were offered for the lack of correlation between the two pretreatment FMISO scans. Some of the reasons included experimental uncertainties such as the variation in the time between tracer injection and imaging, registration errors, and statistical noises; however, these reasons do not completely explain the differences between the serially acquired images. These data strongly suggested that imageable hypoxia, at least that using FMISO as tracer, does fluctuate temporally and spatially without treatment and such fluctuations may be even greater with treatment intervention. This calls into question the feasibility of FMISO-directed dose painting for RT dose escalation to target hypoxia. Until we learn more about the fluctuating nature of hypoxia in human tumors, how these changes are reflected by individual tracers, and how RT may impact such changes, we should rethink the strategy of hypoxia-targeted RT dose painting.

Future Directions of Hypoxia Imaging in Radiation Targeting

Radiobiological modeling suggests that hypoxia is likely to have a much greater impact on the efficacy of a single large fraction than on that of fractionated treatment, because without the opportunity for re-oxygenation, the dose required to control a hypoxic tumor is greater than that required for a normoxic tumor by the multiplicative factor of the full oxygen enhancement ratio. While this has been known for decades, the field of stereotactic radiotherapy has emerged, first for brain tumors and subsequently for extracranial sites, seemingly overlooking this principle. Although early local control outcomes of stereotactic body radiotherapy (SBRT), mostly for lung tumors, have been excellent, longer follow up suggests that the results are poorer for larger tumors, where one would expect to have more hypoxia [24, 46]. Bauman et al. [48] reported that most of their local failures after SBRT were found in larger tumors, as measured by T-stage and GTV. In another phase II study of SBRT for lung cancers, the 3-year local control rate was 78% for small T1 tumors and only 40% for large T2 tumors [49]. For central nervous system (CNS) metastases, lesions with a ring-like pattern of enhancement suggesting central necrosis had a poorer local control rate compared to non-necrotic lesions after single-fraction stereotactic radiosurgery (SRS) [50]. These data suggested that the influence of hypoxia on local control should be investigated for SRS. In the ideal setting, hypoxia imaging can be performed prior to SRS and SBRT, and the intensity as well as extent of hypoxia can be used to correlate with local control in these patients. If hypoxia imaging does correlate with local control then it can be applied in the future to guide patient selection for dose escalation, modification of fractionation, or for concomitant treatment with hypoxic cell radiosensitizers. In addition, in general, SRS involves treating the entirety of a small target to an ablative dose, and dose painting is not applicable. Therefore, the presence of hypoxia is more relevant than its spatial and temporal fluctuation, making hypoxia imaging more applicable in this setting.

Finally, hypoxia has been shown to promote not just radiation resistance but also enhanced systemic metastasis. Our group has shown that lysyl oxidase (LOX), a hypoxia-induced protein, promotes metastases in solid tumors through the regulation of the focal adhesion kinase (FAK) pathway as well as through the formation of the premetastatic niche [14, 15]. Hypoxia imaging can therefore help to identify patients at higher risk for tumor dissemination who may need additional systemic treatment, and it can help to select patients for future systemic hypoxia-specific therapy.

Conclusion

In summary, published data have shown that hypoxia imaging can identify patients at higher risk for recurrence after definitive therapies, specifically those including radiation as a major treatment component. However, both animal and human studies have shown that the hypoxia fluctuates both temporally and spatially within solid tumors, making it impractical to use PET-based hypoxia imaging for dose painting. However, the concept of using hypoxia imaging for guiding patient selection for treatment intensification with either RT dose escalation (specifically with SBRT and SRS), prevention of systemic recurrence, or integration of hypoxia-targeted drugs or hypoxic cell radiosensitizers is quite sound and should be considered in future clinical studies.

Acknowledgments This study was supported by the following grants from the National Institute of Health: 1 R01 CA118582-01 (QTL), PO1- CA67166 (QTL).

References

1. Brown JM, Giaccia AJ (1998) The unique physiology of solid tumors: opportunities (and problems) for cancer therapy. Cancer Res 58: 1408–1416
2. Thomlinson RH, Gray LH (1955) The histological structure of some human lung cancers and the possible implications for radiotherapy. Br J Cancer 9: 539–549
3. Trotter MJ, Chaplin DJ, Durand RE et al (1989) The use of fluorescent probes to identify regions of transient perfusion in murine tumors. Int J Radiat Oncol Biol Phys 16: 931–934
4. Gray LH, Conger AD, Ebert M et al (1953) Concentration of oxygen dissolved in tissues at the time of irradiation as a factor in radiotherapy. Br J Radiol 26: 638–648
5. Hall EA (2000) Radiobiology for the radiobiologist, 5th edn. Lippincott Williams & Wilkins, Philadelphia, PA
6. Pallavicini MG, Laland ME, Miller RG et al (1979) Cell-cycle distribution of chronically hypoxic cells and determination of the clonogenic potential of cells accumulated in G2-M phases after irradiation of a solid tumor. Cancer Res 39: 1891–1897
7. Shen J, Hughes C, Chao C et al (1987) Coinduction of glucose-regulated proteins and doxorubicin resistance in Chinese hamster cells. Proc Natl Acad Sci USA 84: 3278–3282

8. Hughes CS, Shen JW, Subjeck JR (1989) Resistance to etoposide induced by three glucose-regulated stresses in Chinese hamster ovary cells. Cancer Res 49: 4452–4454
9. Murphy BJ, Laderoute KR, Chin RJ et al (1994) Metallothionein IIA is up-regulated by hypoxia in human A431 squamous carcinoma cells. Cancer Res 54: 5808–5810
10. Graeber TG, Osmanian C, Jacks T et al (1996) Hypoxia-mediated selection of cells with diminished apoptotic potential in solid tumours. Nature 379: 88–91
11. Koch WM, Brennan JA, Zahurak M et al (1996) p53 mutation and locoregional treatment failure in head and neck squamous cell carcinoma. J Natl Cancer Inst 88: 1580–1586
12. Denko NC, Fontana LA, Hudson KM et al (2003) Investigating hypoxic tumor physiology through gene expression patterns. Oncogene 22: 5907–5914
13. Le QT, Denko NC, Giaccia AJ (2004) Hypoxic gene expression and metastasis. Cancer Metastasis Rev 23: 293–310
14. Erler JT, Bennewith KL, Nicolau M et al (2006) Lysyl oxidase is essential for hypoxia-induced metastasis. Nature 440: 1222–1226
15. Erler JT, Bennewith KL, Cox TR et al (2009) Hypoxia-induced lysyl oxidase is a critical mediator of bone marrow cell recruitment to form the premetastatic niche. Cancer Cell 15: 35–44
16. Overgaard J (2007) Hypoxic radiosensitization: adored and ignored. J Clin Oncol 25: 4066–4074
17. Krohn KA, Link JM, Mason RP (2008) Molecular imaging of hypoxia. J Nucl Med 49 (Suppl 2):129S-48S
18. Le QT, Courter D (2008) Clinical biomarkers for hypoxia targeting. Cancer Metastasis Rev 27: 351–362
19. Tatum JL, Kelloff GJ, Gillies RJ et al (2006) Hypoxia: importance in tumor biology, non-invasive measurement by imaging, and value of its measurement in the management of cancer therapy. Int J Radiat Biol 82: 699–757
20. Gagel B, Reinartz P, Dimartino E et al (2004) PO_2 polarography versus positron emission tomography ([(18)F] fluoromisonidazole, [(18)F]-2-fluoro-2′-deoxyglucose). An appraisal of radiotherapeutically relevant hypoxia. Strahlenther Onkol 180: 616–622
21. Gagel B, Piroth M, Pinkawa M et al (2007) PO_2 polarography, contrast enhanced color duplex sonography (CDS), [18F] fluoromisonidazole and [18F] fluorodeoxyglucose positron emission tomography: validated methods for the evaluation of therapy-relevant tumor oxygenation or only bricks in the puzzle of tumor hypoxia? BMC Cancer 7: 113–122
22. Bentzen L, Keiding S, Nordsmark M et al (2003) Tumour oxygenation assessed by 18F-fluoromisonidazole PET and polarographic needle electrodes in human soft tissue tumours. Radiother Oncol 67: 339–344
23. Grigsby PW, Malyapa RS, Higashikubo R et al (2007) Comparison of molecular markers of hypoxia and imaging with (60)Cu-ATSM in cancer of the uterine cervix. Mol Imaging Biol 9: 278–283
24. Hu M, Yu JM, Sun X et al (2008) The valuation of [18F] FETNIM PET/CT imaging for detecting tumor hypoxia in non-small cell lung cancer. J Clin Oncol 26: abstract 7504
25. Cherk MH, Foo SS, Poon AM et al (2006) Lack of correlation of hypoxic cell fraction and angiogenesis with glucose metabolic rate in non-small cell lung cancer assessed by 18F-fluoromisonidazole and 18F-FDG PET. J Nucl Med 47: 1921–1926
26. Lee ST, Scott AM (2007) Hypoxia positron emission tomography imaging with 18F-fluoromisonidazole. Semin Nucl Med 37: 451–461
27. Rajendran JG, Schwartz DL, O'Sullivan J et al (2006) Tumor hypoxia imaging with [F-18] fluoromisonidazole positron emission tomography in head and neck cancer. Clin Cancer Res 12: 5435–5441
28. Thorwarth D, Eschmann SM, Holzner F et al (2006) Combined uptake of [18F] FDG and [18F] FMISO correlates with radiation therapy outcome in head-and-neck cancer patients. Radiother Oncol 80: 151–156

29. Hicks RJ, Rischin D, Fisher R et al (2005) Utility of FMISO PET in advanced head and neck cancer treated with chemoradiation incorporating a hypoxia-targeting chemotherapy agent. Eur J Nucl Med Mol Imaging 32: 1384–1391
30. Rischin D, Hicks RJ, Fisher R et al (2006) Prognostic significance of [18F]-misonidazole positron emission tomography-detected tumor hypoxia in patients with advanced head and neck cancer randomly assigned to chemoradiation with or without tirapazamine: a substudy of Trans-Tasman Radiation Oncology Group Study 98.02. J Clin Oncol 24: 2098–2104
31. Cher LM, Murone C, Lawrentschuk N et al (2006) Correlation of hypoxic cell fraction and angiogenesis with glucose metabolic rate in gliomas using 18F-fluoromisonidazole, 18F-FDG PET, and immunohistochemical studies. J Nucl Med 47: 410–418
32. Eschmann SM, Paulsen F, Reimold M et al (2005) Prognostic impact of hypoxia imaging with 18F-misonidazole PET in non-small cell lung cancer and head and neck cancer before radiotherapy. J Nucl Med 46: 253–260
33. Lee N, Nehmeh S, Schoder H et al (2009) Prospective trial incorporating pre-/mid-treatment [(18)F]-misonidazole positron emission tomography for head-and-neck cancer patients undergoing concurrent chemoradiotherapy. Int J Radiat Oncol Biol Phys; e-published before print version
34. Dehdashti F, Mintun MA, Lewis JS et al (2003) In vivo assessment of tumor hypoxia in lung cancer with (60)Cu-ATSM. Eur J Nucl Med Mol Imaging 30: 844–850
35. Dietz DW, Dehdashti F, Grigsby PW et al (2008) Tumor hypoxia detected by positron emission tomography with 60Cu-ATSM as a predictor of response and survival in patients undergoing neoadjuvant chemoradiotherapy for rectal carcinoma: a pilot study. Dis Colon Rectum 51: 1641–1648
36. Komar G, Seppanen M, Eskola O et al (2008) 18F-EF5: a new PET tracer for imaging hypoxia in head and neck cancer. J Nucl Med 49: 1944–1951
37. Evans SM, Fraker D, Hahn SM et al (2006) EF5 binding and clinical outcome in human soft tissue sarcomas. Int J Radiat Oncol Biol Phys 64: 922–927
38. Evans SM, Judy KD, Dunphy I et al (2004) Hypoxia is important in the biology and aggression of human glial brain tumors. Clin Cancer Res 10: 8177–8184
39. Evans SM, Du KL, Chalian AA et al (2007) Patterns and levels of hypoxia in head and neck squamous cell carcinomas and their relationship to patient outcome. Int J Radiat Oncol Biol Phys 69: 1024–1031
40. Lee NY, Mechalakos JG, Nehmeh S et al (2007) Fluorine-18-labeled fluoromisonidazole positron emission and computed tomography-guided intensity-modulated radiotherapy for head and neck cancer: a feasibility study. Int J Radiat Oncol Biol Phys 70: 2–13
41. Thorwarth D, Eschmann SM, Paulsen F et al (2007) Hypoxia dose painting by numbers: a planning study. Int J Radiat Oncol Biol Phys 68: 291–300
42. Grosu AL, Souvatzoglou M, Roper B et al (2007) Hypoxia imaging with FAZA-PET and theoretical considerations with regard to dose painting for individualization of radiotherapy in patients with head and neck cancer. Int J Radiat Oncol Biol Phys 69: 541–551
43. Dewhirst MW, Braun RD, Lanzen JL (1998) Temporal changes in PO2 of R3230AC tumors in Fischer-344 rats. Int J Radiat Oncol Biol Phys 42: 723–726
44. Raleigh JA, Chou SC, Arteel GE et al (1999) Comparisons among pimonidazole binding, oxygen electrode measurements, and radiation response in C3H mouse tumors. Radiat Res 151: 580–589
45. Evans SM, Koch CJ (2003) Prognostic significance of tumor oxygenation in humans. Cancer Lett 195: 1–16
46. Nehmeh SA, Lee NY, Schroder H et al (2008) Reproducibility of intratumor distribution of (18)F-fluoromisonidazole in head and neck cancer. Int J Radiat Oncol Biol Phys 70: 235–242

47. Lin Z, Mechalakos J, Nehmeh S et al (2008) The influence of changes in tumor hypoxia on dose-painting treatment plans based on 18F-FMISO positron emission tomography. Int J Radiat Oncol Biol Phys 70: 1219–1228
48. Baumann P, Nyman J, Lax I et al (2006) Factors important for efficacy of stereotactic body radiotherapy of medically inoperable stage I lung cancer. A retrospective analysis of patients treated in the Nordic countries. Acta Oncol 45: 787–795
49. Koto M, Takai Y, Ogawa Y et al (2007) A phase II study on stereotactic body radiotherapy for stage I non-small cell lung cancer. Radiother Oncol 85: 429–434
50. Goodman KA, Sneed PK, McDermott MW et al (2001) Relationship between pattern of enhancement and local control of brain metastases after radiosurgery. Int J Radiat Oncol Biol Phys 50: 139–146

Molecular Imaging for the Assessment of Tumor Malignancy and Response to Therapy

Yuji Kuge[1,2], Songji Zhao[1,3], Toshiki Takei[3], and Nagara Tamaki[3]

Summary

Fluorine-18-2-deoxy-2-fluoro-D-glucose (^{18}F-FDG) is a useful positron emission tomography (PET) probe for differentiating tumors from benign lesions and for evaluating therapeutic effects. However, limitations of FDG-PET have recently been noted and the development of post-FDG molecular probes is strongly required. Here, we describe our basic studies on post-FDG molecular probes for the assessment of tumor malignancy and response to therapy, focusing on cell proliferation and apoptosis.

We evaluated the usefulness of FDG, 3′-deoxy-3′-^{18}F-fluorothymidine (FLT; a thymidine analog), and L-^{11}C-methionine (MET) for differentiating tumors (hepatomas) from granulomas in experimental rat models. FDG and FLT levels in the granulomas were comparable to those in the hepatomas. MET uptake in the granulomas was significantly lower than that in the tumors. These results were further confirmed by animal PET studies. Thus, MET could have a potential value in accurately differentiating malignant tumors from benign lesions. Next, we compared the potentials of FDG and FLT for assessing response to molecular-targeting therapy with a tyrosine kinase inhibitor, gefitinib. Two days after the treatment, FLT uptake levels in the tumor were significantly reduced, but those of FDG were not. Ki-67 expression and phosphorylated epidermal growth factor receptor (P-EGFR) in the tumor were suppressed after gefitinib therapy, while there was no definite change in the expression levels of EGFR. In addition, there was no statistically significant change in the tumor size in comparisons of before and after the treatment in any groups. Accordingly, FLT can detect the antiproliferative effects of molecular-targeting therapy with gefitinib early, before there are significant changes in tumor size.

[1]Department of Tracer Kinetics and Bioanalysis, Hokkaido University Graduate School of Medicine, Kita 15 Nishi 7, Kita-ku, Sapporo 060-8638, Japan
[2]Department of Molecular Imaging, Hokkaido University Graduate School of Medicine, Kita 15 Nishi 7, Kita-ku, Sapporo 060-8638, Japan
[3]Department of Nuclear Medicine, Hokkaido University Graduate School of Medicine, Kita 15 Nishi 7, Kita-ku, Sapporo 060-8638, Japan

N. Tamaki, Y. Kuge (eds.)
Molecular Imaging for Integrated Medical Therapy and Drug Development.
DOI: 10.1007/978-4-431-98074-2_3, ©Springer 2010

In animal-PET studies, granulomas and tumors were produced by inoculating *Rhodococcus* (3×10^8 CFU/rat) and allogeneic glioma cells (C6; 2×10^6 cells/rat), respectively, into the left and right calf muscles to generate a rat model bearing both a granuloma and a tumor. Ten days after *Rhodococcus* and C6 inoculation, rats were fasted overnight to receive i.v. injection of ^{18}F-FDG and ^{11}C-MET. After the tracer injection, PET scans (list-mode acquisition) were performed for 120 min using a small-animal PET (Inveon; Siemens Medical Solutions USA Inc., Knoxville, TN) in rats bearing the granuloma and C6 glioma ($n = 4$). The time-activity curves, static images (30–40 min for MET, 110–120 min for FDG), and standardized uptake value (SUV) in the lesions were calculated.

For histological evaluation, formalin-fixed, paraffin-embedded tissue sections (tumor and granuloma) of 3-μm thickness were stained with hematoxylin and eosin (H&E). Immunohistochemical staining of an immune-associated antigen (Ia) was also performed, using a monoclonal antibody (mAb; mouse IgG) that recognizes a monomorphic determinant of rat Ia, MHC class II present on B lymphocytes, dendritic cells, some macrophages, and certain epithelial cells, as previously described [28].

Results and Discussion

In the intramuscular granuloma induced by BCG, mature epithelioid cell granuloma formation and massive lymphocyte infiltration were observed; features that were histologically similar to those of sarcoidosis. Immunohistochemical staining also showed the accumulation of Ia-positive (Ia$^+$) macrophages and Ia$^+$ lymphocytes in the periphery of the granuloma. In the intramuscular tumor induced by KDH-8 cells, massive amounts of viable and proliferating cancer cells were observed by H&E staining.

Figure 1 shows the tracer uptake levels in rats bearing the tumors and granulomas. The granulomas showed a high ^{18}F-FDG uptake level, comparable to that in the hepatomas (group I; 8.18 ± 2.40 DUR vs 9.13 ± 1.52 DUR; $P = \text{NS}$ and group II; 8.43 ± 1.45 DUR vs 8.91 ± 2.32 DUR; $P = \text{NS}$). The ^{14}C-MET uptake level in the granuloma was significantly lower than that in the hepatoma (Fig. 1B; 1.31 ± 0.22 DUR vs 2.47 ± 0.60 DUR; $P < 0.01$), whereas the ^{3}H-FLT uptake level in the granuloma was comparable to that in the hepatoma (1.98 ± 0.70 DUR vs 2.30 ± 0.67 DUR; $P = \text{NS}$). Figure 2 shows the representative animal PET images of ^{18}F-FDG and ^{11}C-MET. ^{11}C-MET uptake in the granuloma showed a slow exponential clearance after the initial distribution, while the uptake in the tumor gradually increased with time (data not shown). The SUV in the granuloma (1.53 ± 0.04) was significantly lower than that in the tumor (1.82 ± 0.10; $P < 0.01$). The dynamic patterns, static images, and SUVs of FDG in the granuloma were similar to those in the tumor ($P = \text{NS}$).

The extensive ^{18}F-FDG uptake in the granuloma is related to the active infiltration of inflammatory cells, including lymphocytes, macrophages, and epithelioid cells. The present results indicate that ^{18}F-FDG might reflect active granulomatous

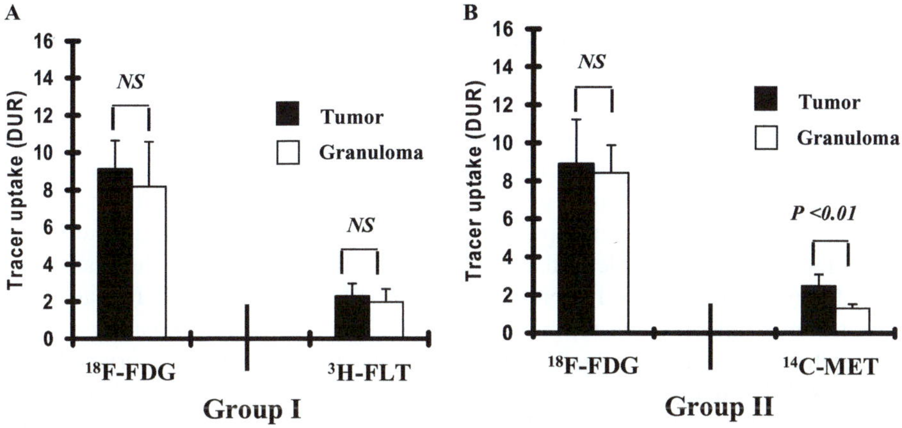

Fig. 1A,B. Fluorine-18-2-deoxy-2-fluoro-D-glucose (*^{18}F-FDG*), [methyl-^{3}H (N)]-3′-fluoro 3′-deoxythymidine (*^{3}H-FLT*) and L-[methyl-^{14}C] methionine (*^{14}C-MET*) uptake levels in the tumor and granuloma. **A** Group I, ^{18}F-FDG and ^{3}H-FLT uptake levels in the tumor and granuloma. **B** group II, ^{18}F-FDG and ^{14}C-MET uptake levels in the tumor and granuloma. Values given are means ± SD. Statistical analyses were performed using a nonparametric Mann-Whitney *U*-test. *NS*, not significant; *DUR*, differential uptake ratio (Reproduction from reference 15 with permission)

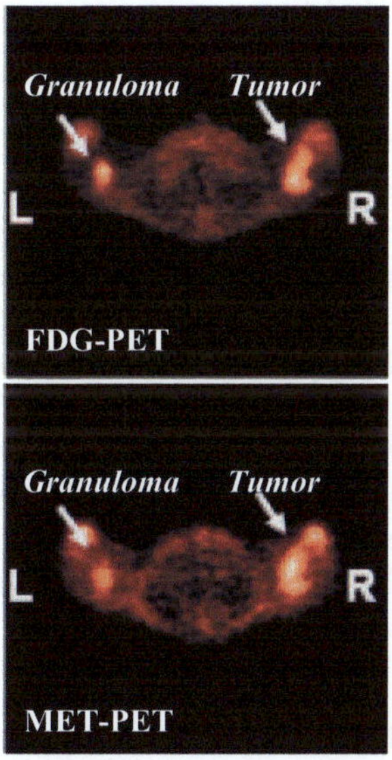

Fig. 2. Representative animal positron emission tomography (*PET*) images of ^{18}F-FDG and ^{11}C-MET in a rat bearing *Rhodococcus* granuloma and C6 glioma. *L*, left; *R*, right

processes, as activated inflammatory cells show markedly increased glycolysis [3–7]. In the present study, we demonstrated that MET uptake in the granuloma was significantly lower than that in the tumor. These results indicate that MET has potential value in accurately differentiating malignant tumors from benign lesions, especially granulomatous lesions, in clinical PET studies. Kubota et al. [29] have also reported that uptake of ^{14}C-MET is proportional to the amount of viable tumor cells, but is low in macrophages and other non-neoplastic cellular components. On the other hand, the present results showed that ^{3}H-FLT uptake in the granuloma was comparable to that in the tumor. ^{3}H-FLT uptake in the granuloma may be related to massive lymphocyte infiltration around the granuloma, although several investigators have reported that ^{18}F-FLT uptake in inflammatory cells is lower than that in tumors, due to the low mitotic activity of inflammatory cells [8].

Thus, MET uptake level in the granuloma was significantly lower than that in the tumor, whereas FDG and FLT were not able to differentiate granulomas from tumors. These results were further confirmed by animal PET studies. Consequently, MET could have potential value in accurately differentiating malignant tumors from benign lesions.

Comparison of FDG and FLT for Assessing Response to Molecular-Targeting Therapy with a Tyrosine Kinase Inhibitor, Gefitinib [16, 17]

Drugs that target molecular abnormalities; namely, molecular-targeting agents, are now expected to hold the greatest promise for cancer therapy. It is useful to clarify the therapeutic effects of these molecular-targeting agents early, for selecting the clinically optimum treatment strategy, and for reducing the side effects. Thus, PET evaluation with various probes could play important roles in assessing early response to the treatment. Gefitinib, a molecular-targeting agent, inhibits epidermal growth factor receptor (EGFR) tyrosine kinase. It is of great importance to clarify the effects of molecular-targeting therapy with gefitinib early, because of the drawbacks of this drug (development of fatal side effects, including interstitial pneumonia) and because of individual differences in its efficacy. Accordingly, we evaluated whether FDG and FLT are useful in monitoring the early therapeutic effects of gefitinib in mice bearing a human tumor xenograft with high EGFR expression.

Materials and Methods

All experimental protocols were approved by the Laboratory Animal Care and Use Committee of Hokkaido University. Female BALB/c athymic (nu+/nu+) mice were used. [Methyl-^{3}H (N)]-3′-fluoro 3′-deoxythymidine (^{3}H-FLT) was purchased from

Moravek Biochemicals. ^{18}F-FDG, synthesized by standard procedures, was obtained from Hokkaido University Hospital Cyclotron Facility.

In dual-tracer tissue distribution studies, an EGFR-dependent human tumor xenograft model (A431) was established with female BALB/c athymic mice, and the mice were assigned to three groups: a control and two treatment groups. Mice in the two treatment groups were treated with a partial regression dose of gefitinib (100 mg/kg per day) and the maximum tolerated dose of gefitinib (200 mg/kg per day), respectively, once daily for 2 days. Tumor size was measured during the experimental period. Biodistribution studies of ^{18}F-FDG and ^{3}H-FLT were performed 2 days after the last treatment. One hour after the tracer injection, radioactivity (percentage of injected dose per g × kg [%ID/g per kg]) in the tumor was determined.

In histological studies, tumor cell proliferative activity was determined with the Ki-67 labeling index. Immunohistochemical staining of EGFR and measurement of phosphorylated EGFR (P-EGFR) were also performed.

Results and Discussion

High expressions of EGFR and Ki-67 were observed in the A431 tumors of control mice. The expression of Ki-67 and P-EGFR in the tumor was remarkably suppressed after the treatment with gefitinib, while there was no definite change in the expression levels of EGFR. There was no statistically significant change in the tumor size between before and after the treatment in any groups. Two days after the treatment with 100 and 200 mg/kg doses of gefitinib, uptake levels of ^{3}H-FLT in the tumor were significantly reduced, to 68% and 58% of the control value, respectively (0.41 ± 0.09 %ID/g per kg) for 100 mg/kg; 0.35 ± 0.06 for 200 mg/kg; 0.60 ± 0.13 for control; $P < 0.05$), but the uptake levels of ^{18}F-FDG were not reduced (110% and 107% of the control value for 100 and 200 mg/kg, respectively).

Our results indicate that FLT can detect the effects of molecular-targeting therapy with gefitinib before significant changes occur in the tumor size. This can be ascribed to the potential of FLT to accurately assess cellular proliferation, leading to the early detection of the antiproliferative effects of gefitinib. FLT reflects phosphorylation by the enzyme thymidine kinase 1 (TK1), which increases during DNA synthesis [8, 12, 22]. On the other hand, FDG could not detect the effects of gefitinib. Unfortunately, the reasons for the discrepancy remain unclear, and further investigations are needed to clarify this point. Autoradiographic studies of FDG and FLT in comparison with histological ones may help us to understand the tumor accumulation mechanisms of these tracers and the responses to molecular-targeting therapy.

In our mouse model, FLT uptake was dose-dependently decreased 2 days after the treatment with gefitinib, whereas FDG uptake was not decreased. FLT has potential for the early detection of the antiproliferative effects of molecular-targeting therapy.

Assessment of Apoptotic Tumor Response with ^{99m}tc-Annexin A5, Following a Single Dose of Chemotherapy, in Comparison with FDG [18–21]

Successful chemotherapy or radiotherapy induces apoptosis in neoplastic cells and indicates tumor response to the therapy [30–32]. Determining baseline levels of apoptosis and the increase of apoptosis induced by therapy can serve as useful prognostic markers [33, 34]. Annexin A5 (annexin V) is an endogenous protein that binds with high affinity and specificity to PS, which is presented on the cell surface in the early process of apoptosis. Accordingly, apoptotic cells can be detected in vivo using annexin A5 labeled with radionuclides, such as ^{99m}Tc [14, 18, 20, 21]. Accordingly, we applied apoptotic imaging with ^{99m}Tc-annexin A5 to the assessment of apoptotic tumor response following a single dose of chemotherapy, and compared it with FDG in rats with implanted hepatomas.

Materials and Methods

All experimental protocols were approved by the Laboratory Animal Care and Use Committee of Hokkaido University. Male WKAH rats were used. Annexin A5 was derivatized with hydrazinonicotinamide (HYNIC) and then labeled with ^{99m}Tc with tricine as a coligand as described [14]. ^{18}F-FDG, synthesized by standard procedures, was obtained from Hokkaido University Hospital Cyclotron Facility.

Rats inoculated with allogeneic hepatoma cells (KDH-8) were randomly divided into three groups: rats in group I ($n = 7$) were untreated and served as the control, rats in group II ($n = 7$) received a single dose of 2′, 2′-difluoro-2′-deoxycytidine (gemcitabine, 90 mg/kg, i.v.), rats in group III ($n = 8$) received cyclophosphamide (150 mg/kg, i.p.). Dual-tracer tissue distribution studies using ^{18}F-FDG and ^{99m}Tc-annexin A5 were performed 48 h after the chemotherapy. Briefly, ^{99m}Tc-annexin A5 was first injected and 5 h later ^{18}F-FDG was injected. Six hours after the ^{99m}Tc-annexin A5 injection (1 h after ^{18}F-FDG), the rats were killed and the organs, including the tumor tissues, were excised to determine the radioactivities of ^{18}F-FDG and ^{99m}Tc-annexin A5 (%ID/g per kg).

Histological evaluation by the terminal deoxynucleotidyl transferase-mediated deoxyuridine triphosphate nick-end labeling (TUNEL) method was performed to obtain the indices of apoptosis.

Results and Discussion

Figure 3 shows the uptake levels of ^{99m}Tc-annexin A5 and ^{18}F-FDG in the tumor after a single dose of chemotherapy using gemcitabine or cyclophosphamide. The uptake levels of ^{99m}Tc-annexin A5 in the tumor were significantly increased by the

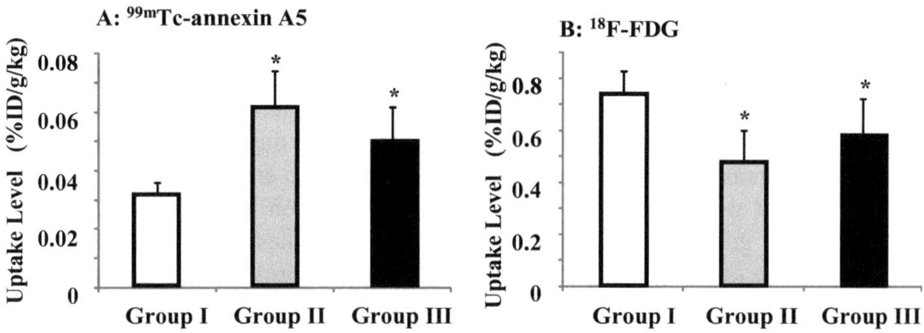

Fig. 3A,B. Uptake levels of **A** ^{99m}Tc-hydrazinonicotinamide (HYNIC)-annexin A5 and **B** ^{18}F-FDG in the tumor after a single dose of chemotherapy using gemcitabine or cyclophosphamide. *Group I*, untreated control; *group II*, gemcitabine (90 mg/kg, i.v.); *group III*, cyclophosphamide (150 mg/kg, i.p.). $^{*}P < 0.01$ vs control. *%ID/g/kg*, percentage of injected dose per gram × kg

chemotherapy (0.031 ± 0.005 %ID/g per kg in group I (control), 0.062 ± 0.012 %ID/g per kg in group II (gemcitabine), 0.050 ± 0.012 %ID/g per kg in group III (cyclophosphamide); $P < 0.01$). On the other hand, the uptake levels of ^{18}F-FDG in the tumor were significantly decreased by the chemotherapy (0.743 ± 0.084 %ID/g per kg in group I, 0.483 ± 0.118 %ID/g per kg in group II, 0.583 ± 0.142 %ID/g per kg in group III; $P < 0.01$). ^{18}F-FDG uptake levels in the tumor were negatively correlated with the ^{99m}Tc-annexin A5 uptake levels ($r = -0.75$; $P < 0.01$). In the gemcitabine- and cyclophosphamide-treated groups, the rates of TUNEL-positively stained cells were significantly higher than that in the control group (10.2% ± 1.7% and 8.0% ± 1.5% vs 5.2% ± 1.5%; $P < 0.01$).

This study demonstrated that the enhanced apoptotic reaction correlated with suppressed tumor glucose utilization after cytotoxic chemotherapy. The results of the tracer experiments were confirmed by histological evaluation. Cancer chemotherapy with agents such as gemcitabine and cyclophosphamide not only arrests metabolism and induces DNA alkylation, respectively but also enhances apoptosis [20, 35]. These molecular events occurred before the actual tumor regression. Thus, the increase in ^{99m}Tc-annexin A5 uptake in tumors, as well as the decrease in ^{18}F-FDG uptake, can be useful markers for predicting therapeutic outcomes and for prognosis at an early stage of chemotherapy.

Conclusions

FDG is a useful PET probe for differentiating tumors from benign lesions and evaluating therapeutic effects. However, limitations of FDG-PET have recently been noted, and the development of post-FDG molecular probes is strongly required. Here, we described our basic studies of post-FDG molecular probes for the

Development of a Prototype 3D PET Scanner Using Semiconductor Detectors and Depth of Interaction Information

Yuichi Morimoto[1], Yuichirou Ueno[1], Shinichi Kojima[1], Wataru Takeuchi[1], Takafumi Ishitsu[1], Kazuki Matsuzaki[1], Kikuo Umegaki[1], Naoki Kubo[2], Chietsugu Katoh[2], Songji. Zhao[3], Tohru Shiga[3], and Nagara Tamaki[3]

Summary

A prototype brain positron emission tomography (PET) scanner using semiconductor detectors and depth of interaction (DOI) information has been developed to achieve high spatial resolution and reduced scatter fraction. At the first step of the development, we created a two-dimensional prototype PET scanner composed of a single-slice full-ring detector unit to confirm the feasibility of the basic technologies that are necessary to realize a semiconductor PET scanner. Through phantom and small-animal studies, the feasibility of the semiconductor PET was confirmed and the results showed that the semiconductor PET could produce quantitative imaging with high spatial resolution. Based on these achievements, a prototype brain PET scanner was developed to demonstrate the high spatial resolution and quantitative imaging capability required in human imaging.

The prototype brain PET scanner has a patient port diameter of 350 mm and an axial field of view of 250 mm without interplane septa. It consists of 18 detector units arranged around the patient port. Each detector unit has 22 detector boards arranged along an axial field of view. The detector board has cadmium telluride (CdTe) detectors on each side. Signals are read by three-layer DOI. The total number of detector channels is about 80 000, and signals from each detector are processed with "individual channel readout" technology. Physical performance indicators, such as spatial resolution, sensitivity, and noise equivalent count rates (NECRs) were measured with National Electrical Manufacturers Association (NEMA) standards.

[1]Central Research Laboratory, Hitachi, Ltd., 7-2-1 Omika-cho, Hitachi, Ibaraki 319-1221, Japan
[2]Faculty of Health Sciences, Hokkaido University, North 12 West 5, Kita-ku, Sapporo 060-0812, Japan
[3]Department of Nuclear Medicine, Hokkaido University Graduate School of Medicine, North 15, West 7, Kita-ku, Sapporo, 060-8638, Japan

N. Tamaki, Y. Kuge (eds.)
Molecular Imaging for Integrated Medical Therapy and Drug Development.
DOI: 10.1007/978-4-431-98074-2_4, ©Springer 2010

The transverse and axial resolutions near the center are 2.3 mm and 5.1 mm, respectively. The absolute sensitivity and the scatter fraction of the scanner evaluated with the NEMA NU2-1994 phantom are 650 kcps/μCi/cc and 23%, respectively. These estimates are for a lower energy threshold of 450 keV. The NECR is 41 kcps at 0.2 μCi/cc.

These results show that the PET scanner based on semiconductor detectors is feasible for clinical use, with high spatial resolution, and with good potential for providing the high spatial resolution and quantitative imaging capability required in the brain and the head-and-neck region.

Key words CdTe · Semiconductor detector · DOI · PET

Introduction

Recently, semiconductor detectors have emerged as a new technology in nuclear medicine imaging, especially for use with gamma cameras and single photon emission computed tomography (SPECT) [1–5]. High-energy resolution and flexibility in both the sizing and the fine arrangement of the detectors are expected to improve image quality. These inherent features of semiconductor detectors may also lead to improved positron emission tomography (PET) images, because their high-energy resolution may offer a reduction in scattered events just as they are expected to be for gamma camera and SPECT uses. Flexible sizing and arrangement of detectors can provide the capability not only to increase spatial resolution but also to obtain depth of interaction (DOI) information [6–7]. For PET application, the feasibility of semiconductor detectors, including those made of silicon [8], and cadmium zinc telluride (CZT) [9] has been investigated. Very high spatial resolution in the submillimeter range has been demonstrated with a Compton camera and the DOI method using silicon detectors [8]. Encouraged by these features of semiconductor detectors, we started the development of a semiconductor PET scanner.

To realize a semiconductor PET scanner, we developed a method to implement the detectors as well as a new application specific integrated circuit (ASIC) for downstream signal processing. We used these developments for CdTe semiconductor detectors [10–11] to build a prototype PET scanner.

At the first stage of our development, a two-dimensional (2D) prototype PET scanner composed of a single-slice full-ring detector unit was developed to confirm the feasibility of the basic technologies that are necessary to realize a semiconductor PET scanner. Through phantom and small-animal studies, the feasibility of the semiconductor PET was confirmed and the results showed that the semiconductor PET could produce quantitative imaging with high spatial resolution [12]. Based on these achievements, the prototype brain PET scanner was developed to demonstrate the high spatial resolution and quantitative imaging capability required in human imaging.

2D Prototype Pet Scanner

Materials and Methods

System Description

Figure 1 shows the developed PET scanner. This system comprised six detector units. Each unit (Fig. 2) consisted of 96 CdTe semiconductor detector modules; signal processing circuit elements, including the ASIC and field programmable gate array (FPGA); and a circuit board on which the detectors and circuit elements were mounted. The transaxial field of view (FOV) of this system was 82 mm in diameter, while the axial FOV was 3.8 mm. Side-shield lead collimators were set in front of the detectors. The gaps between detector units were very large in this system, and we could not correct the gap effect in images. Therefore, we rotated the test objects

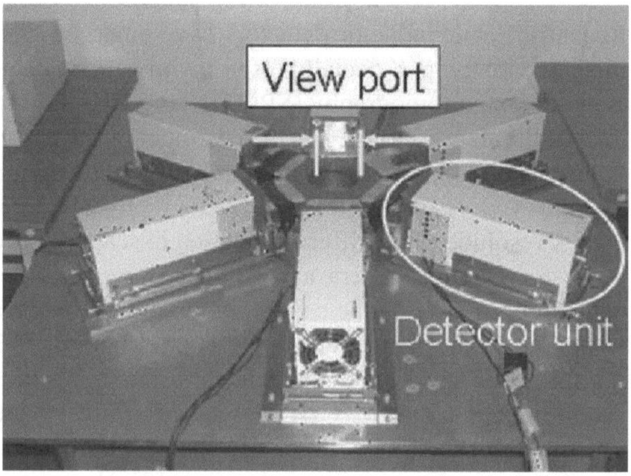

Fig. 1. Two-dimensional (2D) prototype single-slice positron emission tomography (PET) scanner with CdTe semiconductor detector units

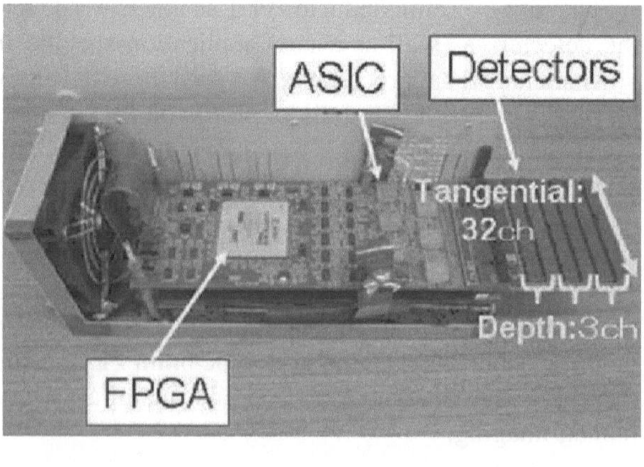

Fig. 2. Inner structure of a detector unit. Each unit has one circuit board on which CdTe modules and signal processing circuit elements are mounted. *ASIC*, application specific integrated circuit; *FPGA*, field programmable gate array; *ch*, channel

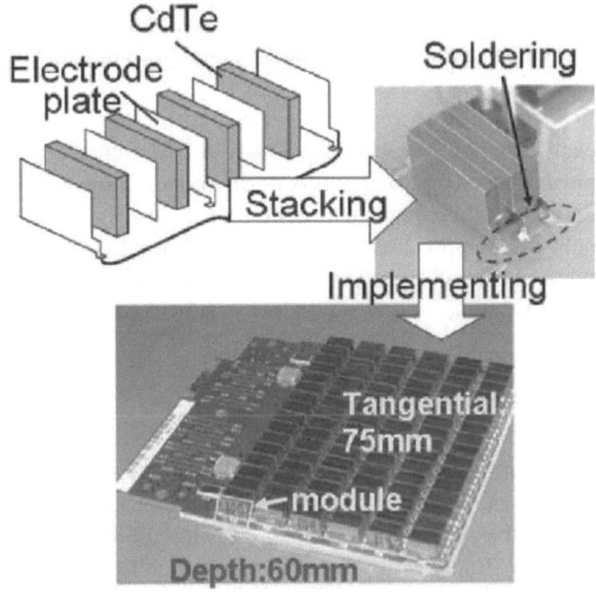

Fig. 3. CdTe module structure. Each module consists of four stacked CdTe elements, with 96 modules (16 × 6) on one circuit board

to reduce sensitivity fluctuation and the effect of the gaps between the detector units.

Figure 3 shows the CdTe module structure. Each module has four stacked CdTe elements (1-mm-thick) between electrode plates. These modules are densely mounted on a circuit board by soldering. This module structure maintains the volume of the detector elements, and a high voltage (500 V) can be applied to them so that sensitivity, energy resolution, timing resolution, and spatial resolution are improved. This is because the detection position accuracy and the transfer time of charged-carriers and loss of charged-carriers by recombination or trapping are improved when the crystal is thin. In contrast, making the crystals thicker is an effective way to improve sensitivity. To balance this trade-off, we fabricated the detector by stacking thin crystals. One detector channel is not limited to one module. We can change one channel by circuit wiring. In this system, the dimensions of a single detector channel were 2.3 mm (two elements of 2 mm and gap 0.3 mm) in the tangential direction, 17.5 mm (two modules of 15 mm and gap 2.5 mm) in the radial direction, and 4 mm (element height) in the axial direction. A circuit board had 96 detector channels arranged in three layers in the radial direction, i.e., 32 detector channels in each layer.

To achieve the expected features of semiconductor detectors, we developed an analog ASIC. This ASIC consists of 32-channel circuits that amplify minute charge signals from detectors with low noise. From the output signals of the ASIC, the FPGA in the detector unit creates a digital data packet whose energy, timing, and position data correspond to a detected gamma ray. When a gamma ray is detected, the ASIC emits a trigger signal and a pulse whose peak height is proportional to the energy of the detected gamma ray, and the FPGA creates a digital time stamp from the trigger signal and energy data from the pulse. These data packets are stored in a PC.

From these stored data, we can correct an offset and a gain of pulse height at each channel. We had previously made a correction table of offset and gain by measuring two pulse heights of different radioisotopes. We had also checked offset delays of all channels using the timing pulses and found that the delays of all channels were under 2 ns. Therefore, we did not need to correct the offset delay. From these corrected data, coincidence detection was performed with a timing window of 14 ns and an energy window of 450–530 keV.

Scanner Performance

We examined the basic performance of this scanner focusing on energy resolution, timing resolution, and spatial resolution. The energy resolution and timing resolution of the system were measured using a 68Ge/68Ga point source at the center of the FOV. Energy resolution was measured from the stored data without coincidence detection. Timing resolution was measured from the data that we stored with coincidence detection of a 128-ns time window, which was sufficiently longer than the timing resolution of the CdTe detectors.

To evaluate spatial resolution, three acrylic capillary tubes (outside diameter, 6 mm; inside diameter, 1 mm) were set on a rotating disk at 15-mm intervals from the center along a radius. The tubes were filled with 860 MBq/ml 18F-fluorodeoxyglucose (FDG). Data were collected at 1° intervals while the phantom was rotated 60°. Data were converted into a sinogram data set of 410 × 600 pixels and a pixel size of 0.2 mm × 0.6°. Images were reconstructed with filtered back projection without attenuation, scatter, or random correction. The reconstruction filter was a ramp filter with a cutoff at the Nyquist frequency. These rotation and reconstruction methods were also used for the following tests using images. Acquisition time of the spatial resolution test was 10 s in each position, so total time was 600 s.

Imaging Experiment

Imaging studies of rat myocardium and a tumor were performed with 18F-FDG. The experiments were performed in accordance with the guidelines for the care and use of laboratory animals of the Hokkaido University Graduate School of Medicine. In the rat myocardium imaging test, 150 MBq of 18F-FDG was injected into a rat weighing 240 g at 1 h after full anesthetic injection. At 4 h and 10 min after 18F-FDG injection, we began to store the data. The acquisition time at each rotating step was 1 min, so total acquisition time was 60 min. In the rat tumor imaging test, 100 MBq of 18F-FDG was injected into a rat weighing 240 g at 1 h after full anesthetic injection. At 1 h and 30 min after 18F-FDG injection, we began to store the data. The acquisition time was the same as in the myocardium test.

Results and Discussion

Scanner Performance

The energy resolution of the system (average of all detectors) was 5.4% full width half maximum (FWHM), which is one of the highest resolutions among conventional clinical PET scanners. The energy window of 450–530 keV was set by considering the trade-off between scatter fraction (SF) and sensitivity. We set the energy window to cover the photo-peak so that the SF would be smaller but the sensitivity would not decrease drastically. Timing resolution was 6.0 ns FWHM, which is not as good as in conventional PET scanners. However, we think it is possible to improve the timing resolution by reducing noise in the ASIC.

The spatial resolution was measured from an image of the point source phantom. The number of coincidences was 2.5×10^5. Figure 4 plots the spatial resolution profile. Figure 4a shows that FWHM was uniform across the FOV, both with and without DOI. A DOI effect appeared in full width tenth maximum (FWTM) (Fig. 4b). In the common interpretation, the DOI effect appears only at the edge of the FOV, but Fig. 4b shows that DOI made a difference even at the center of the FOV. This is because the detector ring diameter was small as compared with length of the detectors, and the number of polygonal detectors was only six in the developed system.

Imaging Experiment

Figure 5 shows images of a rat myocardium. Because the CT image was acquired with a different CT scanner, using a different rat, the fusion image shows a mismatch between the vertebral regions of the PET and CT images. In the PET image, the lung and vertebra can be distinguished, and about a 10-mm-diameter circular image of the myocardium is clearly shown.

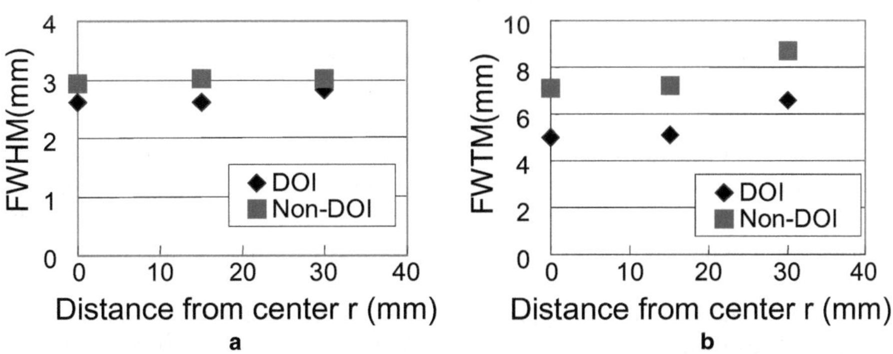

Fig. 4a,b. Spatial resolution of **a** full-width half maximum (*FWHM*) and **b** full width tenth maximum (*FWTM*). *DOI*, depth of interaction; *r*, radius

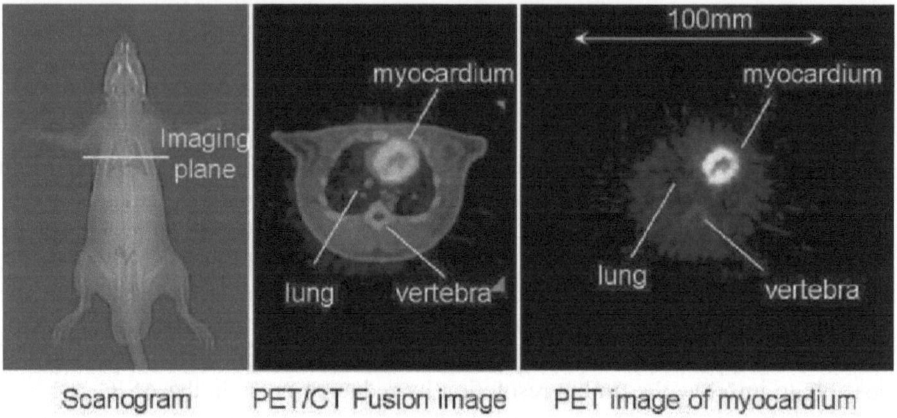

Fig. 5. Myocardium of rat: scanogram, PET/computed tomography (*CT*) fusion image, and PET image. The CT image was acquired using another rat and a different CT scanner

In Fig. 6, images of a tumor in a rat are shown. This 3-mm tumor was clearly imaged, and we checked that the PET tumor image size was equal to that of the dissected tumor.

3D Prototype PET Scanner for Human Brain

Materials and Methods

System Description

Figure 7 shows the structure of the detector unit. The dimensions of the unit package are 100 × 400 × 340 mm. In the unit, the detector boards are arranged in parallel, and the detectors are mounted on both sides of each board. A detector board has 96 detectors on each side (i.e., 192 detectors in total) and signal processors. These processors include ASICs mounted along the incident direction of the gamma rays. Signals are read by a three-layer DOI. Each unit has 22 boards and 4224 detectors. The whole system is cooled by forced air.

To realize dense implementation of detectors in the 3D PET scanner, the ASIC has been improved in terms of the following points. Power consumption for each channel was decreased to 50% of the ASIC in the 2D prototype. The new ASIC consists of 48-channel circuits, and noise level was decreased to one-third of that in the prototype ASIC.

Figure 8 shows our prototype 3D PET scanner, dedicated to human brain imaging. The diameter of the patient port is 350 mm, the transaxial FOV is 310 mm, and the axial FOV is 250 mm. There are 18 detector units radially arranged around the patient port.

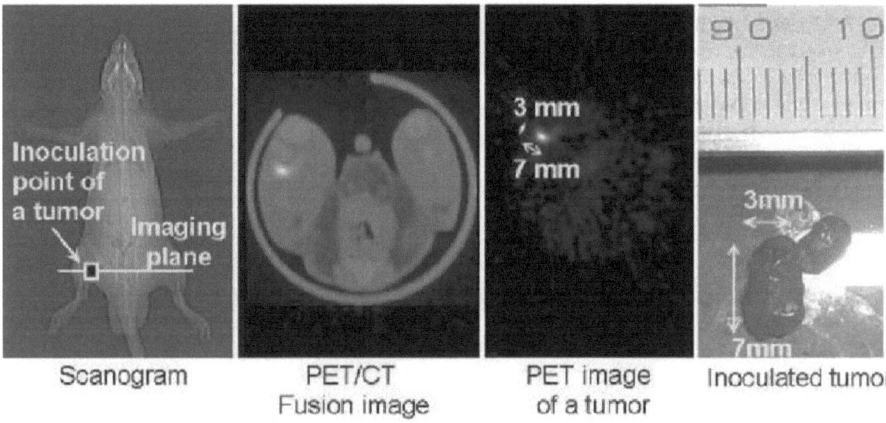

Fig. 6. Tumor of rat: scanogram, PET/CT fusion image, PET image, and inoculated tumor. The CT image was acquired using another rat and a different CT scanner

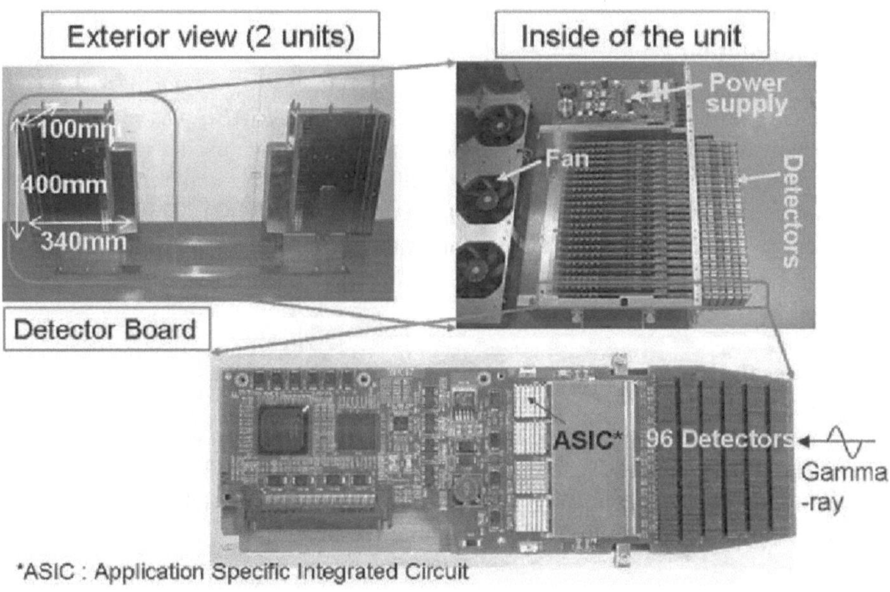

Fig. 7. Detector unit structure of 3D brain PET scanner. The dimensions of the unit package are $100 \times 400 \times 340$ mm. The detector board has 96 detectors on each side, and signal processors

Fig. 8. View of the 3D brain PET prototype that uses semiconductor detectors and depth of interaction information

Performance Indicators

Timing Resolution

A 68Ge-Ga point radiation source with an activity of 3.7 MBq was placed at the FOV center. Measurements were taken for 5 min. When we used a test pulse, the timing resolution was 1.7 ns (FWHM) without detectors, and when we used gamma-ray irradiation with detectors, it was 6.8 ns (FWHM). This was almost equivalent to our 2D prototype PET scanner; therefore, we set the time window at 14 ns.

Energy Resolution

Experimental conditions were the same as for the measurement of the timing resolution. The average energy resolution of all detectors was 4.1% (FWHM). This is superior to the energy resolution obtained using available scintillation detectors and this means the scanner could produce quantitative imaging.

Spatial Resolution

Spatial resolution was evaluated under National Electrical Manufacturers Association (NEMA) standards. The transverse and axial resolutions near the center were 2.3 mm and 5.1 mm, respectively.

A hot rods phantom including 18F-water with an activity of 149 MBq was placed at the FOV center. Measurements were taken for 15 min. Figure 9 shows a reconstructed hot rods phantom image. The diameters are between 1.5 and 4.0 mm.

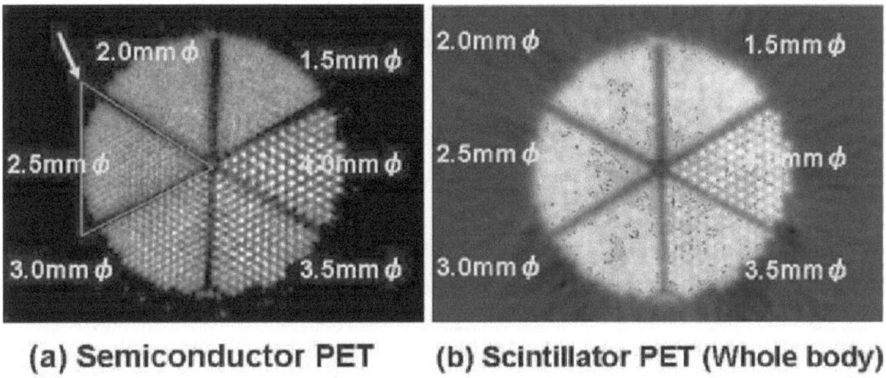

Fig. 9. Reconstructed hot rods phantom image. This image was reconstructed using Fourier rebinning and filtered back projection

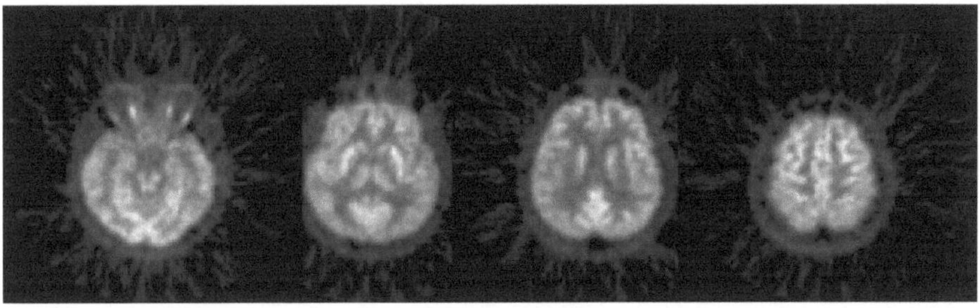

Fig. 10. Reconstructed brain images from a healthy volunteer. These images were reconstructed using an iterative image reconstruction method with scatter, random, and attenuation correction

Hot rods down to 2.5 mm in diameter were visually separated. This means that the scanner could offer high spatial resolution.

Sensitivity

The absolute sensitivity and the SF of the scanner evaluated with the NEMA NU2–1994 phantom were 650 kcps/μCi per cc and 23%, respectively. These estimates were for a lower energy threshold of 450 keV. Noise equivalent count rate (NECR) was 41 kcps at 0.2 μCi/cc.

Clinical Evaluation

We have started clinical evaluations and are working on improving image reconstruction. Figure 10 shows reconstructed brain images. The images are superior to images taken using conventional scanners because of the improved delineation of the cerebral cortex and the better contrast between the white and gray matter.

The first clinical experience in which the high spatial resolution capability of the PET scanner was utilized is reported in the chapter by T. Shiga, in this volume.

Conclusion

We have built a prototype single-slice PET scanner using CdTe semiconductor detectors. Using this system, we evaluated the basic physical performance indicators and obtained phantom and rat images in order to test the effectiveness of the semiconductor PET scanner. Our findings confirmed that the semiconductor PET scanner has high energy resolution and DOI measurement capability, features which are expected to improve the quantitative accuracy of PET measurements.

Based on these results, a prototype brain PET scanner using CdTe semiconductor detectors and DOI information was developed and its performance was evaluated using NEMA standards. These results show that the PET scanner based on semiconductor detectors is feasible for clinical use, with high spatial resolution, and it has good potential for providing the high spatial resolution and quantitative imaging capability required in the brain and the head-and-neck region.

References

1. Kubo N, Zhao S, Fujiki Y et al (2005) Evaluating performance of a pixel array semiconductor SPECT system for small animal imaging, Ann Nucl Med 19: 633–639.
2. Yokoi K, Morimoto Y, Kitaguchi H et al (2006) Development of a large FOV semiconductor gamma camera with finely pixilated CdTe detector. Proc Sci abstracts J Nucl Med 53rd Annual Meeting, San Diego, CA, p 54
3. Yokoi K, Kawaguchi T, Kawasaki Y et al (2007) Study of CdTe-SPECT inspection for subendocardial infarction model. Proc Sci abstracts J Nucl Med 54th Annual Meeting, Washington, DC, p 427
4. Tsuchiya K, Kawaguchi T, Takahashi I et al (2007) Evaluation of the suitable parallel-hole collimator for the finely pixelated CdTe gamma camera. Proc Sci abstracts J Nucl Med 54rd Annual Meeting, Washington, DC, p 430
5. Scheiber C (2000) CdTe and CdZnTe detectors in nuclear medicine. Nucl Instr and Meth A, 448: 513–524
6. Murayama H, Ishibashi H, Uchida H et al (1998) Depth encoding multicrystal detectors for PET. IEEE Trans Nucl Sci 45: 1152–1157
7. Murayama H, Ishibashi H, Uchida H et al (2000) Design of a depth of interaction detector with a PS-PMT for PET. IEEE Trans. Nucl Sci 47: 1045–1050
8. Park S-J, Rogers WL, Huh S et al (2007) A prototype of very high-resolution small animal PET scanner using silicon pad detectors, Nucl Instr and Meth A 570: 543–555
9. Zhang F, He Z, Seifert CE, A prototype three-dimensional position sensitive CdZnTe detector array. IEEE Trans Nucl Sci 54: 843–848
10. Seino T, Takahashi I (2007), CdTe detector characteristics at 30C and 35C when using the periodic bias reset technique. IEEE Trans. Nucl Sci 54: 777–781

11. Kominami S, Yokoi K, Tsuchiya K et al (2006) Effects of gamma-ray damage on energy spectra and polarization of a CdTe detector with the schottky barrier for imaging equipment. IEEE Trans. Nucl Sci 53: 3041–3048
12. Ueno Y, Morimoto Y, Tsuchiya K et al (2009) Basic performance test of a prototype PET scanner using CdTe semiconductor detectors. IEEE Trans. Nucl Sci 56: 24–28

A New PET Scanner with Semiconductor Detector Reveals Intratumoral Inhomogeneous Cell Activity with High Spatial and Energy Resolution

Tohru Shiga[1], Naoki Kubo[1], Yuichi Morimoto[2], Norio Katoh[1], Chietsugu Katoh[1], Yuichirou Ueno[2], Kenji Hirata[1], Reiko Usui[1], Keiji Kobayashi[2], Wataru Takeuchi[2], Hiroki Shirato[1], and Nagara Tamaki[1]

Summary

Objective: Intensity-modulated radiation therapy, an advanced mode of high-precision radiotherapy developed to deliver precise radiation doses to specific areas within a tumor, has become popular in the clinical situation. With this modality, it has become particularly important in the clinical situation to estimate accurate cell activity with positron emission tomography (PET) scanner. We developed a new PET scanner with a semiconductor detector. Phantom images and [18F] fluorodeoxyglucose ([18F] FDG)-PET images of patients were measured to evaluate this new scanner's capacity to identify intratumoral inhomogeneous cell activity.
Methods: The semiconductor PET scanner uses CdTe detectors. A cold spot phantom study was performed with 6-mm-diameter cold sphenoid defects. An "H"-shaped phantom, a dual-cylinder phantom, was also studied. These phantoms were surrounded with water. The phantom images and FDG-PET images of eight patients suffering with nasopharyngeal cancer were compared with images from a conventional BGO PET scanner (Siemens HR+). Profile curves of the phantom were measured in the phantom study. Contrast was defined as the peak-to-valley ratio. Intratumoral inhomogeneity and tumor edge sharpness were evaluated on the images of the patients.
Results: The contrast obtained with the semiconductor PET scanner (1.53) was 28% higher than that obtained with the HR+ (1.20) for the 6-mm-diameter cold sphenoid phantom. The contrast obtained with the semiconductor PET scanner (1.43) was 27% higher than that obtained with the HR+ (1.13) for the dual-cylinder phantom. Similarly, the 2-mm cold region between 1-mm hot rods in the "H"-shaped phantom was identified only by the new PET scanner and not by the HR+. The new PET scanner identified intratumoral inhomogeneity in more detail than the HR+ in six of the eight patients. The tumor edge was sharper

[1]Department of Nuclear Medicine, Hokkaido University, Kita-15 Nishi-5, Kita-ku, Sapporo 060-8638, Japan
[2]Hitachi Ltd., Central Research Laboratory, Hitachi, Ibaraki, Japan

N. Tamaki, Y. Kuge (eds.)
Molecular Imaging for Integrated Medical Therapy and Drug Development.
DOI: 10.1007/978-4-431-98074-2_5, ©Springer 2010

on the images obtained with the new PET scanner than on those obtained with the HR+.
Conclusion: The new PET scanner has the potential for better identification of intratumoral inhomogeneous cell activity, with high spatial resolution and contrast.

Key words Semiconductor · PET · FDG · Head and neck cancer · High resolution

Introduction

[18F] Fluorodeoxyglucose positron emission tomography ([18F] FDG-PET) has been widely used for detection and evaluation in oncology studies. A high-resolution PET camera permits the precise evaluation of both tumor localization and treatment effects. Recently, CdTe semiconductors have been used for the direct conversion of gamma-rays without scintillator material [1, 2]. High-energy resolution and flexibility in both the sizing and fine arrangement of detectors are expected to improve image quality. These characteristics of semiconductor detectors may also lead to improved PET images because their high-energy resolution offers a reduction in scattered noise [1, 3].

Intensity-modulated radiation therapy (IMRT) is an advanced mode of high-precision radiotherapy that utilizes computer-controlled X-ray accelerators to deliver precise radiation doses to a malignant tumor or specific areas within the tumor; IMRT has become widely used in the clinical setting [4–6]. We have reported that image fusion between 18-FDG-PET and magnetic resonance imaging (MRI)/computed tomography (CT) was useful in the determination of gross tumor volume and clinical target volume in conformal radiation therapy, and several studies have revealed the clinical impact of FDG-PET/CT fusion on radiation planning [6–9]. In radiation therapy, it is increasingly important to estimate intratumoral inhomogeneity for more accurate radiation planning.

We developed a new PET system using CdTe semiconductor detectors with a depth of interaction (DOI) system. Phantom images and patients' images were both analyzed to evaluate intratumoral inhomogeneity compared with conventional BGO PET.

Materials and Methods

Study of Phantoms

Two different phantom studies were performed. The first phantom, a 136 × 188-mm elliptical phantom, consisted of four small compartments filled with water. The first compartment was a uniform cylinder in the middle, the second

Results

Study of Phantoms

The images of the phantom showed a 25-mm cold spot clearly for both the semiconductor PET and HR+. However, the profile curve of the second phantom indicated a better contrast for the 6-mm sphenoid cold phantom, as well as the dual-cylinder phantom, on images obtained with the semiconductor PET (Fig. 1a) [11]. The contrast with the semiconductor PET (1.53) increased by 27% as compared to the HR+ (1.20). The contrast with the semiconductor PET (1.43) increased by 27% as compared to the HR+ (1.13) in the ducal cylinder phantom.

The 1-mm hot rods on the "H"-shaped hot phantom were detected only by the semiconductor PET, but not by HR+. Similarly, the 2-mm cold region between the 1-mm hot rods was also identified only by the semiconductor PET (Fig. 1b) [11]. The profile curves confirmed the detection of the small hot rods and the cold spot in between because of the increased contrast of the small structures by the semiconductor PET.

Study of Patients

For qualitative analysis, the semiconductor PET identified intratumoral inhomogeneous glucose metabolism in more detail than the HR+ in six of the eight patients. In all six of these patients, intratumoral inhomogeneity or small structures in the tumor were confirmed by MRI.

The profile curves were sharper for the semiconductor PET than for the HR+ in both brain mode and whole-body mode in all the images of the eight patients who showed abnormal uptake. Figures 2 and 3 show typical cases [11].

Discussion

Because the PET scanner we have described provides high resolution and low scatter noise, it has the potential to provide better delineation of intratumoral inhomogeneity in the study of nasopharyngeal cancer. The phantom study confirmed a higher contrast of small cold and hot spots obtained by the semiconductor PET, with uniformity similar to that obtained by the conventional PET (HR+). In the study of our specially designed "H"-shaped hot-rod phantom, a 2-mm defect between 1-mm hot rods was identified by the semiconductor PET. In the clinical setting, intratumoral inhomogeneity and edge sharpness were better delineated by the semiconductor PET than by the HR+. This better delineation of intratumoral

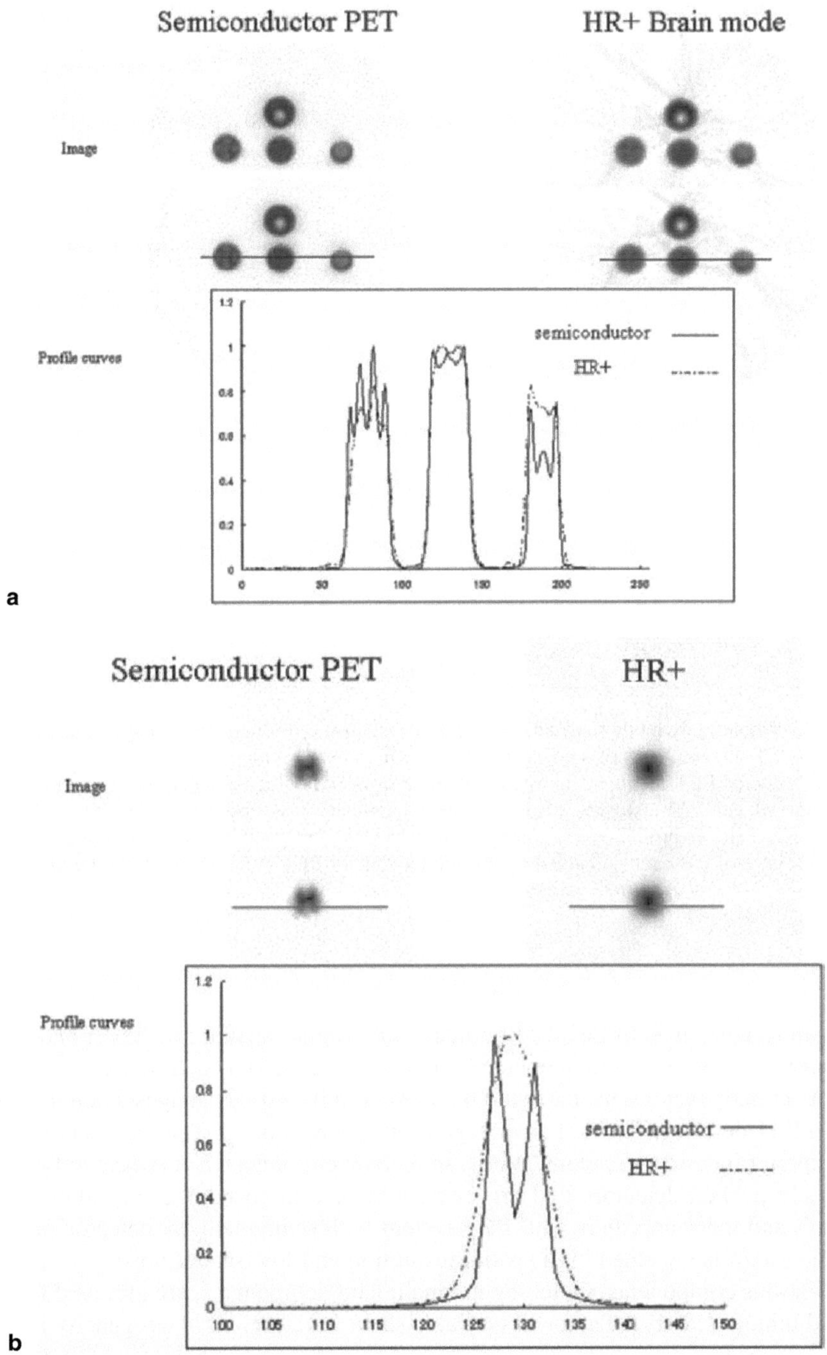

Fig. 1a,b. Phantom images and profile curves obtained by semiconductor positron emission tomography (PET) scanner (*semiconductor PET*) and Siemens EXACT HR+ (Asahi-Siemens; *HR+*). **a** Tumor phantom; **b** "H"-shaped hot phantom. The *upper panels* of **a** and **b** show PET images. The *middle panels* show the positions of the profile curves. The *lower panels* show the profile curves. The profile curves were normalized by maximum counts

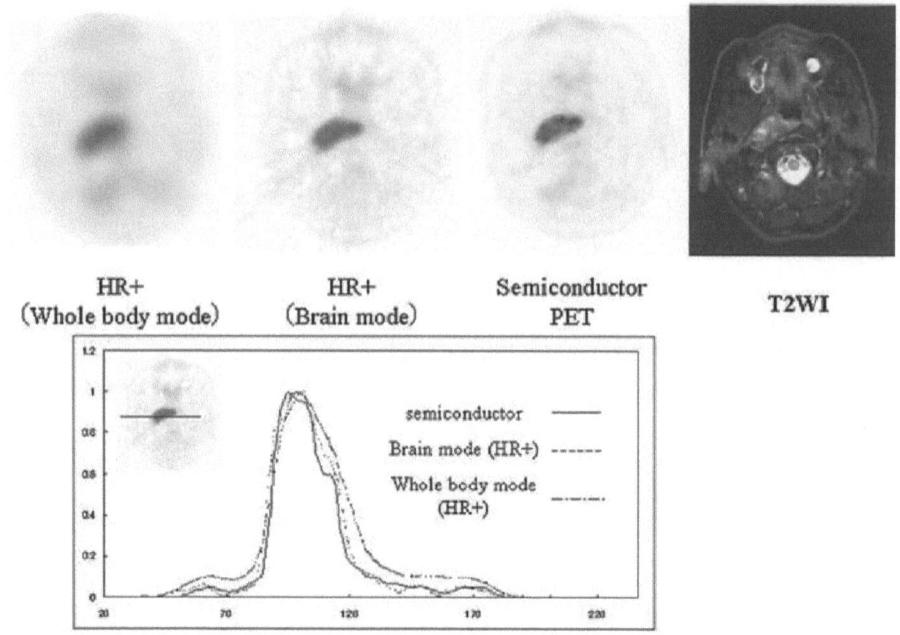

Fig. 2. Whole-body mode (*left*) and brain mode (*middle*) scintillator PET images, and semiconductor PET image (*right*) of a 61-year-old man with nasopharyngeal squamous cell cancer. The semiconductor PET identified intratumoral inhomogeneous glucose metabolism in more detail than the HR+. T2WI also shows inhomogeneous intensity in the primary lesion (*far right*). The *lower panel* shows profile curves. The profile curves were normalized by maximum counts. The profile curve of the tumor obtained by the semiconductor PET is the sharpest among these three images

inhomogeneity is considered to be due to the higher spatial resolution and lower scatter noise of our semiconductor PET, not to higher statistical noise.

A semiconductor detector could have two significant advantages when it is used for a PET detector. The first advantage is energy resolution. The second advantage is superior individual readout. With a semiconductor detector, it is easy to fabricate densely packed detectors that are composed of a large number of small crystal arrays and their amplifiers. Our PET system with semiconductor detectors and the DOI system has yielded high spatial resolution and low scatter noise.

Various components, including nonmalignant components, are involved in most solid tumors [12–19]. Tumor tissues also show intratumoral heterogeneity in their various properties, which may originate from diverse phenotypic properties of tumor cells or may be induced by their metabolic microenvironment. In this regard, intratumoral heterogeneity assessed by FDG distribution has been well demonstrated by autoradiography [14–19]. Intratumoral 18F-FDG distribution has been described at the cellular level by several investigators. Kubota et al. [16] showed,

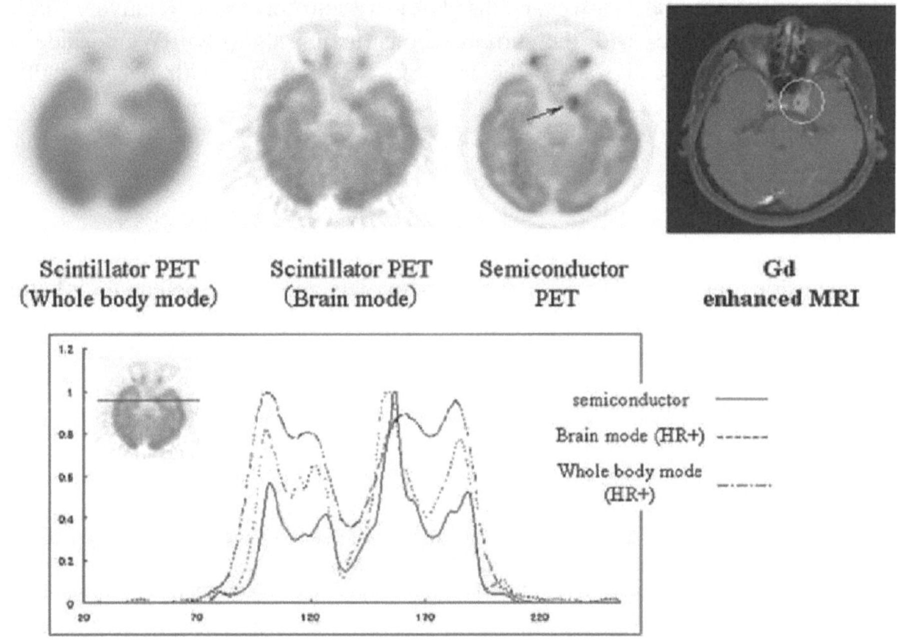

Fig. 3. Whole-body mode (*left*) and brain mode (*middle*) scintillator PET images and semiconductor PET image (*right*) of a 35-year-old woman with nasopharyngeal squamous cell cancer. The semiconductor PET scanner revealed a low uptake in the tumor, which corresponded to the internal carotid artery (*arrow*). Gd-enhanced magnetic resonance imaging (*MRI*) showed that the left internal carotid artery had penetrated the tumor (*far right*). The *lower panel* shows profile curves. The profile curves were normalized by maximum counts. The profile curve of the tumor obtained by the semiconductor PET scanner is the sharpest among these three images

using a malignant tumor mouse model, that 18F-FDG preferentially accumulated in macrophages and young granulation tissues surrounding necrotic foci, rather than in tumor cells. In contrast, Brown et al. [17] observed relatively less 3H-FDG accumulation in necrotic/inflammatory infiltrations compared with that in tumor cells. The semiconductor PET with the DOI system may reveal intratumoral heterogeneity even in the clinical setting.

One of the major limitations of the semiconductor PET is the relatively low sensitivity of the semiconductor detector. This low sensitivity mainly arises from the narrow energy window setting (490–530 keV) and low interaction of 511-keV gamma-rays and CdTe material. However, our uniformity phantom study confirmed that uniformity was well maintained and uniformity values were similar for the semiconductor PET and conventional PET.

Another limitation is that we did not compare our new PET results with a state-of-the-art PET system, but compared the results with those obtained with a relatively old camera, the Siemens HR+ system, which has a standard OSEM and DIFT reconstruction method. However, the HR+ system provides relatively high-resolution PET images with the current reconstruction algorithm. We consider it necessary to compare a state-of-the-art LSO PET scanner with our new PET scanner in the future.

Conclusion

In conclusion, our new PET with semiconductor detector has the potential for better identification of intratumoral inhomogeneous cell activity, with high spatial resolution and contrast.

References

1. Kubo N, Zhao S, Fujiki Y et al (2005) Evaluating performance of a pixel array semiconductor SPECT system for small animal imaging. Ann Nucl Med 19: 633–639
2. Kiyono Y, Kuge Y, Katada Y et al (2007) Applicability of a high-resolution small semiconductor gamma camera to small animal imaging. Nucl Med Commun 28: 736–741
3. Scheiber C (2000) CdTe and CdZnTe detectors in nuclear medicine. Nucl Instr and Meth A 448: 513–524
4. Shirato H, Oita M, Fujita K et al (2004) Feasibility of synchronization of real-time tumor-tracking radiotherapy and intensity-modulated radiotherapy from viewpoint of excessive dose from fluoroscopy. Int J Radiat Oncol Biol Phys 60: 335–341
5. Lee N, Puri DR, Blanco AI et al (2007) Intensity-modulated radiation therapy in head and neck cancers: an update. Head Neck 29: 387–400
6. Schwartz DL, Ford EC, Rajendran J et al (2005) FDG-PET/CT-guided intensity modulated head and neck radiotherapy: a pilot investigation. Head Neck 27: 478–487
7. Wang D, Schultz CJ, Jursinic PA et al (2006) Initial experience of FDG-PET/CT guided IMRT of head-and-neck carcinoma. Int J Radiat Oncol Biol Phys 65: 143–151
8. Vanderstraeten B, Duthoy W, De Gersem W et al (2006) [18F] Fluoro-deoxy-glucose positron emission tomography ([18F] FDG-PET) voxel intensity-based intensity-modulated radiation therapy (IMRT) for head and neck cancer. Radiother Oncol 79: 249–258
9. Nishioka T, Shiga T, Shirato H et al (2002) Image fusion between 18FDG-PET and MRI/CT for radiotherapy planning of oropharyngeal and nasopharyngeal carcinomas. Int J Radiat Oncol Biol Phys 53: 1051–1057
10. Takeuchi W, Morimoto Y, Suzuki A et al (2008) Iterative reconstruction method using prior information and point spread function for high resolution CdTe PET scanner (Meeting abstract). J Nucl Med 49 (suppl 1): 388
11. Shiga T, Morimoto Y, Kubo N et al (2009) A new PET scanner with semiconductor detectors enables better identification of intratumoral inhomogeneity. J Nucl Med 50: 148–155
12. Tovi M, Hartman M, Lilja A et al (1994) MR imaging in cerebral gliomas. Tissue component analysis in correlation with histopathology of whole-brain specimens. Acta Radiol 35: 495–505
13. Dachman AH, Pakter RL, Ros PR et al (1987) Hepatoblastoma: radiologic-pathologic correlation in 50 cases. Radiology 164: 15–19
14. Zhao S, Kuge Y, Mochizuki T et al (2005) Biologic correlates of intratumoral heterogeneity in 18F-FDG distribution with regional expression of glucose transporters and hexokinase-II in experimental tumor. J Nucl Med 46: 675–682

15. Kubota R, Yamada S, Kubota K et al (1992) Intratumoral distribution of fluorine-18-fluorodeoxyglucose in vivo: high accumulation in macrophages and granulation tissues studied by microautoradiography. J Nucl Med 33: 1972–1980
16. Kubota R, Kubota K, Yamada S et al (1994) Microautoradiographic study for the differentiation of intratumoral macrophages, granulation tissues and cancer cells by the dynamics of fluorine-18-fluorodeoxyglucose uptake. J Nucl Med 35: 104–112
17. Brown RS, Leung JY, Fisher SJ et al (1995) Intratumoral distribution of tritiated fluorodeoxyglucose in breast carcinoma. I. Are inflammatory cells important? J Nucl Med 36: 1854–1861
18. Kubota R, Kubota K, Yamada S et al (1994) Active and passive mechanisms of [fluorine-18] fluorodeoxyglucose uptake by proliferating and prenecrotic cancer cells in vivo: a microautoradiographic study. J Nucl Med 35: 1067–1075
19. Kubota R, Kubota K, Yamada S et al (1995) Methionine uptake by tumor tissue: a microautoradiographic comparison with FDG. J Nucl Med 36: 484–492

A New Positron Emission Tomography (PET) Scanner with Semiconductor Detectors for Target Volume Delineation and Radiotherapy Treatment Planning in Patients with Nasopharyngeal Carcinoma

Norio Katoh[1], Koichi Yasuda[1], Tohru Shiga[2], Masakazu Hasegawa[1], Rikiya Onimaru[1], Shinichi Shimizu[1], Gerard Bengua[3], Masayori Ishikawa[3], Nagara Tamaki[2], and Hiroki Shirato[1]

Summary

Purpose: Reducing noise in medical imaging has improved dose distribution in the treatment planning of external radiotherapy since the advent of computed tomography (CT). We compared treatment planning of stereotactic boost for nasopharyngeal carcinoma (NPC) between using conventional scintillator positron emission tomography (PETconv) and using a new semiconductor PET scanner (PETnew), which has been shown to have a better signal-to-noise ratio in phantom studies.

Methods: The subjects considered in this study were six patients with NPC. [F-18] Fluorodeoxyglucose (^{18}F-FDG)-PET images were acquired with both the semiconductor PET (PETnew) and a scintillator PET (PETconv) on the same day. CT and two PET data sets were transferred to a treatment planning system, and the PETconv and PETnew images were co-registered with the same set of CT images. The window width and level for all PET images were fixed. The gross tumor volume (GTV) was visually delineated on PET images using either conventional scintillator PET images (GTVconv) or semiconductor PET images (GTVnew). Assuming a stereotactic radiotherapy boost of seven ports, the prescribed dose delivered to 95% of the planning target volume (PTV) was set at 2000 cGy in four fractions.

Results: The average absolute volume of GTVnew was 16.6 ml (SD, 12.3), and that of GTVconv was 29.9 ml (SD, 20.2). The average of GTVnew was significantly

---

[1]Department of Radiology, Hokkaido University Graduate School of Medicine, North-15 West-7, Kita-ku, Sapporo 060-8638, Japan

[2]Department of Nuclear Medicine, Hokkaido University Graduate School of Medicine, North-15 West-7, Kita-ku, Sapporo 060-8638, Japan

[3]Department of Medical Physics, Hokkaido University Graduate School of Medicine, North-15 West-7, Kita-ku, Sapporo 060-8638, Japan

N. Tamaki, Y. Kuge (eds.)
Molecular Imaging for Integrated Medical Therapy and Drug Development.
DOI: 10.1007/978-4-431-98074-2_6, ©Springer 2010

smaller than that of GTVconv ($P = 0.028$). There was no statistically significant difference between PTVnew and PTVconv in the maximum dose ($P = 0.173$) or the mean dose ($p = 0.249$). The radiotherapy treatment plan based on GTVnew, defined as PLANnew, significantly reduced the maximum dose to cerebrum and cerebellum ($P = 0.028$) and brain stem ($P = 0.028$).
Conclusions: The present study suggests that the new semiconductor PET scanner can provide more accurate tumor delineation than scintillator PET scanners and will be an important tool for functional and molecular radiotherapy treatment planning.

Key words Semiconductor · Positron emission tomography · Radiotherapy planning · Target volume delineation · Nasopharyngeal carcinoma

Introduction

Since the advent of computed tomography (CT), sophisticated techniques in radiation treatment, such as three-dimensional conformal radiotherapy, stereotactic radiotherapy, and intensity-modulated radiotherapy, have been developed in order to focus and escalate the radiation dose to the tumor while sparing normal tissues. With these techniques, it is important to precisely determine the tumor volume. With their high anatomic resolution, CT and magnetic resonance imaging (MRI) have been primarily used for target volume delineation in radiotherapy treatment planning. However, when delineating the target volume, it is sometimes difficult to distinguish between tumor and nontumor tissues using anatomical imaging alone. In the past 10 years, positron emission tomography (PET) with [F-18] fluorodeoxyglucose (^{18}F-FDG), which is able to visualize molecular information for the tumor, has been widely used in oncology for the diagnosis and staging of various cancers. This functional imaging has been adopted in radiotherapy, and several studies have examined the clinical impact of PET on radiotherapy planning [1–5]. IIowever, because PET is not an inherently accurate test, with a spatial resolution of approximately 4 to 7 mm [6], it is difficult to determine tumor boundaries on conventional scintillator PET images. In 2007, a novel PET scanner with semiconductor detectors, the first in the world, was developed with Hitachi (Hitachi, Japan), and installed at our institute. Phantom studies have revealed that this new device has the potential to give high-quality images with lower scatter noise and a spatial resolution of 2.3 mm [7]. The difference in the signal-to-noise ratio between the new semiconductor PET scanner and a conventional scintillator PET scanner in a phantom was impressive; allowing radiation oncologists to expect precise delineation with the new machine (Fig. 1). When we used the two PETs for ^{18}F-FDG examinations of the same patient with nasopharyngeal carcinoma (NPC), it was apparent that the semiconductor PET image showed a different image from that shown by the scintillator PET scanner (Fig. 2). However, there are many questions to find the reason for the difference behind the images. Is the difference due to

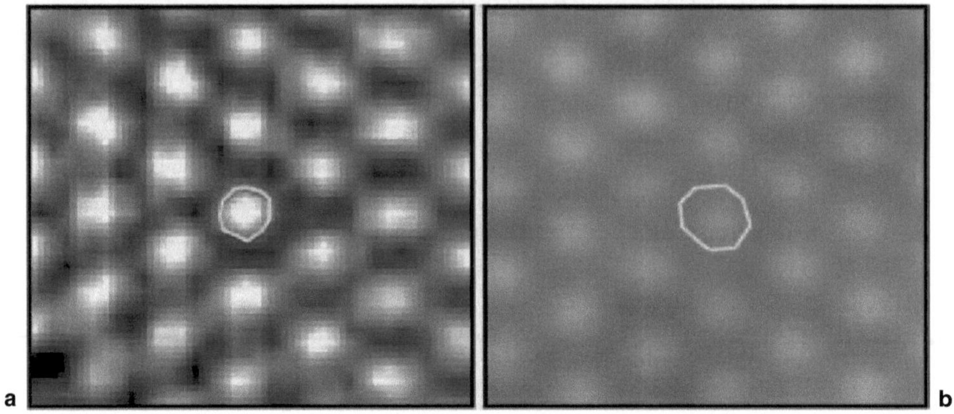

Fig. 1a,b. Illustration of the difference in delineation of a tumor (*yellow outlines*) in a radiotherapy treatment plan based on **a** semiconductor positron emission tomography (PET) images (gross tumor volume [GTV] new), and **b** a radiotherapy treatment plan based on conventional scintillator PET images (GTVconv) for phantom data. It is apparent that it is better to use the semiconductor PET scanner for tumor delineation

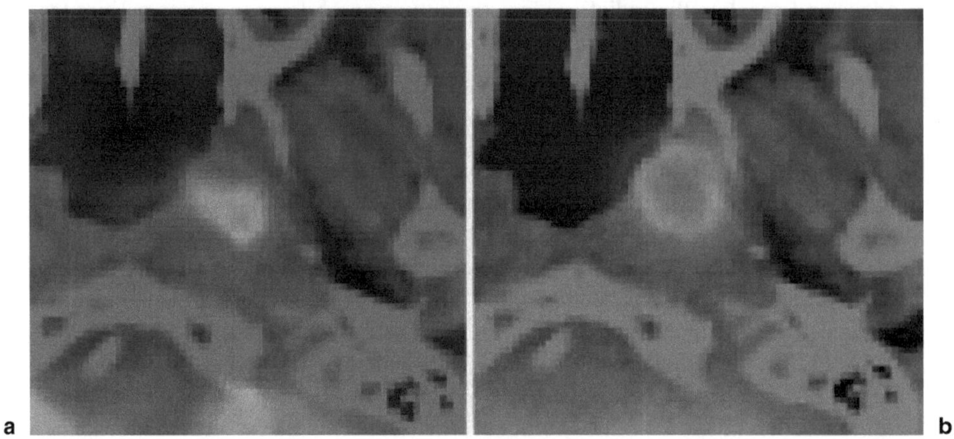

Fig. 2a,b. The two PET examinations of the same nasopharyngeal carcinoma (NPC) patient with T1N2 disease. **a** Semiconductor PET image; **b** scintillator PET image

time of examination? Or is it due to fluctuation of the uptake of the FDG? Or is it because of the better signal-to-noise ratio of the semiconductor PET scanner? The purpose of this study was to compare the new semiconductor PET scanner and a conventional scintillator PET scanner with regard to tumor volume delineation, using treatment planning for NPC to find the answers to these questions.

Patients, Materials, and Methods

Patients

The subjects considered in this study were six NPC patients who had been newly diagnosed between July 2007 and January 2008. There were five men and a woman, and their median age was 57 years, with a range of 30 to 76 years. The distribution of the TNM tumor stage was stage T1N1–T3N3b (2002 International Union Against Cancer [UICC] stage IIB–IVB). Written informed consent for the study was obtained from all patients.

Image Acquisition/Target Volume Delineation

CT with a slice thickness of 2–5 mm was performed. ^{18}F-FDG-PET images were acquired with the semiconductor PET scanner (PETnew) and with a scintillator PET scanner (EXACT HR+; Asahi-Siemens, Tokyo, Japan; PETconv) on the same day. Three patients were examined first with PETnew and the other three patients were examined first with PETconv. Because the new PET scanner with semiconductor detectors is dedicated to brain imaging, PETnew images for NPC were limited to the primary tumors and retropharyngeal lymph nodes. The CT and two PET data sets were transferred to the Pinnacle^3 treatment planning system (version 8.0d; Philips Medical Systems, Fitchburg, WI, USA) for image registration, target-volume delineation, and volume analysis. The PETconv and PETnew images were co-registered with the same set of CT images. In the Pinnacle^3 treatment planning system, the window width and level in all PET images were fixed. The gross tumor volume (GTV) was visually delineated on PET images alone by an experienced nuclear medicine physician and a radiation oncologist in consensus. When drawing the GTV contour, CT images were not used. GTVconv was determined using conventional scintillator PET images, while GTVnew was determined using semiconductor PET images. There was an interval of approximately a week between the delineation of GTVconv and GTVnew.

Radiotherapy Treatment Planning Simulation

The clinical target volume (CTV) was defined as the GTV with a 2-mm margin three-dimensionally. The planning target volume (PTV) was defined as the CTV plus a 3-mm margin three-dimensionally. Assuming a stereotactic radiotherapy boost of seven ports, the prescribed dose delivered to 95% of the PTV was set at 2000 cGy in four fractions. A radiotherapy treatment plan was prepared for GTVnew and GTVconv. Dose-volume histograms (DVHs) were calculated for the PTV, the cerebrum and cerebellum, and the brain stem in both plans.

Statistical Analysis

Absolute volumes of GTV and DVH parameters were compared. The difference was evaluated using the Wilcoxon signed-rank test. $P < 0.05$ was considered statistically significant.

Results

The results are summarized in Table 1. The average absolute volume of GTVnew was 16.6 ml (range, 4.9–31.6 ml), and that of GTVconv was 29.9 ml (range, 10.6–63.0 ml). The average absolute volume of GTVnew was significantly smaller than that of GTVconv (P = 0.028). There was no statistically significant difference between PTVnew and PTVconv in the maximum dose (P = 0.173) or the mean dose (P = 0.249). The radiotherapy treatment plan based on GTVnew was defined as PLANnew, and the plan based on GTVconv was defined as PLANconv. In PLANnew, the average maximum dose to cerebrum and cerebellum (CC) was 2077 cGy (range, 1860–2260 cGy) and that to brain stem (BS) was 1475 cGy (range, 586–2186 cGy). In PLANconv, the average maximum dose to CC was

Table 1. Summary of the results

	Average	SD	*P* value
Absolute GTV			
GTVnew (ml)	16.6	12.3	0.028
GTVconv (ml)	29.9	20.2	
The maximum PTV dose			
PLANnew (cGy)	2289	60	0.173
PLANconv (cGy)	2306	76	
The mean PTV dose			
PLANnew (cGy)	2136	19	0.249
PLANconv (cGy)	2146	24	
The maximum dose to cerebrum and cerebellum			
PLANnew (cGy)	2077	175	0.028
PLANconv (cGy)	2251	85	
The maximum dose to brain stem			
PLANnew (cGy)	1475	682	0.028
PLANconv (cGy)	1728	602	

GTV, gross tumor volume; GTVnew, using semiconductor positron emission tomography (PET) images; GTVconv, using conventional scintillator PET images; PTV, planning target volume; PLANnew, radiotherapy treatment plan based on GTVnew; PLANconv, radiotherapy treatment plan based on GTVconv

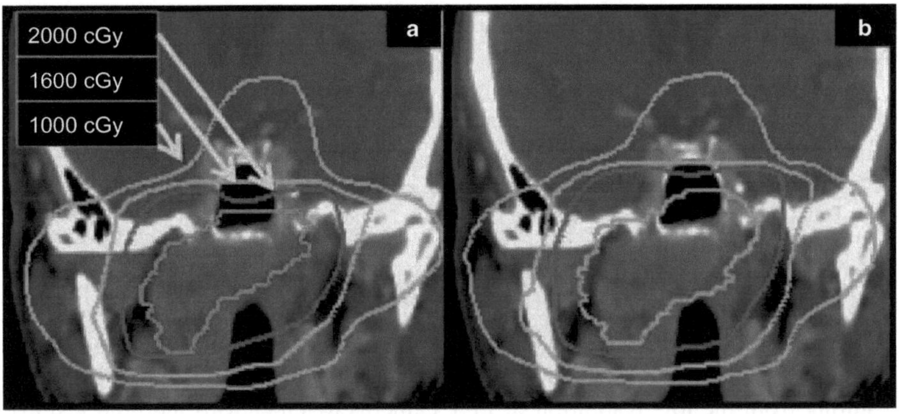

Fig. 3. **a** Radiotherapy treatment plan based on GTVnew (PLANnew) and **b** radiotherapy treatment plan based on GTVconv (PLANconv) in the same NPC patient with T3N3bM0 disease. *Blue*, *aqua*, and *orange lines* show 2000-cGy, 1600-cGy, and 1000-cGy isodose lines, respectively. The *red line* indicates GTVnew, while the *green line* indicates GTVconv

2251 cGy (range, 2163–2340 cGy) and that to BS was 1728 cGy (range, 664–2191 cGy). Compared with PLANconv, PLANnew significantly reduced the maximum dose to CC ($P = 0.028$) and BS ($P = 0.028$). An example of PLANnew and PLANconv is shown in Fig. 3.

There were no significant differences between those who were examined by PETnew first and those who were examined by PETconv first. The GTVnew was always smaller than GTVconv for all six patients.

Discussion

Although PET offers better identification of tumor localization than anatomical imaging modalities because of its higher contrast resolution, tumor boundaries are not accurate on conventional scintillator PET, due to its relatively low spatial resolution. In the present study, the absolute GTV volumes on the semiconductor PET scanner with high spatial resolution and low scatter noise were smaller than those on a conventional PET scanner. Because there were no differences between those who were examined by PETnew first and those who were examined by PETconv first, and the GTVnew was always smaller than the GTVconv, it seems that the smaller size of the GTV on semiconductor PET was not likely to have been due to the time of examination or the fluctuation of uptake. In the radiotherapy treatment planning, this target volume reduction resulted in a decrease in the radiation dose to organs at risk. Although we did not compare the pathologic specimens to the target volumes on PET images, and it is unclear whether the semiconductor PET-based GTV accurately reflected the true tumor volume, we consider the reduction of absolute GTV to be primarily due to the tumor edge being more clearly defined on the semiconductor PET scanner.

We adopted a visual interpretation method for the delineation of GTV. This method is commonly used [8–12], but is influenced by the display windowing and is dependent on the operator. Therefore, several objective methods for contouring PET images have been developed, including isocontouring, based on a fixed threshold of a standardized uptake value [1, 12–15]; a fixed threshold of 40% to 50% of the maximum activity [12, 15–17]; and a threshold adapted to the signal-to-background ratios [12, 18]. However, the appropriate standardized technique for the segmentation of PET images is still under investigation in the head and neck region [6, 19–21]. From a methodological point of view, the use of threshold-based methods was mainly motivated by the intrinsically low quality of PET images resulting from the low resolution of PET cameras together with the high level of noise [21]. As such, semiconductor PET images could lead to a new standardized segmentation method, and we consider it necessary to evaluate the interobserver variability of the target delineation and to compare objective segmentation methods for the semiconductor PET images.

We have previously reported that the semiconductor PET scanner has the potential to provide better identification of intratumoral inhomogeneity [7]. It is possible that, with the use of the semiconductor PET scanner, intensity-modulated radiotherapy can accurately deliver a higher dose to a lesion, with higher intratumoral uptake.

Conclusions

Although the number of patients in the present study was too small to draw definitive conclusions, our results suggest that the new semiconductor PET scanner can provide more accurate tumor delineation than scintillator PET scanners and thereby it will be an important tool for functional and molecular radiotherapy treatment planning.

References

1. Wang D, Schultz CJ, Jursinic PA et al (2006) Initial experience of FDG-PET/CT guided IMRT of head-and-neck carcinoma. Int J Radiat Oncol Biol Phys 65: 143–151
2. Paulsen F, Scheiderbauer J, Eschmann SM et al (2006) First experiences of radiation treatment planning with PET/CT. Strahlenther Onkol 182: 369–375
3. Vernon MR, Maheshwari M, Schultz CJ et al (2008) Clinical outcomes of patients receiving integrated PET/CT-guided radiotherapy for head and neck carcinoma. Int J Radiat Oncol Biol Phys 70: 678–684
4. Erdi YE, Rosenzweig K, Erdi AK et al (2002) Radiotherapy treatment planning for patients with non-small cell lung cancer using positron emission tomography (PET). Radiother Oncol 62: 51–60

5. Bradley JD, Dehdashti F, Mintun MA et al (2004) Positron emission tomography in limited-stage small-cell lung cancer: a prospective study. J Clin Oncol 22: 3248–3254
6. Ahn PH, Garg MK (2008) Positron emission tomography/computed tomography for target delineation in head and neck cancers. Semin Nucl Med 38: 141–148.
7. Shiga T, Morimoto Y, Kubo N et al (2009) A new PET scanner with semiconductor detectors enables better identification of intratumoral inhomogeneity. J Nucl Med 50: 148–155
8. Heron DE, Andrade RS, Flickinger J et al (2004) Hybrid PET-CT simulation for radiation treatment planning in head-and-neck cancers: a brief technical report. Int J Radiat Oncol Biol Phys 60: 1419–1424
9. Riegel AC, Berson AM, Destian S et al (2006) Variability of gross tumor volume delineation in head-and-neck cancer using CT and PET/CT fusion. Int J Radiat Oncol Biol Phys 65: 726–732
10. Nishioka T, Shiga T, Shirato H et al (2002) Image fusion between 18FDG-PET and MRI/CT for radiotherapy planning of oropharyngeal and nasopharyngeal carcinomas. Int J Radiat Oncol Biol Phys 53: 1051–1057
11. Ciernik IF, Dizendorf E, Baumert BG et al (2003) Radiation treatment planning with an integrated positron emission and computer tomography (PET/CT): a feasibility study. Int J Radiat Oncol Biol Phys 57: 853–863
12. Schinagl DA, Vogel WV, Hoffmann AL et al (2007) Comparison of five segmentation tools for 18F-fluoro-deoxy-glucose-positron emission tomography-based target volume definition in head and neck cancer. Int J Radiat Oncol Biol Phys 69: 1282–1289
13. Nestle U, Kremp S, Schaefer-Schuler A et al (2005) Comparison of different methods for delineation of 18F-FDG PET-positive tissue for target volume definition in radiotherapy of patients with non-small cell lung cancer. J Nucl Med 46: 1342–1348
14. Hong R, Halama J, Bova D et al (2007) Correlation of PET standard uptake value and CT window-level thresholds for target delineation in CT-based radiation treatment planning. Int J Radiat Oncol Biol Phys 67: 720–726
15. Burri RJ, Rangaswamy B, Kostakoglu L et al (2008) Correlation of positron emission tomography standard uptake value and pathologic specimen size in cancer of the head and neck. Int J Radiat Oncol Biol Phys 71: 682–688
16. Koshy M, Paulino AC, Howell R et al (2005) F-18 FDG PET-CT fusion in radiotherapy treatment planning for head and neck cancer. Head Neck 27: 494–502
17. Bradley J, Thorstad WL, Mutic S et al (2004) Impact of FDG-PET on radiation therapy volume delineation in non-small-cell lung cancer. Int J Radiat Oncol Biol Phys 59: 78–86
18. Daisne JF, Duprez T, Weynand B et al (2004) Tumor volume in pharyngolaryngeal squamous cell carcinoma: comparison of CT, MR imaging, and FDG PET and validation with surgical specimen. Radiology 233: 93–100
19. Rembielak A, Price P (2008) The role of PET in target localization for radiotherapy treatment planning. Onkologie 31: 57–62
20. Gregoire V, Haustermans K, Geets X et al (2007) PET-based treatment planning in radiotherapy: a new standard? J Nucl Med 48 (Suppl 1): 68S–77S
21. Gregoire V, Bol A, Geets X et al (2006) Is PET-based treatment planning the new standard in modern radiotherapy? The head and neck paradigm Semin Radiat Oncol 16: 232–238

Initial Performance Measurement of an Integrated PET/SPECT/CT System for Small-Animal Imaging

Keiichi Magota[1,2], Naoki Kubo[3], Kenji Narihiro[4], Kotaro Suzuki[1], Kenichi Nishijima[4], Songji Zhao[4], Yuji Kuge[4], and Nagara Tamaki[1]

Summary

Objectives: Recently, an integrated positron emission tomography (PET)/single photon emission computed tomography (SPECT)/computed tomography (CT)·system was developed for the purpose of studying molecular imaging of small animals in vivo. With PET/CT or SPECT/CT, lesions visualized by functional imaging can be correlated with anatomic structures. We aimed to validate the basic characterization and performance of the PET and SPECT components.
Methods: We characterized the performance of this system for both PET and SPECT components in terms of spatial resolution and sensitivity. For the PET component, the count rate performance was also evaluated using a cylindrical phantom (25 mmϕ).
Results: PET: the spatial resolutions were 1.60 mm the full width at the half maximum (FWHM) at the center of the field of view (FOV) and 2.66 mm FWHM at 32 mm offset. The absolute sensitivity was 2.1%. The peak trues of the count rate performance reached 1.1 Mcps at 200 MBq. SPECT: the spatial resolutions at the center of the FOV in the 0.5-, 1.0-, 2.0-, and 3.0-mm pinhole apertures were 0.77, 1.02, 1.81, and 2.76 mm, respectively. The system sensitivities in the 0.5-, 1.0-, 2.0-, and 3.0-mm pinhole apertures were 11.84, 24.97, 80.16, and 169.1 cps/MBq, respectively.
Conclusions: This integrated PET/SPECT/CT system provided high system performance with excellent spatial resolution for both PET and SPECT. This system holds promise for multitracer imaging and fusion images with high-resolution morphology information in small animals.

---

[1]Department of Nuclear Medicine, Hokkaido University Graduate School of Medicine, Kita 15 Nishi 7, Kita-ku, Sapporo 060-8638, Japan
[2]Department of Radiology, Hokkaido University Hospital, Kita 14 Nishi 5, Kita-ku, Sapporo 060-8648, Japan
[3]Faculty of Health Sciences, Hokkaido University, Kita 12 Nishi 5, Kita-ku, Sapporo 060-0812, Japan
[4]Department of Tracer Kinetics and Bioanalysis, Hokkaido University Graduate School of Medicine, Kita 15 Nishi 7, Kita-ku, Sapporo 060-8638, Japan

N. Tamaki, Y. Kuge (eds.)
Molecular Imaging for Integrated Medical Therapy and Drug Development.
DOI: 10.1007/978-4-431-98074-2_7, ©Springer 2010

Key words Positron emission tomography (PET) · Single photon emission computed tomography (SPECT) · Computed tomography (CT) · Small animal · Molecular imaging

Introduction

Functional imaging of small animals using positron emission tomography (PET), single photon emission tomography (SPECT), and computed tomography (CT) is becoming a valuable tool for studying animal models of human disease [1–12]. The combination of PET, SPECT, and CT techniques for small animal studies could offer the significant possibility of developing new and interesting protocols for the investigation of many biological phenomena more effectively than with PET, SPECT, or CT modalities alone [5]. Furthermore, with PET/CT or SPECT/CT, lesions visualized by functional imaging can be correlated with anatomic structures. The Inveon (Siemens Medical Solutions, Knoxville, TN, USA) is an integrated PET/SPECT/CT scanner developed for small animal imaging [6]. In this study, we aimed to validate the basic characterization and performance of its PET and SPECT components.

Materials and Methods

The Inveon is one of the latest commercial small-animal PET/SPECT/CT scanners [6]. The system, developed by Siemens Medical Solutions, is described in Table 1 and is shown in Figs. 1 and 2. The PET, SPECT, and CT components were combined by placing the SPECT and CT at the front and the PET at the rear in same gantry.

We have characterized the performance of the Inveon in regard to the PET and SPECT components in terms of spatial resolution and sensitivity. For PET, the count rate performance was also evaluated using a cylindrical phantom.

PET

The coincidence timing window and energy window were 3.432 ns and 350–650 keV, respectively. Images could be reconstructed using a three-dimensional filtered back-projection (FBP) algorithm using a ramp filter cutoff at the Nyquist frequency. Reconstructed voxel size was $0.5 \times 0.5 \times 0.5$ mm^3. Attenuation and scatter corrections were performed using a CT-based method.

Table 1. Inveon system description

Modality	Parameter	Description and value
PET	Crystal material	LSO
	Crystal element size	1.5 mm × 1.5 mm × 10 mm
	Crystal array	20 × 20
	Total number of crystal elements	25,600
	Transaxial FOV	12.7 cm
	Axial FOV	10 cm
SPECT	Number of detector heads	2
	Active area per detector head	15 cm × 15 cm
	Scintillator crystal material	NaI (Tl)
	Crystal element size	2 mm × 2 mm × 10 mm
	Number of crystal elements per detector head	4624
	Aperture size of single pinhole collimator	0.5, 1.0, 2.0, 3.0 mm
CT	Number of detector pixels	2048 × 3072
	Detector element size	32 μm
	Maximum FOV	5.5 cm × 8.4 cm
	Focal spot size	50 μm
	Voltage range	35–80 kVp
	Maximum anode current	500 μA

PET, positron emission tomography; SPECT, single photon emission computed tomography; CT, computed tomography; LSO, lutetium oxyorthosilicate; FOV, field of view

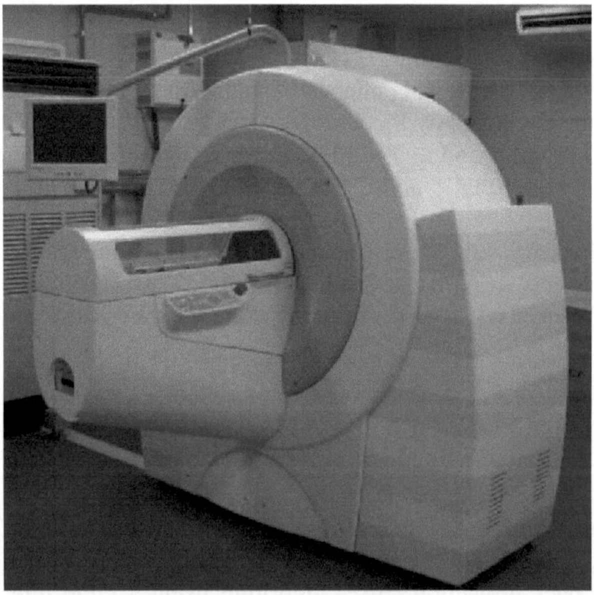

Fig. 1. Photograph of the Inveon scanner (Siemens Medical Solutions, Knoxville, TN, USA)

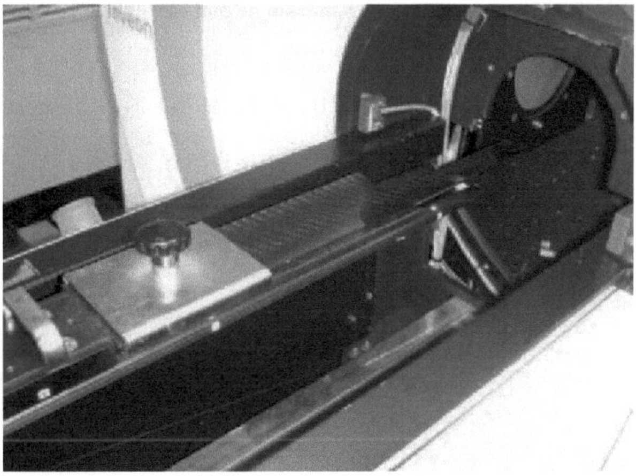

Fig. 2. Photograph of the animal bed of the Inveon

Spatial Resolution

All spatial resolutions were measured using a ^{22}Na point source. The point source was attached to the animal bed, positioned at 0, 16, and 32 mm radially from the center. The count profiles were plotted in the radial, tangential, and axial directions, and the full width at the half maximum (FWHM) and full width at tenth maximum (FWTM) were determined.

Absolute Sensitivity

Absolute sensitivity was measured using a 7.5-cm glass tube with inner diameter of 0.85 mm, which was filled with 370 kBq of ^{18}F-fluoro deoxy-glucose (^{18}F-FDG). The glass line source was sealed with a set of five concentric aluminum sleeves. The data were acquired for each decreasing number of sleeves. The sensitivity was extrapolated to zero wall thickness from an exponential fit to the slice sensitivity plotted as a function of the number of sleeves.

Count Rate Performance

A plexiglass phantom was used to estimate the prompt, true, and random counts to be expected in a mouse that measured 90 mm in length and 25 mm in diameter on the exterior. This phantom was filled with a uniform solution of ^{11}C with 400 MBq initially. As the activity decayed over this period, the prompt, true, and random counts were recorded as a function of time.

Fig. 3. Photograph of the 0.5 mm single pinhole collimator of the Inveon

SPECT

In the spatial resolution investigation, and phantom and animal imaging, the projection data were acquired in step-and-shoot mode with 60 views over 360°at 60 s/view. The photopeak was centered at 140 keV with a ± 10%. The reconstruction method used was a three-dimensional ordered subset expectation maximization (3D-OSEM) algorithm (iteration time, 2; subsets, 6). The image matrix voxel size was $0.4 \times 0.4 \times 0.4$ mm^3. Neither attenuation nor scatter correction were performed. The aperture size of the single pinhole collimator was 0.5, 1.0, 2.0, or 3.0 mm [7–12]. Fig. 3 is the photograph of the 0.5 mm single pinhole collimator.

System Spatial Resolution

The FWHMs of the count profile were measured in an axial image of SPECT using a ^{99m}Tc line source placed at 30 mm distant from the pinhole collimator.

System Sensitivity

The system sensitivity on the central axis was measured using a ^{99m}Tc point source placed at 30 mm distant from the pinhole collimator. This point source was created by drawing a small fluid column into the tip of a 0.85-mm (inside diameter) capillary tube; taping this tube to the animal bed perpendicular to, and on, the axis of the scanner, and imaging this source as the bed was stepped in small increments along the central axis.

Results and Discussion

PET

Spatial Resolution

At the center of the field of view (FOV), radial, tangential, and axial spatial resolutions were 1.60, 1.67, and 1.54 mm FWHM, respectively (Table 2). Radial resolution degraded at larger distances from the center of the FOV, and in particular, at 32 mm, the radial component of the spatial resolution was 2.66 mm FWHM. The radial component degrades due to depth of interaction. This value reported here is similar to values reported by Visser et al. [6].

Sensitivity

The absolute sensitivity at the center of the FOV was 2.1%. This value was less than that in a previous report using same scanner [6], which reported sensitivity of 6.8% with a ^{22}Na point source.

Count Rate Performance

Figure 4 shows the behavior of the prompt, true, and random count rates as a function of the activity in the mouse-size phantom. The trues saturate at a rate of

Table 2. Spatial resolution of PET component

	FWHM (mm)	FWTM (mm)
	Center of FOV	
Radial	1.60	3.03
Tangential	1.67	3.15
Axial	1.54	3.04
	16-mm offset	
Radial	2.29	4.25
Tangential	2.07	3.85
Axial	2.06	3.89
	32-mm offset	
Radial	2.66	4.91
Tangential	2.46	4.54
Axial	3.04	5.75

FWHM, full width at the half maximum; FWTM, full width at tenth maximum

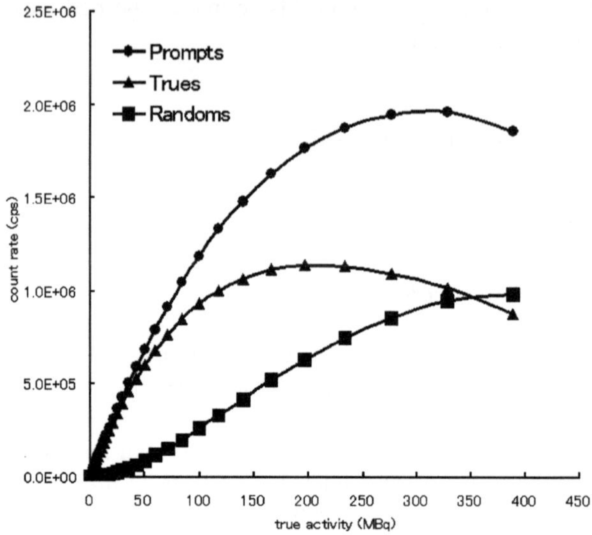

Fig. 4. Plot of the count rate performance obtained with a cylindrical phantom (25 ϕ). The *plots* show the behavior of the prompt, true, and random count rates versus the activity in the phantom

Table 3. Spatial resolution and system sensitivity of SPECT component

Aperture size (mm)	0.5	1.0	2.0	3.0
Spatial resolution (mm)	0.77	1.02	1.81	2.76
System sensitivity (cps/MBq)	11.84	24.97	80.16	169.1

about 1.1 Mcps. The energy window of 350–650 keV and the timing window of 3.432 ns provided a peak trues of 200 MBq, corresponding to 4525 kBq/ml.

SPECT

Spatial Resolution

Table 3 (top) summarizes the spatial resolution at the center of the FOV in each pinhole aperture. FWHMs at the center of the FOV in the 0.5-, 1.0-, 2.0-, and 3.0-mm pinhole diameters were 0.77, 1.02, 1.81, and 2.76 mm, respectively.

Sensitivity

Table 3 (bottom) summarizes the system sensitivities at a pinhole-to-source diameter distance of 30 mm in each pinhole aperture. The system sensitivities in the 0.5-, 1.0-, 2.0-, and 3.0-mm pinhole diameters were 11.84, 24.97, 80.16, and 169.1 cps/MBq, respectively. Figure 5 shows the normalized sensitivity profile as

Fig. 5. The normalized sensitivity profile as a function of axis position (z-direction)

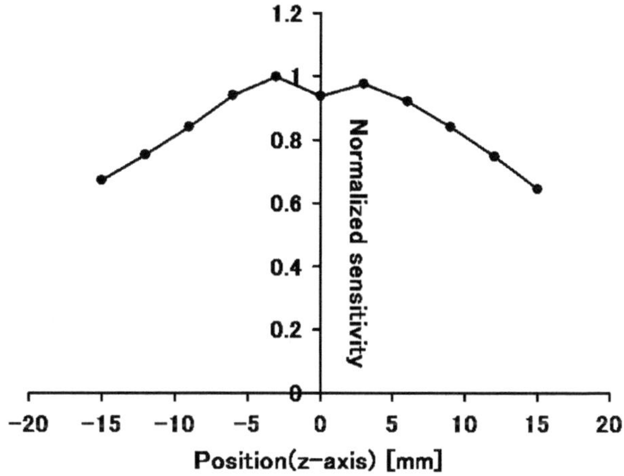

a function of axis position. This profile was low around the center of the collimator. We think that this phenomenon may have occurred because of a gap in the NaI(Tl) pixelated scintillator array [11].

Conclusion

This integrated PET/SPECT/CT system provided high system performance with excellent spatial resolution for both PET and SPECT. This system holds promise for multitracer imaging and fusion images with high-resolution morphology information in small animals.

References

1. Tai YC, Chatziioannou A, Siegel S et al (2001) Performance evaluation of the microPET P4: a PET system dedicated to animal imaging. Phys Med Biol 46: 1845–1862
2. Knoess C, Siegel S, Smith A et al (2003) Performance evaluation of the microPET R4 PET scanner for rodents. Eur J Nucl Med Mol Imaging 30: 737–747
3. Kim JS, Lee JS, Im KC et al (2007) Performance measurement of the microPET Focus 120 Scanner. J Nucl Med 48: 1527–1535
4. Tai YC, Chatziioannou A, Yang Y et al (2003) MicroPET II: design, development and initial performance of an improved microPET scanner for small-animal imaging. Phys Med Biol 48: 1519–1537
5. Guerra AD, Bartoli A, Belcari N et al (2006) Performance evaluation of the fully engineered yap-(s)pet scanner for small animal imaging. IEEE Trans Nucl Sci 53: 1078–1083
6. Visser EP, Disselhorst JA, Brom M et al (2009) Spatial resolution and sensitivity of the Inveon small-animal PET scanner. J Nucl Med 50: 139–147

7. Weber DA, Ivanovic M, Franceschi D et al (1994) Pinhole SPECT: an approach to in vivo high resolution SPECT imaging in small laboratory animals. J Nucl Med 35: 342–348
8. McElroy DP, MacDonald LR, Beekman FJ et al (2002) Performance evaluation of A-SPECT: a high resolution desktop pinhole SPECT system for imaging small animals. IEEE Trans Nucl Sci 49: 2139–2147
9. Song TY, Choi Y, Jung JH et al (2005) Performance amelioration for small animal SPECT using optimized pinhole collimator and image correction technique. IEEE Trans Nucl Sci 52(5): 1396–1400
10. Beekman FJ, van der Have F, Vastenhouw B et al (2005) U-SPECT-I: a novel system for submillimeter-resolution tomography with radiolabeled molecules in mice. J Nucl Med 46: 1194–1200
11. Zeniya T, Watabe H, Aoi T et al (2006) Use of a compact pixellated gamma camera for small animal pinhole SPECT imaging. Ann Nucl Med 20: 409–416
12. Kubo N, Zhao S, Fujiki Y et al (2005) Evaluating performance of a pixel array semiconductor SPECT system for small animal imaging. Ann Nucl Med 19: 633–639

Molecular Imaging of Atherosclerotic Plaque Vulnerability: Comparison between ^{18}F-FDG and ^{99m}Tc-Annexin A5

Yan Zhao[1,2], Yuji Kuge[1,3], Songji Zhao[1,2], H. William Strauss[4], Francis G. Blankenberg[5], and Nagara Tamaki[2]

Summary

Macrophage infiltration, apoptosis, and lipid deposition greatly contribute to plaque vulnerability. Nuclear imaging provides at least two potential indicators to identify vulnerable plaques: ^{18}F-fluorodeoxyglucose (FDG) as a marker of intraplaque inflammation and ^{99m}Tc-annexin A5 as a marker of apoptosis. In this article we summarize our recent study on the comparison of ^{18}F-FDG and ^{99m}Tc-annexin A5 uptake in atherosclerotic lesions of the apolipoprotein E knockout (ApoE-/-) mouse, and briefly review previous studies on molecular imaging of atherosclerotic plaques using ^{18}F-FDG and ^{99m}Tc-annexin A5.

The intralesional distribution of ^{18}F-FDG and ^{99m}Tc-annexin A5 was determined in male ApoE-/- mice. The mice were maintained on a high-fat diet after the age of 5 weeks. At 25 weeks, ^{18}F-FDG or ^{99m}Tc-annexin A5 was injected and the aortas were harvested. Regional radioactivity distribution was compared in relation to Oil Red O staining. Both ^{18}F-FDG and ^{99m}Tc-annexin A5 showed preferential uptakes into atherosclerotic lesions, with higher uptake levels for ^{18}F-FDG than ^{99m}Tc-annexin A5. The regional uptake levels of these tracers were significantly correlated with the Oil Red O staining score. The uptake ratios of advanced lesions to early lesions were significantly higher for ^{99m}Tc-annexin A5 than for ^{18}F-FDG.

Previous studies have indicated that ^{18}F-FDG and ^{99m}Tc-annexin A5 are useful for the detection of vulnerable atherosclerotic lesions. Our findings could further characterize these tracers as follows: the high aortic uptake level of ^{18}F-FDG may offer higher sensitivity in lesion detection, whereas the preferential uptake of ^{99m}Tc-

[1]Department of Tracer Kinetics and Bioanalysis, Hokkaido University Graduate School of Medicine, Kita 15 Nishi 7, Kita-ku, Sapporo 060-8638, Japan
[2]Department of Nuclear Medicine, Hokkaido University Graduate School of Medicine, Kita 15 Nishi 7, Kita-ku, Sapporo 060-8638, Japan
[3]Department of Molecular Imaging, Hokkaido University Graduate School of Medicine, Kita 15 Nishi 7, Kita-ku, Sapporo 060-8638, Japan
[4]Department of Nuclear Medicine, Memorial Sloan-Kettering Cancer Center, 1275 York Avenue, New York, NY 10065, USA
[5]Department of Pediatric Radiology, Stanford University School of Medicine, 725 Welch Road, RM 1671 MC 5654, Palo Alto, CA 94304, USA

N. Tamaki, Y. Kuge (eds.)
Molecular Imaging for Integrated Medical Therapy and Drug Development.
DOI: 10.1007/978-4-431-98074-2_8, ©Springer 2010

annexin A5 in advanced lesions suggests its potential for assessing the vulnerability of atherosclerotic plaques.

Key words Atherosclerosis · Molecular imaging · ^{18}F-FDG·^{99m}Tc-annexin A5

Introduction

Because the rupture of vulnerable atherosclerotic plaques and subsequent thrombus formation are the major cause of ischemic diseases, such as cerebral and myocardial infarction [1, 2], the detection of vulnerable atherosclerotic plaques is clinically important for the early selection and administration of appropriate therapy. Various invasive and noninvasive imaging methods are used to obtain morphological and functional information necessary for the identification of atherosclerotic plaques with a risk of rupture (vulnerable plaques). Molecular imaging methods, including positron emission tomography (PET) and single photon emission computed tomography (SPECT), are diagnostic technologies that noninvasively and sensitively visualize and quantify important determinants of plaque vulnerability, based on specific cellular or biochemical changes in vivo, utilizing various probes labeled with radioisotopes (RIs).

Inflammation and apoptosis are important components of plaque vulnerability. In this regard, nuclear imaging provides at least two potential indicators to identify vulnerable plaques: ^{18}F-fluorodeoxyglucose (FDG) and ^{99m}Tc-annexin A5 [3, 4]. Because ^{18}F-FDG is a glucose derivative, and PET using this marker can measure cellular energy metabolism, ^{18}F-FDG-PET has been used for the detection of active inflammation in arterial walls. One of the earliest events in apoptosis is the externalization of phosphatidylserine (PS), a membrane phospholipid normally restricted to the inner leaflet of the lipid bilayer. Annexin A5 (annexin V), a human protein with a high affinity for membrane-bound PS, can be labeled with ^{99m}Tc for in vivo imaging of apoptosis. Accordingly, the vulnerability of atherosclerotic lesions may be assessed using radiolabeled annexin A5 [4].

In this article, we summarize our recent study on the comparison of ^{18}F-FDG and ^{99m}Tc-annexin A5 uptake in atherosclerotic lesions of the apolipoprotein E knockout (ApoE-/-) mouse, a rodent model of severe hypercholesterolemia and spontaneous atherosclerosis [5], and we briefly review previous studies on the molecular imaging of atherosclerotic plaques using ^{18}F-FDG and ^{99m}Tc-annexin A5.

Comparison of ^{18}F-FDG and ^{99m}Tc-Annexin A5 in ApoE-/- mice

Several previous studies demonstrated the detection of plaque instability using ^{18}F-FDG or ^{99m}Tc-annexin A5 in human and animal models of atherosclerosis [6–9]. However, it remains to be elucidated which tracer is more useful for detecting atherosclerotic lesions and for determining lesion vulnerability. Thus, we compared the intralesional distribution of ^{99m}Tc-annexin A5 and ^{18}F-FDG in relation to lesion development in ApoE-/- mice [10]. ApoE-/- mice have been widely used in studies of atherosclerosis, because they spontaneously develop hypercholesterolemia and atherosclerotic lesions similar to those found in humans [11, 12]. These data suggest that ApoE-/- mice could be a useful model for the evaluation of tracers for detecting unstable atherosclerotic lesions.

Materials and Methods

ApoE-/- mice ($n = 12$–14/group) and wild-type mice (C57BL/6J: $n = 11$–15/group) were maintained on a high-fat diet (21% fat, 0.15% cholesterol) from the age of 5 weeks. At the age of 25 weeks the animals were fasted overnight, anesthetized, and intravenously injected with ^{99m}Tc-annexin A5 or ^{18}F-FDG (18.5 MBq/mouse). Two hours after ^{99m}Tc-annexin A5 injection or 1 h after ^{18}F-FDG injection, the aortas were collected and mounted on a glass slide after longitudinal incision.

The aorta specimens were exposed to phosphor imaging plates and images were analyzed using a computerized imaging analysis system [13]. Regions of interest (ROIs; 0.36 mm^2) were placed over the autoradiographic images (ARGs). The radioactivity in each ROI was calculated as percentage injected dose per m^2 of aortic tissue surface area (%ID/m^2) and normalized by animal body weight (%ID $\times$ kg/m^2). To determine the relative tracer uptake level in advanced lesions to early lesions, relative uptake values (RUVs) were defined as the ratio of tracer uptake in each ROI group to that in the ROI group with Oil Red O staining scores of 0–0.5 (see below).

The same aortic tissues that were used for ARG were stained with Oil Red O to coordinate with the ARG. The Oil Red O-stained specimens were microscopically examined, and the intensity of staining and percentage of positively stained area were determined in each ROI. The intensity of staining was graded from 0 to 3 (0, not stained; 1, slightly stained; 2, intensely stained;, 3, very intensely stained). The atherosclerosis severity levels were assessed semiquantitatively using the product of these scores (intensity $\times$ % positive area; Oil Red O staining score). On the basis of the Oil Red O staining score, ROIs were divided into three ROI groups (Oil Red O staining scores: 0 to 0.5, 0.5 to 1, and more than 1).

Results

In the ApoE-/- mice, various atherosclerotic changes, from early to advanced lesions, were observed throughout the aortic tissues. No obvious atherosclerotic changes were observed in the aortas of control mice. Both ^{18}F-FDG and ^{99m}Tc-annexin A5 uptake levels were significantly higher, by nearly twofold, in

ApoE-/- mice than in the corresponding control animals (Fig. 1; 56.07 ± 29.94 %ID × kg/m^2 vs 29.42 ± 19.07 for ^{18}F-FDG groups, and 10.38 ± 3.16 %ID × kg/m^2 vs 5.28 ± 1.19 for ^{99m}Tc-annexin A5 groups). The ^{18}F-FDG uptake levels in the aortic tissues were markedly higher than those of ^{99m}Tc-annexin A5 in both strains (Fig. 1). ^{18}F-FDG and ^{99m}Tc-annexin A5 were preferentially accumulated into the regions stained with Oil Red O in ApoE -/- mice (Fig. 2). For both tracers, the regional tracer uptake levels in the aortic tissues were significantly correlated with the sever-

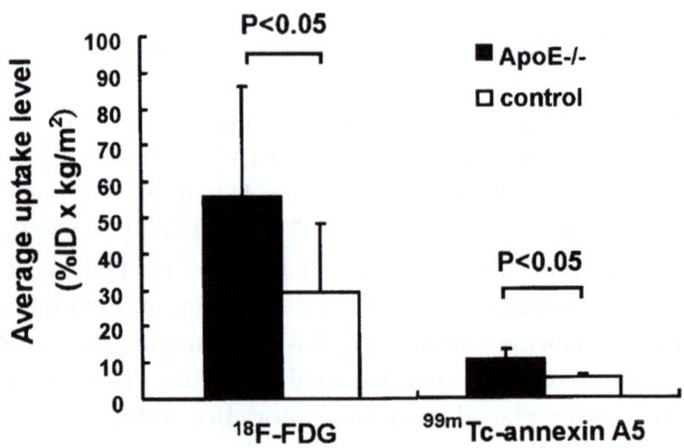

Fig. 1. Uptake levels of ^{18}F-fluorodeoxyglucose (*^{18}F-FDG*) and ^{99m}Tc-annexin A5 in aortic tissues of apolipoprotein E knockout (*ApoE-/-*) and control mice. Both ^{18}F-FDG and ^{99m}Tc-annexin A5 showed higher uptake levels (% injected dose [*ID*] × kg/m^2) in ApoE-/- mice than in control mice. ^{18}F-FDG showed higher absolute uptake levels than ^{99m}Tc-annexin A5 in both strains (Reproduced with permission from [10])

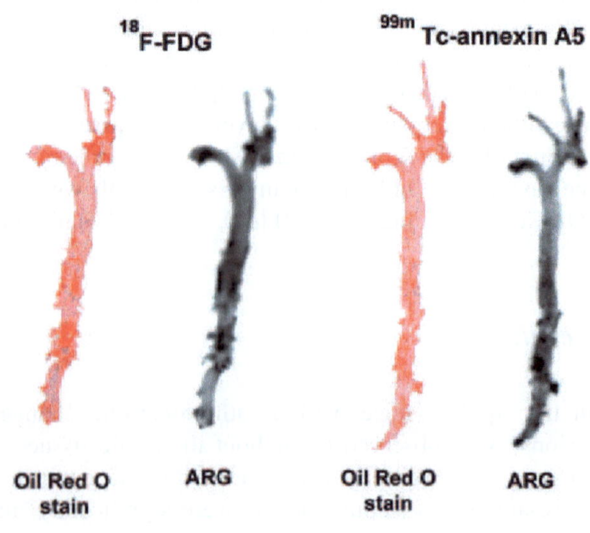

Fig. 2. Autoradiograms (*ARGs*) and corresponding Oil Red O staining in the aortas of ApoE-/- mice. The regions stained red with Oil Red O reveal the presence of atherosclerotic lesions. Corresponding autoradiography showed tracer accumulation in the atherosclerotic lesions (Reproduced with permission from [10])

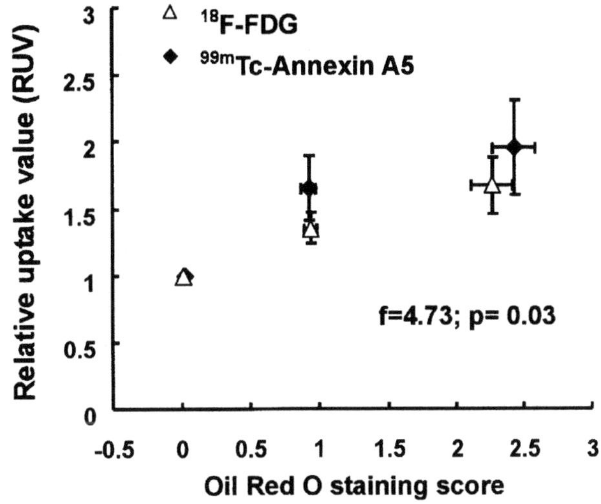

Fig. 3. Relative uptake values (*RUVs*) of tracers in aortic tissues of ApoE-/- mice. The RUVs were defined as the ratio of tracer uptake in each region of interest (ROI) group to that in the ROI group with Oil Red O staining scores of 0–0.5. The uptake ratios of advanced lesions (score, >0.5) to early lesions (score, <0.5) were significantly higher for ^{99m}Tc-annexin A5 than for ^{18}F-FDG (f = 4.73, $P = 0.03$) (Reproduced with permission from [10])

ity of atherosclerotic changes determined by the Oil Red O staining. ^{99m}Tc-annexin A5 accumulation showed better correlation with the severity of atherosclerotic changes ($r = 0.64$; $P < 0.05$) than that of ^{18}F-FDG ($r = 0.56$, $P < 0.05$). The tracer uptake ratios of advanced to early lesions are shown in Fig. 3. ^{18}F-FDG showed relatively low uptake ratios for advanced lesions with higher Oil Red O staining scores (more than 0.5), because of a relatively high uptake in early lesions scored 0 to 0.5. The uptake ratios of ^{99m}Tc-annexin A5 for advanced lesions scored 0.5 to 3 were significantly higher than those of ^{18}F-FDG (f = 4.73; $P = 0.03$).

Discussion

From our comparison between the two tracers, ^{99m}Tc-annexin A5 showed higher correlation between regional tracer uptake level and atherosclerotic severity ($r = 0.65$; $P < 0.05$) than that of ^{18}F-FDG ($r = 0.56$, $P < 0.05$). More importantly, significantly higher uptake levels of ^{99m}Tc-annexin A5 were observed in advanced lesions (data not shown; see reference [10]), indicating the potential of this tracer to identify plaques with higher risk of rupture. The uptake ratios of advanced to early lesions were significantly higher for ^{99m}Tc-annexin A5 than for ^{18}F-FDG (Fig. 3). On the other hand, ^{18}F-FDG showed significantly higher absolute uptake levels in the aortic tissues than those of ^{99m}Tc-annexin A5 (Fig. 1), suggesting that ^{18}F-FDG may be more sensitive for detecting plaques of different severities. Thus, our study using ApoE-/- mice clearly characterized the potential of these tracers.

To further clarify the histological basis of the nuclear imaging on plaque detection with ^{18}F-FDG and ^{99m}Tc-annexin A5, and directly compare the two tracers, we preliminarily performed a dual-tracer autoradiographic study with ^{14}C-FDG and ^{99m}Tc-annexin A5 in ApoE-/- mice [14]. After 20- to 30-week high-cholesterol feeding, tracer uptakes were evaluated in relation to the morphological lesion characteristics [American Heart Association (AHA) lesion phenotypes] and degree of macrophage infiltration. In ApoE-/- mice, plaque size and intimal thickness were increased with lesion progression, while macrophage infiltration was increased from early (type I-II) to atheromatous lesions (type III-IV), but decreased in fibroatheromatous lesions (type V). Higher ^{99m}Tc-annexin A5 uptake was found in atheromatous lesions (0.49 ± 0.20 %ID × kg/mm^2) compared to early (0.40 ± 0.17 %ID × kg/mm^2) or fibroatheromatous (0.30 ± 0.13 %ID × kg/mm^2) lesions, being concordant with macrophage infiltration. ^{14}C-FDG uptake was also the highest in atheromatous lesions (3.1 ± 1.5%ID × kg/mm^2 in atheromatous, 2.8 ± 1.5 %ID × kg/mm^2 in early, and 2.3 ± 1.4 %ID × kg/mm^2 in fibroatheromatous lesions). The uptakes of both ^{99m}Tc-annexin A5 ($r = 0.65$, $P < 0.01$) and ^{14}C-FDG ($r = 0.44$, $P < 0.001$) were positively correlated with the degrees of macrophage infiltration.

Detection of Increased Metabolism of Inflammatory Cells with ^{18}F-FDG

Recently, Rudd et al. [8] investigated the detection of atherosclerotic lesions in the internal carotid artery by ^{18}F-FDG PET in patients, and found that ^{18}F-FDG was markedly accumulated in symptomatic plaques, but not in contralateral asymptomatic plaques. In addition, they incubated ^{3}H-FDG with human atherosclerotic plaques in vitro, and detected high ^{3}H-FDG uptake by macrophages. Ogawa et al. [3] reported that ^{18}F-FDG accumulation in atherosclerotic lesions was correlated with the number of macrophages in the lesions in rabbits with spontaneous atherosclerosis. Tawakol et al. [6] found a high correlation between the accumulations of ^{18}F-FDG and macrophages in 17 patients with severe carotid arterial stenosis who underwent carotid endarterectomy (CEA) within 1 month after ^{18}F-FDG PET. Graebe et al. [15] reported that ^{18}F-FDG uptake in carotid plaques was correlated to the gene expression of CD68 (a macrophage-specific marker) and other known molecular markers of inflammation and plaque vulnerability (interleukin [IL]-18, matrix metalloproteinase [MMP]-9). As data that demonstrated the high reproducibility of ^{18}F-FDG PET on plaque detection, Rudd et al. [16] imaged 20 patients twice, 14 days apart, and found ^{18}F-FDG uptake [standardized uptake value (SUV)] max and plaque-to-background ratio) was highly reproducible in carotid, iliac, and femoral arteries. These findings suggested that ^{18}F-FDG PET allows the quantitative evaluation of inflammatory cells in plaques, particularly macrophages, which may enable the selective imaging of vulnerable atherosclerotic plaques.

Tahara et al. [17] and Ogawa et al. [18] recently investigated the application of ^{18}F-FDG PET to monitor the therapeutic effects of lipid-lowering therapy and antioxidant therapy on atherosclerosis, and found it useful. Tahara et al. [17] randomly allocated 43 patients with arterial ^{18}F-FDG accumulation to groups treated with simvastatin and dietary therapy alone and performed ^{18}F-FDG PET after 3 months, and observed significant reductions of low-density lipoprotein (LDL)-

cholesterol and ^{18}F-FDG accumulation in the simvastatin treatment group. Ogawa et al. [18] treated rabbits exhibiting spontaneous atherosclerosis with an antioxidative agent, probucol, and performed ^{18}F-FDG PET before and 1, 3, and 6 months after the initiation of treatment. They found that ^{18}F-FDG accumulation in the arterial wall slowly increased in the control group, but decreased with the reduction of inflammatory reaction in the treatment group, and the accumulation in the arterial wall was undetectable 6 months after the initiation of treatment. These findings indicate the potential of ^{18}FDG PET to evaluate the therapeutic effects of statins and probucol on atherosclerosis, based on the reduction of inflammatory reactions in aortic tissues.

Imaging of Ongoing Apoptosis with ^{99m}Tc-Annexin A5

The possibility of imaging vulnerable plaques using ^{99m}Tc-annexin A5 has been extensively investigated in animals and patients by Narula's group (Kologdie et al. [9], Johnson et al. [19], and Kietsaler et al. [20]). Recently, we compared ^{99m}Tc-annexin A5 accumulation in plaques with the histological findings in a rabbit model of spontaneous atherosclerosis (myocardial infarction-prone Watanabe heritable hyperlipidemic rabbits [WHHLMI rabbits]), which develop lesions similar to human atherosclerosis [4], to further investigate the relationship between ^{99m}Tc-annexin A5 accumulation and plaque vulnerability. ^{99m}Tc-annexin A5 was injected into the auricular vein in WHHLMI and control rabbits, and SPECT was performed. The aorta in WHHLMI was clearly imaged 3 h after administration, and a significantly higher level of ^{99m}Tc-annexin A5 (5.6 times) was detected in the excised aorta of WHHLMI than in that of the control ($P < 0.0001$). The excised tissues were subjected to autoradiography. Radioactivity accumulation in the aortic sections was detected in the intima, and the accumulation was positively correlated with the density of terminal deoxynucleotidyl transferase-mediated deoxyuridine triphosphate nick-end labeling (TUNEL)-positive cells in the sections ($r = 0.76$, $P < 0.0001$). Furthermore, we classified atherosclerotic lesions into the following four types based on the pathological findings: (1) neointimal lesions showing early adaptive thickening of the intima, (2) vulnerable-like atheromatous lesions, (3) stable fibroatheromatous lesions covered with a thick fibrous cap, and (4) collagen-rich lesions consisting of smooth muscle cells and collagen fibers, and investigated the radioactivity accumulation in each type. The highest radioactivity accumulation was detected in atheromatous lesions, which were considered to be vulnerable-like plaques. These findings suggest the possibility of the selective imaging of vulnerable atherosclerotic plaques using ^{99m}Tc-annexin A5 targeting apoptosis in plaques.

Recently, Hartung et al. [21] treated rabbits that had undergone balloon de-endothelialization with statin and/or diet modification. Quantitative ^{99m}Tc-annexin A5 uptake in the diseased aorta was determined after 3 months of high-cholesterol feeding and 1 month of treatment. They found that ^{99m}Tc-annexin A5 accumulation in the arterial wall was substantially reduced in the diet modification and statin therapy groups, which both showed histological characteristics of stabilization, including the resolution of macrophage infiltration and marked reduction in macrophage apoptosis. These findings indicate the potential of ^{99m}Tc-annexin A5 SPECT for evaluating the therapeutic effects of statin therapy or dietary modification on atherosclerosis, based on the reduction of apoptosis.

Conclusion

Previous studies have indicated that ^{18}F-FDG, a PET probe as a marker of inflammation, and ^{99m}Tc-annexin A5, a SPECT probe as a marker of ongoing apoptotic cell death, are useful for the detection of vulnerable atherosclerotic lesions. Our findings verified the usefulness of ^{18}F-FDG and ^{99m}Tc-annexin A5 as molecular imaging agents for detecting atherosclerotic lesions. Particularly, our findings clarified which tracer is more useful for detecting atherosclerotic lesions and which is more useful for determining lesion vulnerability, as follows: the high aortic uptake level of ^{18}F-FDG may offer higher sensitivity in lesion detection, whereas the preferential uptake of ^{99m}Tc-annexin A5 in advanced lesions suggests its potential for assessing the vulnerability of atherosclerotic plaques.

References

1. Virmani R, Kolodgie FD, Burke AP et al (2000) Lessons from sudden coronary death: a comprehensive morphological classification scheme for atherosclerotic lesions. Arterioscler Thromb Vasc Biol 20: 1262–1275
2. Virmani R, Burke AP, Kolodgie FD et al (2003) Pathology of the thin-cap fibroatheroma: a type of vulnerable plaque. J Interv Cardiol 16: 267–272
3. Ogawa M, Ishino S, Mukai T et al (2004) ^{18}F-FDG accumulation in atherosclerotic plaques: immunohistochemical and PET imaging study. J Nucl Med 45: 1245–1250
4. Ishino S, Kuge Y, Takai N et al (2007) ^{99m}Tc-annexin A5 for noninvasive characterization of atherosclerotic lesions: imaging and histological studies in myocardial infarction-prone Watanabe heritable hyperlipidemic rabbits. Eur J Nucl Med Mol Imaging 34: 889–899
5. Meir KS, Leitersdorf E (2004) Atherosclerosis in the apolipoprotein E-deficient mouse: a decade of progress. Arterioscler Thromb Vasc Biol 24: 1006–1014
6. Tawakol A, Migrino RQ, Bashian GG et al (2006) In vivo ^{18}F-fluorodeoxyglucose positron emission tomography imaging provides a noninvasive measure of carotid plaque inflammation in patients. J Am Coll Cardiol 48: 1818–1824
7. Lederman RJ, Raylman RR, Fisher SJ et al (2001) Detection of atherosclerosis using a novel positron-sensitive probe and 18-fluorodeoxyglucose (FDG). Nucl Med Commun 22: 747–753
8. Rudd JHF, Warburton EA, Fryer TD et al (2002) Imaging atherosclerotic plaque inflammation with [^{18}F]-fluorodeoxyglucose positron emission tomography. Circulation 105: 2708–2711
9. Kolodgie FD, Petrov A, Virmani R et al (2003) Targeting of apoptotic macrophages and experimental atheroma with radiolabeled annexin V: A technique with potential for noninvasive imaging of vulnerable plaque. Circulation 108: 3134–3139

10. Zhao Y, Kuge Y, Zhao S et al (2007) Comparison of ^{99m}Tc-annexin A5 with ^{18}F-FDG for detecting atherosclerosis in ApoE-/- mice. Eur J Nucl Med Mol Imaging 34(11): 1747–1755
11. Plump AS, Smith JD, Hayek T et al (1992) Severe hypercholesterolemia and atherosclerosis in apolipoprotein E-deficient mice created by homologous recombination in ES cells. Cell 71: 343–353
12. Zhang SH, Reddick RL, Piedrahita JA et al (1992) Spontaneous hypercholesterolemia and arterial lesions in mice lacking apolipoprotein E. Science 258: 468–471
13. Zhao S, Kuge Y, Mochizuki T et al (2005) Biologic correlates of intratumoral heterogeneity in 18F-FDG distribution with regional expression of glucose transporters and hexokinase-II in experimental tumor. J Nucl Med 46: 794–799
14. Zhao Y, Kuge Y, Zhao S et al (2008) Macrophage infiltration and apoptosis can be imaged utilizing ^{18}F-FDG and ^{99m}tc-annexin A5: potential indicators of plaque vulnerability. *Circulation* 118:S-102
15. Græbe M, Pedersen SF, Borgwardt L et al (2009) Molecular pathology in vulnerable carotid plaques: correlation with [18]-fluorodeoxyglucose positron emission tomography (FDG-PET). Eur J Vasc Endovasc Surg 37: 714–721
16. Rudd JH, Myers KS, Bansilal S et al (2008) Atherosclerosis inflammation imaging with 18F-FDG PET: carotid, iliac, and femoral uptake reproducibility, quantification methods, and recommendations. J Nucl Med 49: 871–878
17. Tahara N, Kai H, Ishibashi M et al (2006) Simvastatin attenuates plaque inflammation: evaluation by fluorodeoxyglucose positron emission tomography. J Am Coll Cardiol 48: 1825–1831
18. Ogawa M, Magata Y, Kato T et al (2006) Application of ^{18}F-FDG PET for monitoring the therapeutic effect of antiinflammatory drugs on stabilization of vulnerable atherosclerotic plaques. J Nucl Med 47: 1845–1850
19. Johnson LL, Schofield L, Donahay T et al (2005) ^{99m}Tc-annexin V imaging for in vivo detection of atherosclerotic lesions in porcine coronary arteries. J Nucl Med 46: 1186–1193
20. Kietselaer BL, Reutelingsperger CP, Heidendal GA et al (2004) Noninvasive detection of plaque instability with use of radiolabeled annexin A5 in patients with carotid-artery atherosclerosis. N Engl J Med 350: 1472–1473
21. Hartung D, Sarai M, Petrov A et al (2005) Resolution of apoptosis in atherosclerotic plaque by dietary modification and statin therapy. J Nucl Med 46: 2051–2056

Quantification of Myocardial Blood Flow Using Rubidium-82 PET

Keiichiro Yoshinaga[1], Chietsugu Katoh[2], Osamu Manabe[3], Masanao Naya[4], Robert A. deKemp[5], Ran Klein[5], Rob S.B. Beanlands[5], and Nagara Tamaki[3]

Summary

Background Rubidium-82 (^{82}Rb) is a generator-produced positron emission tomography (PET) myocardial perfusion tracer and is widely used at cardiac imaging centers without access to an on-site cyclotron. Quantitative measurements of myocardial blood flow (MBF) have provided additional information over relative flow measurements and have been applied for risk assessment. However, there have been limited reports on MBF measurements using ^{82}Rb and MBF response during the cold pressor test (CPT), which can evaluate coronary endothelial function. The purpose of this study was to develop a method for the quantitative assessment of MBF using ^{82}Rb PET.

Methods The two-dimensional acquisition mode was used for all PET emission scans. MBF was assessed at rest and during the CPT with ^{82}Rb in nine controls. A one-compartment model was used to estimate MBF.

Results The rate pressure product was significantly increased during CPT (6152.7 ± 831.9 to 9757.8 ± 2258.7; $P < 0.001$). MBF was also significantly increased during CPT (0.62 ± 0.20 to 1.02 ± 0.36 ml/min/g; $P = 0.013$). The percent increase of MBF during CPT was 69.8% ± 52%.

Conclusions ^{82}Rb MBF measurements detected increased flow response during CPT in normal volunteers. Thus, ^{82}Rb has the potential for wide clinical application, including the measurement of MBF.

Key words Coronary · Endothelium · Tomography

---

[1]Department of Molecular Imaging, Hokkaido University Graduate School of Medicine, Kita 15 Nishi 7, Kita-ku, Sapporo 060-8638, Japan
[2]Department of Health Sciences, Hokkaido University Graduate School of Medicine, Kita 15 Nishi 7, Kita-ku, Sapporo 060-8638, Japan
[3]Department of Nuclear Medicine, Hokkaido University Graduate School of Medicine, Kita 15 Nishi 7, Kita-ku, Sapporo 060-8638, Japan
[4]Department of Cardiology, Hokkaido University Graduate School of Medicine, Kita 15 Nishi 7, Kita-ku, Sapporo 060-8638, Japan
[5]Division of Cardiology, University of Ottawa Heart Institute, 40 Ruskin Street, Ottawa, Ontario, K1Y 4W7 Canada

N. Tamaki, Y. Kuge (eds.)
Molecular Imaging for Integrated Medical Therapy and Drug Development.
DOI: 10.1007/978-4-431-98074-2_9, ©Springer 2010

Introduction

Coronary artery disease is considered a major cause of death both in modern industrialized societies and in developing countries. Techniques for evaluating myocardial blood flow (MBF) have an important role to play in the diagnosis of coronary artery disease and prediction of its prognosis [1, 2]. MBF measurements can provide additional significant functional information over conventional relative perfusion images in patients with coronary artery disease [3, 4].

Positron emission tomography (PET) is an advanced nuclear imaging modality, and has been applied for cardiac imaging [5]. PET can noninvasively measure MBF [6] and, combined with the cold pressor test (CPT), can estimate coronary endothelial function [3, 7, 8]. Most PET MBF measurements have been performed using either ^{15}O water or ^{13}N ammonia. Both of these tracers require immediate access to an on-site cyclotron [2, 4, 9, 10]. This makes it difficult to carry out risk assessment or therapeutic evaluations in larger study populations. Rubidium-82 (^{82}Rb) is a PET perfusion tracer produced from a strontium-82 (^{82}Sr)/ ^{82}Rb generator and is widely used for the diagnosis of coronary artery disease at centers without immediate access to a cyclotron [11–13]. Thus, ^{82}Rb perfusion studies may be applied in larger populations [2, 14]. However, the data on MBF quantification using ^{82}Rb are limited [15–17]. There has also been very little data evaluating the MBF response during CPT for the measurement of coronary endothelial function by ^{82}Rb PET [1]. The purpose of the present study was to evaluate MBF at rest and during CPT.

Subjects, Materials, and Methods

Study Population

Nine healthy volunteers participated in the study. All participants had a normal resting electrocardiogram and did not have a history of cardiovascular disease, nor were they taking any cardiac medications.

Study Protocol

Each subject underwent rest and CPT imaging using ^{82}Rb PET (see Fig. 1 for the mechanism of ^{82}Rb accumulation). The study was approved by the Hokkaido University Graduate School of Medicine Human Research Ethics Board. Written informed consent was obtained from all subjects.

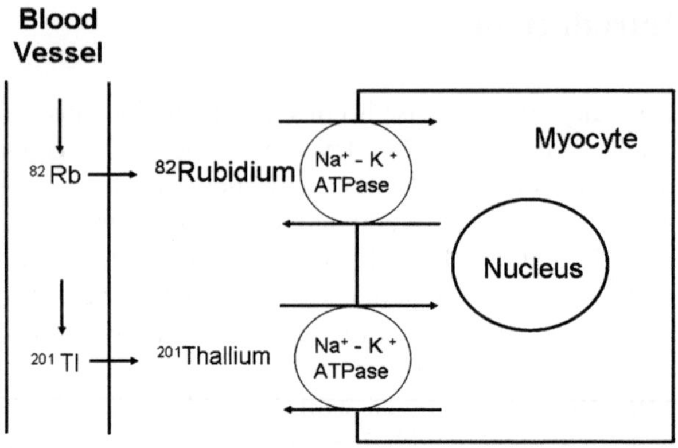

Fig. 1. Mechanism of rubidium-82 (*^{82}Rb*) accumulation. *ATPase*, adenosine triphosphatase

Administration Dose-Finding Study

To determine the appropriate ^{82}Rb administration dose for two-dimensional dynamic image acquisition, one volunteer had a low and a standard dose of ^{82}Rb (480 MBq and 1600 MBq). The low dose (480 MBq) was based on previous three-dimensional ^{82}Rb dynamic studies [4, 15, 18]. The standard dose, 1600 Mbq, was based on the American Society of Nuclear Cardiology guidelines for relative perfusion imaging [9]. The quality of left ventricular myocardium input function was evaluated in both studies.

Data Acquisition

The two-dimensional acquisition mode was used for all emission and transmission scans. Subjects were positioned in a whole-body PET scanner (ECAT HR+; Siemens/CTI, Knoxville, TN, USA).

Immediately following the 6-min transmission scan, 1480 MBq of ^{82}Rb (Bracco Diagnostics, Princeton, NJ, USA; see Fig. 2 for overview of ^{82}Sr/ ^{82}Rb generator) was administered intravenously over 1 min [16]. A 10-min, 24-frame dynamic scan was initiated with increasing frame duration (15 × 6 s, 5 × 12 s, 1 × 30 s, 1 × 60 s, 1 × 120 s, 1 × 240 s) [2, 15, 16, 18].

CPT PET Imaging

The subjects were instructed to abstain from caffeine-containing products for more than 24 h before the PET studies [19–21]. For the CPT, the subject's foot was put into ice water up to the ankle [19, 22]. PET data acquisition started 60 s after the

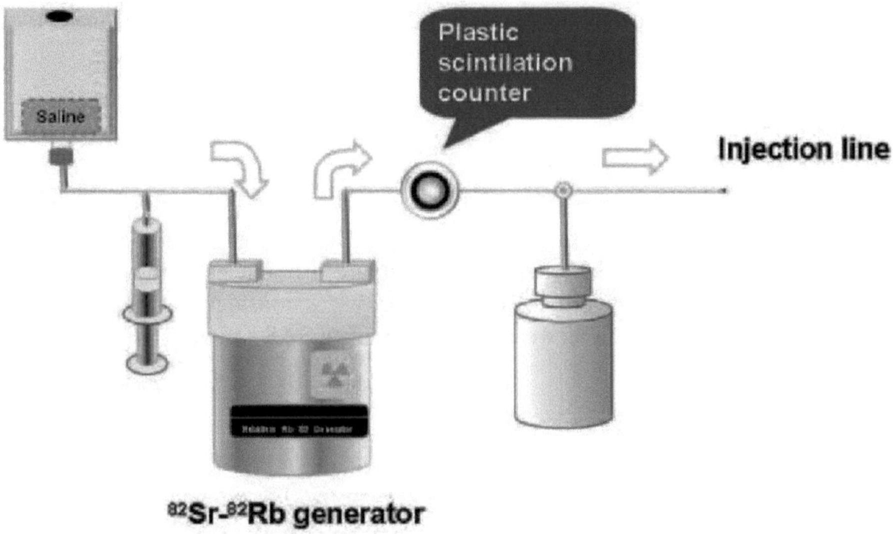

Fig. 2. Overview of ^{82}Sr/ ^{82}Rb generator

beginning of CPT and the CPT continued for 4 min during PET data acquisition [19, 22].

MBF Data Analysis

The left ventricular cavity time-activity curve was used as the input function. The myocardial time-activity curves were fitted by a one-tissue-compartment model [4, 15].

The early-dynamic phase ^{82}Rb images were used to define a region of interest (ROI) in the left ventricle (LV) blood pool. The myocardial uptake images were calculated by adding the late-phase data from 4 to 6 min. We modified a semiautomatic program developed b y investigators at our facility and the LV region of interest (ROI) was set semiautomatically [23].

The following equation used to estimate the inflow rate (K1) of ^{82}Rb into myocardium, Ct(t), was as follows [15]:

$$dCt(t)/dt = K1 \cdot Ca(t) - k2 \cdot Ct(t)$$

where k2 is the outflow rate from myocardium into the blood ,Ca(t).

After obtaining the K1 value, MBF was derived with a modified Renkin-Crone model [15, 16].

MBF at rest and MBF during CPT were corrected for the rate pressure product (RPP) as rest MBF × (normal mean RPP/individual RPP), because these MBF values are closely related to cardiac work [22, 24, 25]. The normal mean RPP at rest in our data base is 8150 [16, 21].

Statistical Analysis

Continuous measures are presented as means ± SDs. For intragroup comparisons, Wilcoxon signed rank tests were used to compare differences within groups for continuous variables. A P value of less than 0.05 was considered statistically significant. Statistical calculations were carried out using SAS software (SAS Institute, Cary, NC, USA).

Results

Subject Characteristics

The baseline characteristics of the nine subjects are shown in Table 1.

^{82}Rb Administration Dose

The time-activity curves of LV blood cavity and LV myocardium are shown in Fig. 3. The standard dose, 1600 MBq, showed lower noise in both LV cavity and LV myocardium than the lower dose (480 MBq).

Hemodynamics

CPT stimulation significantly increased heart rate (54.1 ± 5.0 to 67.6 ± 11.3 beats/min; $P = 0.0039$), systolic blood pressure (114.0 ± 7.7 to 144.8 ± 19.4 mm Hg; $P < 0.001$), and RPP (6152.7 ± 831.9 to 9757.8 ± 2258.7; $P < 0.001$).

Table 1. Baseline characteristics

	Control ($n = 9$)
Age (years)	36.9 ± 11.9*
Gender (M/F)	9/0
BMI	22.7 ± 2.9

BMI, body mass index

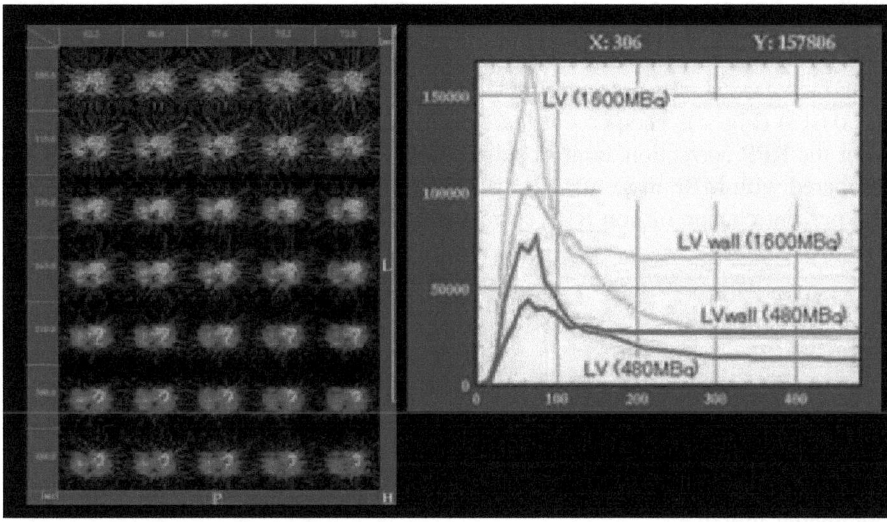

Fig. 3. Time activity curves of ^{82}Rb at lower dose (480 MBq) and standard dose (1600 MBq). *LV*, left ventricle

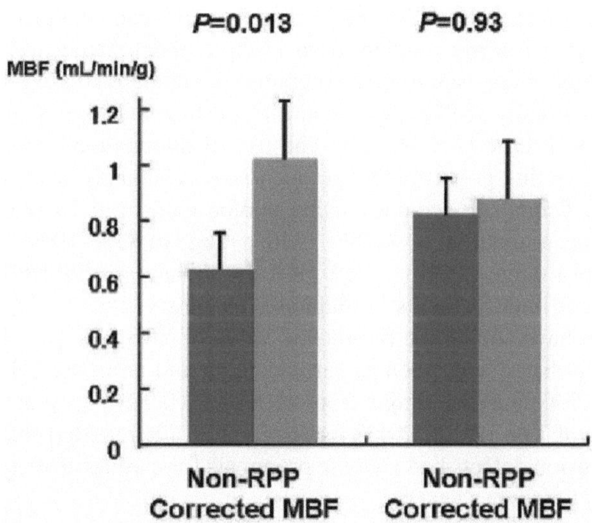

Fig. 4. Myocardial blood flow (*MBF*; ml/min per g; *blue bars*) at rest and during cold pressor test (CPT; *red bars*). *RPP*, rate pressure product

At-Rest MBF and MBF During CPT

MBF increased significantly during CPT (0.62 ± 0.20 to 1.02 ± 0.36 ml/min per g; $P = 0.013$) (Fig. 4). The percent increase in MBF during CPT was 69.8% ± 52%. After the RPP correction, control subjects showed no change in MBF during CPT compared with MBF at rest (0.82 ± 0.22 to 0.0.88 ± 0.31 ml/min per g; $P = 0.93$). The percent change of non-RPP corrected MBF during CPT to the rest MBF was 10.9 ± 37.0%.

Discussion

In the present study, ^{82}Rb MBF measurements detected an increased MBF response during CPT in normal volunteers, demonstrating that generator-produced ^{82}Rb PET MBF measurement combined with the CPT has potential as a marker of coronary endothelial function.

^{82}Rb PET Relative Perfusion Imaging

^{82}Rb is produced from a Sr-82/Rb-82 generator which can be eluted every 10 min. ^{82}Rb PET imaging has been widely used in the clinical setting in North America [9]. ^{82}Rb has high diagnostic ability compared to standard single photon emission computed tomography (SPECT) myocardial perfusion imaging in patients with coronary artery disease [11, 26, 27]. The use of attenuation correction plays an important role in this good diagnostic accuracy. Beanlands et al. [12] evaluated perfusion PET diagnostic accuracy using pooled data from 14 studies with 1460 patients. The mean sensitivity was 89%, with a range of 83%–100%, and specificity was 89%, with a range of 73%–100%. Recent advances in PET/CT are expected to lead to further improvements in diagnostic accuracy.

Recent data have shown the prognostic value of ^{82}Rb PET [2, 14, 26, 28]. The rates of hard cardiac events such as cardiac death and non-fatal MI are low in the normal ^{82}Rb PET imaging range from 0.09% to 0.9% per year depending on the patient population [2, 28, 29]. Yoshinaga et al. [2] reported that patients with ^{82}Rb PET perfusion defects had a worse prognosis for cardiac death (4.3% per year) or hard events (7.0% per year).

Administration Dose-Finding Study

Most previous ^{82}Rb MBF quantification studies used three-dimensional data acquisition [4, 15, 18, 30]. These studies used a relatively low dose (230 MBq) for dynamic data acquisition because ^{82}Rb has high noise, and increasing noise may attenuate MBF. We have developed MBF quantification using oxygen-15 labeled water PET with two-dimensional data acquisition [19, 21, 23]. Two-dimensional data acquisition can reduce noise and can use higher doses of tracer administration compared to three-dimensional data acquisition. Thus, we compared a relatively low dose and the standard dose. The standard dose (1600 MBq) showed less noise with either the LV cavity or the LV myocardium time-activity curve. Therefore, we selected the standard ^{82}Rb administration dose for the present study.

MBF Measurements Using ^{82}Rb PET

^{82}Rb PET has high diagnostic accuracy; however, relative perfusion imaging has its limitations, especially in multivessel disease [13]. Parkash et al. [31] reported reduced MBF in segments without obvious perfusion defects by relative perfusion image in patients with multivessel coronary disease. Yoshinaga et al. [21] reported that coronary flow reserve was reduced in remote segments without coronary artery stenosis in patients with coronary artery disease. These two studies established the clinical importance of MBF quantification. Coronary flow reserve is also reduced in subjects with several coronary risk factors such as smoking, hypertension, diabetes, and hyperlipidemia [1, 3]. MBF quantification also plays an important role in risk assessment and the evaluation of therapeutic intervention.

^{15}O water and ^{13}N ammonia have been used in MBF quantification [1, 3]. However, there are limited data reported on MBF quantification using ^{82}Rb [15–18, 32].

The initial approach for MBF quantification using ^{82}Rb was a simple net retention model [17]. The net retention equals the measured myocardial tracer concentrations at time T divided by the integral of the blood [17] input curve to that certain time. This model is simple; however, the data are dependent on the time at which the model equation is estimated. Lortie et al. [15] applied a single-tissue compartment model to estimate absolute MBF using ^{82}Rb. A mathematical model is applied with parameters describing the flux of radioactivity between the compartments. This approach has been applied for ^{15}O water and is considered the standard approach for MBF quantification [23, 33]. Lortie et al. [15] compared the absolute MBF value with ^{82}Rb and ^{13}N ammonia. They found that ^{82}Rb MBF correlated with MBF measured by ^{13}N ammonia ($r = 0.85$), thus validating their measurements. Manabe et al. [16] evaluated the repeatability of rest and hyperemic MBF measurements using ^{82}Rb. There was no significant difference in repeated at-rest MBF values (0.77 ± 0.25 vs 0.82 ± 0.25 ml/min per g; $P = 0.31$; mean difference, 6.18 ± 12.22%) or repeated hyperemic MBF values (3.35 ± 1.37 vs 3.39 ± 1.37 ml/min per g; $P = 0.81$;, mean difference, 1.17 ± 13.64%). Thus, MBF measurement using ^{82}Rb is highly reproducible. Table 2 summarizes normal MBF values determined with ^{82}Rb in five studies with a total of 62 patients. The at-rest MBF in the present study agrees with the values in the previous studies, indicating the validation of our measurements.

Table 2. Normal MBF values at rest and hyperemic MBF, using rubidium-82 (^{82}Rb) in normal subjects

Reference	Stress agent	Model	Number of subjects	Age (years)	MBF at rest (ml/min/g)	MBF at hyperemia (ml/min/g)	CFR
Lin [39]	Dipy	Two-compartment	11	44	1.15 ± 0.46	2.50 ± 0.54	(-)
Wassenaar [40]	Dipy	Net retention	15	34 ± 6	0.95 ± 0.35	3.0 ± 0.70	3.2 ± 0.8
Lortie [15]	Dipy	One-compartment	14	23–44	0.69 ± 0.14	2.83 ± 0.81	4.25 ± 1.37
Manabe [16]	ATP	One-compartment	15	29 ± 9	0.77 ± 0.25	3.35 ± 1.37	4.47 ± 1.47
Anagnostopoulos [32]	Dipy	Two-compartment	7	56 ± 9	0.9	(-)	2.2
Total	Weighted mean		62		0.877	2.95	3.72
	Mean		62		0.892	2.92	3.53

MBF, myocardial blood flow; CFR, coronary flow reserve; Dipy, dipyridamole

^{82}*Rb PET Endothelial Function Measurements*

We have reported coronary endothelial dysfunction in smokers using ^{15}O water PET [34, 35] and we also evaluated therapeutic interventions [19, 35]. ^{13}N ammonia PET has been widely applied for the measurement of coronary endothelial function [8, 36, 37]. However, there have been limited data on MBF quantification using compartment models with ^{82}Rb, [15, 16] and there are no previous data evaluating the MBF response during the cold pressor test (CPT) for the measurement of coronary endothelial function. In the current study, the non-RPP corrected ^{82}Rb MBF was significantly increased during CPT. This finding agreed with previous reports using ^{15}O-labeled water [22, 34]. After the RPP correction, this significant change disappeared, this result being similar to that in a previous study by Furuyama et al. [22]. Thus, the present data have further extended the previous studies regarding ^{82}Rb MBF measurements in normal subjects and demonstrate the potential of the new endothelial function testing using ^{82}Rb.

The myocardial extraction of ^{82}Rb is similar to that of ^{201}Thallium. The net extraction of ^{82}Rb is around 50% at rest and decreases with increasing MBF [38]. Lortie et al. [15] evaluated ^{82}Rb MBF quantification using a single-compartment model and compared it to ^{13}N ammonia PET. The difference between ^{82}Rb and ^{13}N ammonia was small for at-rest MBF. Therefore, it is considered that, theoretically, ^{82}Rb MBF measurements may be more accurate in the lower-flow range [38]. CPT stress increases MBF by 30% to 50% compared to at-rest MBF, and usually MBF during CPT is less than 2.0 ml/min per g. [1] In the present study, non-RPP corrected MBF for ^{82}Rb at rest and during CPT was less than 1.1 ml/min per g. These MBF values at rest and during CPT are quite suitable for ^{82}Rb MBF measurements. Thus, we expect that ^{82}Rb can accurately estimate MBF during CPT.

Study Limitations

This study assessed the MBF in the whole LV myocardium, but not in regional myocardial segments. Further evaluations are needed to estimate regional MBF during CPT stress in patients with coronary artery disease, where additional regional heterogeneity may be expected.

Conclusions

^{82}Rb MBF measurements detected an increased flow response during CPT in normal volunteers. Thus, ^{82}Rb PET myocardial perfusion imaging may be applicable for risk assessment or for the evaluation of therapeutic interventions in subjects with coronary risk factors.

Acknowledgments The authors thank Sayaka Takamori, RT, Keiichi Magota, RT, Naoko Mizuno, RT, and Hiroshi Arai, RT, for their technical expertise and Mayumi Komoribayashi and Eriko Suzuki for their administrative support of this study.

Sources of Funding

This study was supported in part by grants from the Ministry of Education, Science and Culture (No.19591395) and the Northern Advancement Center for Science & Technology (Sapporo, Japan) (Grant #H19-C-068). Ran Klein was supported by the JSPS and NSERC Summer Program (2008) (Tokyo, Japan, and Ottawa, Ontario, Canada).

Disclosures

None.

References

1. Yoshinaga K, Chow BJ, deKemp RA et al (2005) Application of cardiac molecular imaging using positron emission tomography in evaluation of drug and therapeutics for cardiovascular disorders. Curr Pharm Des 11: 903–932
2. Yoshinaga K, Chow BJ, Williams K et al (2006) What is the prognostic value of myocardial perfusion imaging using rubidium-82 positron emission tomography? J Am Coll Cardiol 48: 1029–1039
3. Camici PG, Crea F (2007) Coronary microvascular dysfunction. N Engl J Med 356: 830–840
4. deKemp RA, Yoshinaga K, Beanlands RS (2007) Will three-dimensional PET-CT enable the routine quantification of myocardial blood flow? J Nucl Cardiol 14: 380–397
5. Hachamovitch R, Di Carli MF (2008) Methods and limitations of assessing new noninvasive tests: part I: Anatomy-based validation of noninvasive testing. Circulation 117: 2684–2690
6. Knuuti J, Bengel FM (2008) Positron emission tomography and molecular imaging. Heart 94: 360–367
7. Bengel FM, Abletshauser C, Neverve J et al (2005) Effects of nateglinide on myocardial microvascular reactivity in type 2 diabetes mellitus–a randomized study using positron emission tomography. Diabet Med 22: 158–163
8. Di Carli MF, Tobes MC, Mangner T et al (1997) Effects of cardiac sympathetic innervation on coronary blood flow. N Engl J Med 336: 1208–1215
9. Machac J (2005) Cardiac positron emission tomography imaging. Semin Nucl Med 35: 17–36
10. Yoshinaga K, Tamaki N, Ruddy T et al (2009) Evaluation of myocardial perfusion. In: Wahl R (ed) Principles and practice of PET and PET/CT, 2nd edn. Lippincott Williams and Wilkins, Philadelphia, PA, pp 541–564
11. Bateman TM, Heller GV, McGhie AI et al (2006) Diagnostic accuracy of rest/stress ECG-gated Rb-82 myocardial perfusion PET: comparison with ECG-gated Tc-99m sestamibi SPECT. J Nucl Cardiol 13: 24–33

12. Beanlands RS, Chow BJ, Dick A et al (2007) CCS/CAR/CANM/CNCS/CanSCMR joint position statement on advanced noninvasive cardiac imaging using positron emission tomography, magnetic resonance imaging and multidetector computed tomographic angiography in the diagnosis and evaluation of ischemic heart disease–executive summary. Can J Cardiol 23: 107–119
13. Sampson UK, Dorbala S, Limaye A et al (2007) Diagnostic accuracy of rubidium-82 myocardial perfusion imaging with hybrid positron emission tomography/computed tomography in the detection of coronary artery disease. J Am Coll Cardiol 49: 1052–1058
14. Lertsburapa K, Ahlberg AW, Bateman TM et al (2008) Independent and incremental prognostic value of left ventricular ejection fraction determined by stress gated rubidium 82 PET imaging in patients with known or suspected coronary artery disease. J Nucl Cardiol 15: 745–753
15. Lortie M, Beanlands RS, Yoshinaga K et al (2007) Quantification of myocardial blood flow with 82Rb dynamic PET imaging. Eur J Nucl Med Mol Imaging 34: 1765–1774
16. Manabe O, Yoshinaga K, Katoh C et al (2009) Repeatability of rest and hyperemic myocardial blood flow measurements with Rb-82 dynamic PET. J Nucl Med 50: 68–71
17. Yoshida K, Mullani N, Gould KL (1996) Coronary flow and flow reserve by PET simplified for clinical applications using rubidium-82 or nitrogen-13-ammonia. J Nucl Med 37: 1701–1712
18. Yoshinaga K, Beanlands RS, Dekemp RA et al (2006) Effect of exercise training on myocardial blood flow in patients with stable coronary artery disease. Am Heart J 151: 1324 e11–1328
19. Naya M, Tsukamoto T, Morita K et al (2007) Olmesartan, but not amlodipine, improves endothelium-dependent coronary dilation in hypertensive patients. J Am Coll Cardiol 50: 1144–1149
20. Siegrist PT, Gaemperli O, Koepfli P et al (2006) Repeatability of cold pressor test-induced flow increase assessed with H(2)(15)O and PET. J Nucl Med 47: 1420–6
21. Yoshinaga K, Katoh C, Noriyasu K et al (2003) Reduction of coronary flow reserve in areas with and without ischemia on stress perfusion imaging in patients with coronary artery disease: a study using oxygen 15-labeled water PET. J Nucl Cardiol 10: 275–283
22. Furuyama H, Odagawa Y, Katoh C et al (2002) Assessment of coronary function in children with a history of Kawasaki disease using (15)O-water positron emission tomography. Circulation 105: 2878–2884
23. Katoh C, Morita K, Shiga T et al (2004) Improvement of algorithm for quantification of regional myocardial blood flow using 15O-water with PET. J Nucl Med 45: 1908–1916
24. Campisi R, Czernin J, Schoder H et al (1998) Effects of long-term smoking on myocardial blood flow, coronary vasomotion, and vasodilator capacity. Circulation 98: 119–125
25. Czernin J, Muller P, Chan S et al (1993) Influence of age and hemodynamics on myocardial blood flow and flow reserve. Circulation 88: 62–69
26. Marwick TH, Go RT, MacIntyre WJ et al (1991) Myocardial perfusion imaging with positron emission tomography and single photon emission computed tomography: frequency and causes of disparate results. Eur Heart J 12: 1064–1069
27. Tamaki N, Ruddy TD, deKemp RA et al (2002) Myocardial perfusion. In: Wahl R (ed) Principles and practice of positron emission tomography. Lippincott Williams and Wilkins, Philadelphia, PA, pp 320–333
28. Chow BJ, Wong JW, Yoshinaga K et al (2005) Prognostic significance of dipyridamole-induced ST depression in patients with normal 82Rb PET myocardial perfusion imaging. J Nucl Med 46: 1095–1101
29. Marwick TH, Shan K, Go RT et al (1995) Use of positron emission tomography for prediction of perioperative and late cardiac events before vascular surgery. Am Heart J 130: 1196–1202
30. deKemp RA, Ruddy TD, Hewitt T et al (2000) Detection of serial changes in absolute myocardial perfusion with 82Rb PET. J Nucl Med 41: 1426–1435

31. Parkash R, deKemp RA, Ruddy Td T et al (2004) Potential utility of rubidium 82 PET quantification in patients with three-vessel coronary artery disease. J Nucl Cardiol 11: 440–449
32. Anagnostopoulos C, Almonacid A, El Fakhri G et al (2008) Quantitative relationship between coronary vasodilator reserve assessed by 82Rb PET imaging and coronary artery stenosis severity. Eur J Nucl Med Mol Imaging 35: 1593–1601
33. Iida H, Kanno I, Takahashi A et al (1988) Measurement of absolute myocardial blood flow with H215O and dynamic positron-emission tomography. Strategy for quantification in relation to the partial-volume effect. Circulation 78: 104–115
34. Iwado Y, Yoshinaga K, Furuyama H et al (2002) Decreased endothelium-dependent coronary vasomotion in healthy young smokers. Eur J Nucl Med Mol Imaging 29: 984–990
35. Morita K, Tsukamoto T, Naya M et al (2006) Smoking cessation normalizes coronary endothelial vasomotor response assessed with 15O-water and PET in healthy young smokers. J Nucl Med 47: 1914–1920
36. Prior JO, Quinones MJ, Hernandez-Pampaloni M et al (2005) Coronary circulatory dysfunction in insulin resistance, impaired glucose tolerance, and type 2 diabetes mellitus. Circulation 111: 2291–2298
37. Schindler TH, Nitzsche EU, Schelbert HR et al (2005) Positron emission tomography-measured abnormal responses of myocardial blood flow to sympathetic stimulation are associated with the risk of developing cardiovascular events. J Am Coll Cardiol 45: 1505–1512
38. Huang SC, Williams BA, Krivokapich J et al (1989) Rabbit myocardial 82Rb kinetics and a compartmental model for blood flow estimation. Am J Physiol 256: H1156–1164
39. Lin JW, Sciacca RR, Chou RL et al (2001) Quantification of myocardial perfusion in human subjects using 82Rb and wavelet-based noise reduction. J Nucl Med 42: 201–208
40. Wassenaar R, Beanlands R, Ruddy T et al (2002) Three dimensional cardiac positron emission tomography. Res Adv Nucl Med 1: 51–60

Part II
Molecular Imaging for Regenerative Treatment

New Concepts for Molecular and Functional Imaging of the Heart: Implications for Regenerative Treatments

Ignasi Carrió[1], Manel Ballester[2], and Albert Flotats[1]

Summary

New advances in the knowledge of myocardial structure and function at genetic, molecular, and microscopic levels bring new opportunities for drug discovery and integrated therapies. The helical ventricular myocardial band is a new concept in our understanding of the functional architecture of the ventricular myocardium. The previous concept of constriction is replaced by the understanding that clockwise and counterclockwise helical motions are necessary to cause the predominant twisting motion. This new model allows for novel understanding of cardiac mechanisms and provides new insights for new treatments such as the restoration or de-novo engineering of heart structures. Recent Fourier analysis of the sequence of ventricular mechanical activation has confirmed a base-to-apex sequence of activation. This is in agreement with recent studies showing the temporal evolution of the three-dimensional strain maps derived from magnetic resonance imaging, which allow mapping of the electromechanical motion of the ventricles. There are several fields in which the impact of these new concepts may be of importance. Imaging online of myocardial fiber dynamics through resonance tensor diffusion imaging, or other imaging means, opens the way to research and the development of more adequate surgical strategies to compensate for cardiac dysfunction. In surgery for heart failure, it seems appropriate to plan surgery in such a way as to spare the helical ventricular anatomy. In cardiac resynchronization, knowledge of the natural sequence of electromechanical ventricular activation should open the way to study the effects of selective stimulation of the band segments on ventricular function, and should provide a rationale for improved ventricular pacing protocols. Coupling cardiac (123)I-metaiodobenzylguanidine (MIBG) studies with functional assessment, taking into account the newly discovered complex mechanics of ventricular activation, should improve the selection of candidates for cardiac resynchronization and the assessment of this treatment. In addition, the eventual site of administration

[1]Department of Nuclear Medicine, Hospital Sant Pau: Autonomous University of Barcelona, Pare Claret 167, 08025-Barcelona, Spain
[2]University of Lleida, Spain

N. Tamaki, Y. Kuge (eds.)
Molecular Imaging for Integrated Medical Therapy and Drug Development.
DOI: 10.1007/978-4-431-98074-2_10, ©Springer 2010

of stem/progenitor cells could be decided on the basis of knowledge of the new model of myocardial fiber distribution. Integrated imaging technologies will bridge the newly discovered structural and functional phenomena with new integrated, targeted, and tailored therapies.

Key words Myocardial band · Cardiac imaging · Regenerative treatments

Introduction

We live in an era of progress in the understanding of myocardial structure and function at genetic, molecular, and microscopic levels, bringing new opportunities for drug discovery and integrated therapies.

The helical ventricular myocardial band is a revolutionary new concept in our understanding of the functional architecture of the ventricular myocardium. Recent investigations have disclosed that the myocardium is formed by a single muscular band that expands from the pulmonary artery to the aorta and spatially adopts a double-loop helicoid structure [1–3]. The muscle fibers originating from the pulmonary artery and the right ventricle can be followed visually to the basal portion of the left ventricle, at which point they descend to the apex in a spiral way and finally ascend to the base of the heart and eventually reach the aorta. In consequence, the previous concept of constriction is replaced by an understanding that clockwise and counterclockwise helical motions are necessary to cause the predominant twisting motion.

Expansion of information from this model allows a novel understanding of cardiac mechanisms, clarifying why a component of ventricular suction involves a systolic event, elucidating septal function, and determining diastolic dysfunction, as well as leading to the establishment of new treatments such as the restoration or de-novo engineering of heart structures. Further, knowledge of the helical structure influences the understanding of atrioventricular and biventricular pacing, and creates novel methods for introducing septal pacing stimuli. In this study, cardiac molecular and functional imaging methods are surveyed to characterize these new pathophysiologic concepts.

The complexity of ventricular myocardial fiber distribution has appealed to investigators for a long time. We now know that the complex structure of the ventricular myocardium is due to the double-loop helical orientation of the single muscular band that extends from the pulmonary artery to the aorta, with a 180° twist in its middle part (Fig. 1) [1–3]. The band is deemed to be a consequence of changes in the circulatory system through evolution, evolving to an extravasation of blood into a self-contained pumping structure that constitutes the ventricles.

The precise contribution of each segment of the band to the myocardial thickness in each plane can be measured (Fig. 2) [3]. In the long-axis section, the right segment (shown in blue in Fig. 2) is well delineated and constitutes 100% of the thickness of the free wall of the right ventricle. The left segment (shown in red in

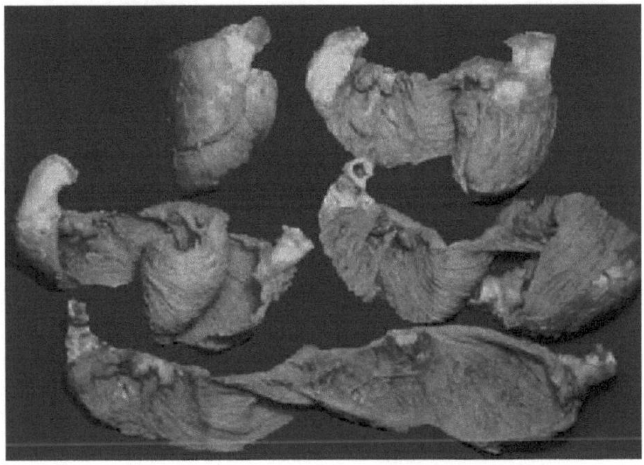

Fig. 1. Stages of the unfolding of the ventricular myocardium into a single myocardial band that has a 180° twist in its middle portion. In the extended myocardial band, the myocardial fibers can be followed from the pulmonary artery to the aorta

Fig. 2) occupies a small area in the most posterobasal portion of the free wall of the left ventricle and overlies the ascendant segment (shown in green in Fig. 2). At the level of the midseptum, the ascendant and descendant segments contribute to septal thickness. In the four-chamber section, the descendant segment occupies the inner surface of the left ventricle and is surrounded by the ascendant segment. In the lateral wall, the proportion occupied by the ascendant segment is in contrast with the small contribution of the basal segment. In the short-axis section, a complex distribution of the band segments is observed. At the base of the heart, the right and left segments are in continuity and enclose both descendant and ascendant segments. In the middle portion of the ventricles, the ascendant segment wraps around the descendant segment almost 360°; at this section, the basal loop is barely seen. At the level of the apex, the segmental distribution is similar to that in the middle portion of the ventricles. Therefore, the left ventricular myocardium is not a homogeneous structure but is formed by different overlapping ventricular segments that constitute the myocardial walls.

Functional Imaging Studies

The unique anatomy of the reciprocal helical configuration of the ventricular myocardial band determines the way the ventricles exert their double function of ejection and suction of the blood. Recent Fourier analysis of the sequence of ventricular mechanical activation has confirmed a base-(right and left)-to-apex sequence of activation (Fig. 3) [4]. At the beginning of the sequence, a signal invariably appears in the most basal portion of the right ventricle, at the level of the pulmonary infundibulum. This signal gradually spreads to involve the entire right ventricular wall. After the right ventricular mechanical motion starts, but is not yet completed, the activity appears in the basal portion of the left ventricle and fully extends to the

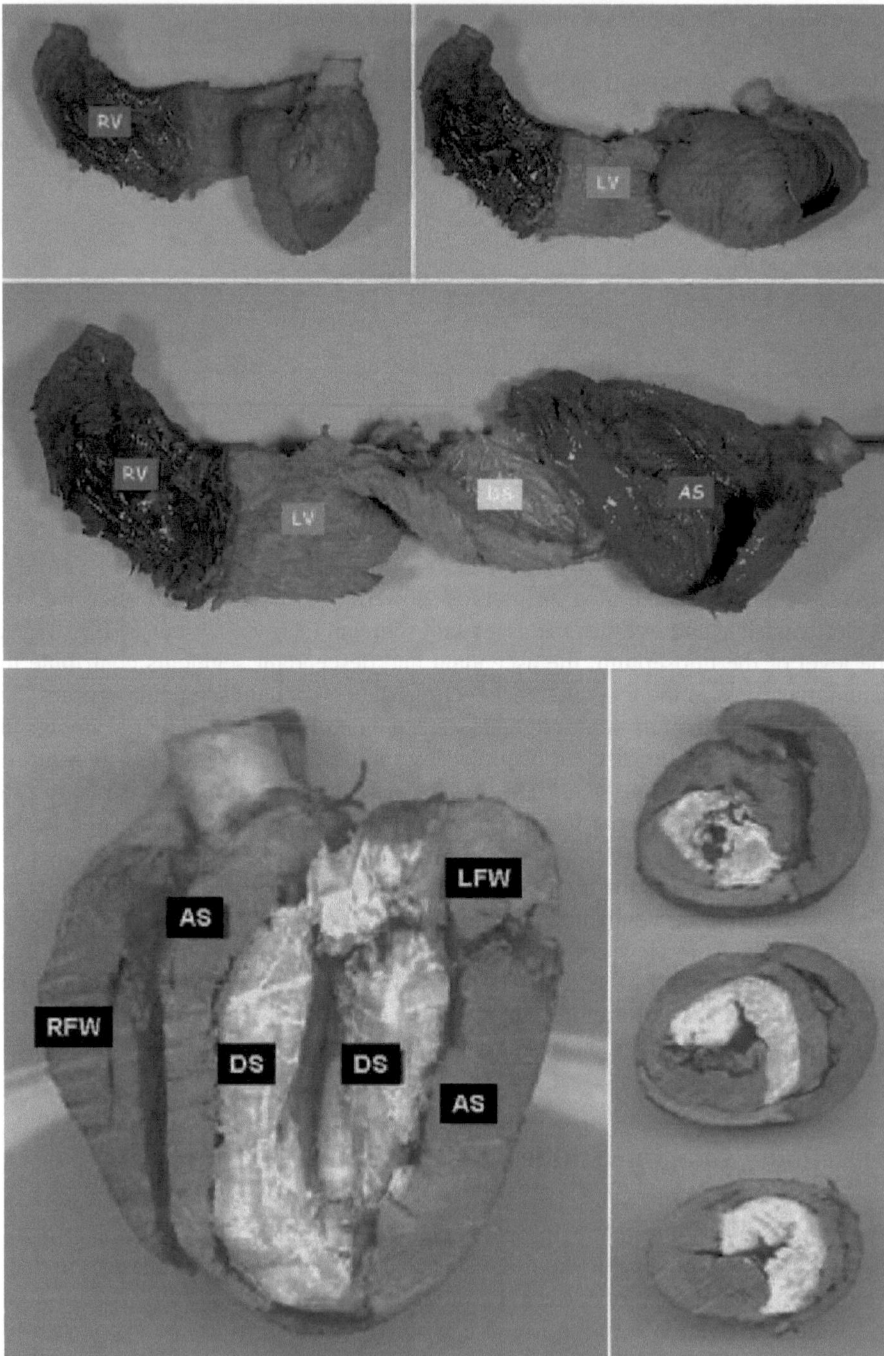

A

B

Fig. 2. **A** Scheme of the colored band segments. **B** The right ventricle (*blue*) and upper portion of the left ventricle (*red*) form the basal loop. The descendant (*yellow*) and ascendant (*green*) segments form the apical loop. *AS*, ascendant segment; *DS*, descendant segment; *LS*, left segment; *RS*, right segment; *RV*, right ventricle; *LV*, left ventricle; *LFW*, left; *RFW*, right

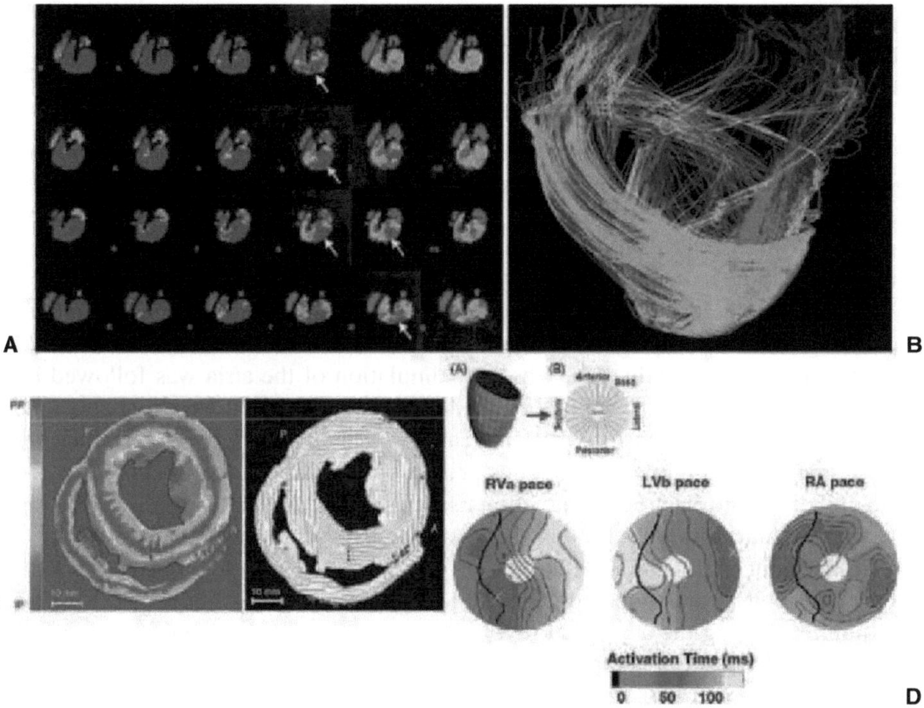

Fig. 3. **A** Fourier analysis in normal individuals. The wave front of ventricular contraction is shown in *yellow*. The earliest activation occurs in the base of the right ventricle (*white arrows*) near the pulmonary infundibulum, extending to the basal portion of the left ventricle; but much before the entire right ventricle is activated, the basal portion of the left begins its mechanical motion. **B** Diffusion tensor magnetic resonance imaging (MRI)-based ventricular reconstruction. The directions of the ventricular fibers closely agree with the myocardial band spiral distribution. **C** Cardiac fiber field in an equatorial slice as measured by MRI-diffusion tensor imaging. The disposition of the different layers closely resembles the myocardial band wrapped anatomy (see [6]). *A*, anterior; *P*, posterior. **D** Temporal evolution of the three-dimensional mechanical activation maps derived from MRI of paced canine hearts and mapped into a bull's eye plot of the ventricle. The *left map*, corresponding to the pacing of the lowest part of the right ventricle, near the septum, shows that the mechanical activity which appears near the right ventricle (*blue*) spreads uniformly from this ventricle to involve the left one. In contrast, *the map in the middle* of the illustration represents the pacing of the coronary sinus, where the basal portion of the left ventricle is first activated (*blue*) and the front wave spreads uniformly in the direction of the right ventricle; finally, the *bull's eye map on the right* of the figure, obtained when the pacing is from the right atrium (*RA*; thus reproducing the physiologic ventricular activation), discloses two opposing initial sites at the ventricular base, in a way similar to the findings revealed by Fourier studies.

entire base of both ventricles. At this stage, no activity is detected in the apical or septal regions of the ventricle. Finally, the mechanical activity extends to involve the apical and septal regions. This sequence of mechanical motion is typically observed when normal individuals are studied [4].

These findings are in agreement with recent studies showing the temporal evolution of three-dimensional strain maps derived from magnetic resonance imaging (MRI), which allow the mapping of the electromechanical motion of the ventricles(Fig. 3D) [5]. These maps illustrate that the onset of motion is observed at the level of the base of the heart at two distinct sites. In a recent study [6], pacing of the right or left ventricle revealed isochronic waves of mechanical activity spreading across the myocardium at a velocity that doubled the ventricular activation via the normal conduction, whereas stimulation of the atria was followed by a simultaneous motion of the base of the heart at two different opposing sites: the right ventricle and the base of the left ventricle. Therefore, these observations are in accordance with the Fourier studies previously described, and could reflect the role of the specific electrical conduction tissue in synchronizing and/or accelerating electromechanical motion through the ventricular myocardium.

These sequences have been further tested in experimental studies using microcrystals placed in the endocardium, along the descendant myofibers, and the epicardium, following the ascendant myofibers, where simultaneous initial activation of both segments but a late prolongation of the activity of the ascendant segment have been observed [6]. In addition, several imaging modalities and experiments support the twist-untwist ventricular mechanics. The first evidence was provided by tagging MR studies [7, 8], where torsion and untorsion dynamics were described, but no interpretation as to the nature of this motion was provided. Recent reports using Doppler tissue imaging have allowed the noninvasive quantification of ventricular torsion, which closely correlated with MRI studies [9]. These results demonstrate that MRI, using velocity recordings for a detailed assessment of myocardial motion and for acceleration tract mapping of structure in healthy volunteers, displays a reproducible and accurate range of myocardial motion, corresponding to contraction/expansion (narrowing/widening) and clockwise/counterclockwise rotation (twisting) in short-axis orientation and shortening/lengthening along the long-axis direction. In addition, such an approach defines myocardial fiber tracking, which correlates with the structural model of the helical ventricular myocardial band. More importantly, these motion images and evaluation of potential fiber orientation improve the recognition of the structure/function relationships of the left ventricular myocardium by interfacing with other imaging modalities that include echocardiography, sonomicrometer crystal measurement of shortening, radionuclide ventriculography, and corrosion cast models of spiral ventricular architecture.

The characteristic global movements documented by MR velocity recordings during the cardiac cycle have traditionally been evident by echocardiography and ventriculograms, and include narrowing during the isometric phase of contraction, shortening during ejection, lengthening during the isovolumetric phase of rapid lengthening in preparation for rapid filling, and widening as passive ventricular filling occurs. The patterning of these findings with other imaging modalities, by using the model of the helical ventricular myocardial band, gives insight into the structural reasons for the global and regional motion dynamics defined by these imaging tools.

A recent modality of ultrasound studies, speckled tracking echocardiography [10, 11], has provided a unique means to tag speckles from the B-mode, which are natural acoustic markers that can be selectively identified and followed in consecutive frames during the heart cycles. This technique has shown that, at an apical level, the left ventricle rotates counterclockwise as viewed from apex, whereas the base rotates clockwise. This gradient of left ventricle rotation between the two levels creates a "wringing" motion and vortex-type hemodynamics within the cavities which can be imaged in vivo using this technique.

Magnetic resonance of myocardial motion and fiber tracking by acceleration velocities in normal volunteers demonstrate a favorable interaction of MR with echocardiography, sonomicrometer crystals, radionuclide ventriculography, and left ventricular corrosion casts, thereby implying that myocyte orientation correlates with the helical ventricular myocardial band. These databases may be useful to evaluate global and regional systolic and diastolic functions in cardiac pathologic processes. Of greater importance, the close agreement among a spectrum of imaging studies provides a very powerful integration that transcends a single look by any one imaging modality. Each part of the integrated global technology of function sees the same structure that has evolved from the underlying spatial architecture. Collaboration between these different imaging methods provides novel explanations to answer questions posed and unanswered by individual imaging tools. These pooled imaging findings imply that the helical ventricular band is the structural basis for the global functional changes observed by most imaging technologies.

Implications for New Integrated Treatments

There are several fields in which the impact of these new concepts is presumed to be of importance.

In the assessment of systolic ventricular function, the angle between the descendant fibers and the ascendant ones seems to be crucial for efficient systolic function [12, 13]. In a similar way that a wider helical angle of a screw is more efficient for . . . the wider the angle between the descendant fibers and the ascendant ones, the more efficient are the torsion dynamics and ejection. In dilated cardiomyopathy, the more horizontal angle of the fibers due to ventricular remodelling favors the subsequent impairment of cardiac function; therefore, in the surgical treatment of heart failure, not only the size of the ventricle but also the fiber angle should be taken into account. The near-future possibility of imaging myocardial fiber dynamics online, through resonance tensor diffusion imaging or other imaging means, opens the way to research and the development of more adequate surgical strategies to compensate for cardiac dysfunction.

Over the past 20 years, the noninvasive evaluation of diastolic function has mainly relied on echo-Doppler analysis of how blood volume enters the ventricle, which largely depends on its loading conditions. One approach is to evaluate the degree of ventricular untwist that occurs during diastole [14], either by speckle tracking echocardiography [10] or by velocity vector imaging, a new ultrasound technique that is angle-independent and thus provides an avenue to evaluate the short-axis mechanics of the left ventricle [15]. Sampling of the descendant myocardial band segment responsible for systole and of the ascendant band responsible for diastole has recently been performed by Vannan and colleagues [16]. The results revealed that normal and hypertensive patients disclosed a similar peak twist velocity (1.9 +/– 0.2 versus 2.3 +/– 0.2 cm/s), whereas untwist was significantly less in hypertensive patients (1.7 +/– 0.1 versus 1.1 +/– 0.1 cm/s), suggesting that early diastolic dysfunction can be ascribed to the generation of less force by the ascendant segment. This finding may have important implications in terms of the early diagnosis of diastolic dysfunction and therapy. In this respect, the anatomy of the band may explain why diastolic dysfunction occurs before systolic alteration of function.

Furthermore, in surgery for heart failure, it seems appropriate, once the helical conformation of the ventricle is known, to plan surgery in such a way as to spare the helical ventricular anatomy. In an initial study by Torrent-Guasp and colleagues [1] and Ballester et al. [3], four theoretical possibilities of surgically reducing left ventricular volume without impairing the helical structure of the ventricles were considered. The most logical approach seems to be to spare the apical loop (small volume and the true motor of the ventricle) and reduce the basal loop; however, new initiatives are needed to convert these theoretical possibilities into practical surgical approaches.

In resynchronization therapy, an explanation of why some patients respond to device implantation whereas others fail to improve is not forthcoming. The issue is not that the left ventricle should contract simultaneously with (most of the indexes developed are aimed at the detection of this phenomenon); rather, what is crucial is the sequence of contraction of the descendant and ascendant segments. In fact, knowledge of the natural sequence of electromechanical ventricular activation [4, 17] should open the way to study the effects of selective stimulation of the band segments on ventricular function, and should provide a rationale for ventricular pacing protocols. In this respect, two recent studies using microcrystals selectively placed in the descendant and ascendant segments of the band have shown the hemodynamic disruption caused by right or left ventricular pacing [18]. This disruption is not solved by atrio-biventricular pacing, which compares less favorably (QRS complex, systolic and diastolic function, and cardiac index) with a more physiologic high septal pacing. With the latter mode of stimulation, the dynamics between descendant and ascendant segments are close to normal [19]. In the same way, the cardiac cycle should be explained in the light of the new anatomy and function and should include the twist and untwist motion and its relationship with the mechanical and acoustic events. Especially important is a description of muscular movements during isovolumic contraction and relaxation. The paradigmatic change in ventricular anatomy and function will prompt research in every field of cardiology.

Several studies show that cardiac resynchronization therapy induces a reduction in cardiac sympathetic nerve activity in responders that parallels an improvement in left ventricular ejection fraction, whereas nonresponders do not show any significant changes [20, 21]. Typically, imaging with (123)I-metaiodobenzylguanidine (MIBG) is used for the assessment of cardiac sympathetic activity. When cardiac sympathetic activity is analyzed before and after cardiac resynchronization therapy, and these findings are examined in relation to treatment response, improvement in cardiac sympathetic activity correlates with a positive response to resynchronization. Lower MIBG uptake before therapy is usually associated with nonresponse to resynchronization. Therefore, the assessment of cardiac sympathetic activity could be helpful in selecting patients for cardiac resynchronization. However, about a third of patients who receive cardiac resynchronization therapy do not improve. It seems that coupling cardiac MIBG studies with functional assessment, taking into account the newly discovered complex mechanics of ventricular activation, could improve the selection of candidates and the assessment of cardiac resynchronization.

Stem-cell, tissue-engineering, and regenerative sciences are currently undergoing vigorous laboratory studies and early clinical trials [22]. In the field of cardiothoracic surgery, such studies may lead to new therapeutic approaches, such as stem-cell implantation to repair damaged myocardium and improve cardiac function, a procedure known as cellular cardiomyoplasty; tissue-engineered cardiac valves and conduits constructed using autologous cells populated on biodegradable scaffolds, which will require neither anticoagulation therapy nor repeated replacements to adapt to the growing infant or child recipient; and regeneration of tissues and organs such as the limbs and heart. Therefore, cell transplantation may preserve or even restore contractile function to infarcted hearts. A typical human infarct involves the loss of approximately 1 billion cardiomyocytes, and, therefore, many investigators have sought to identify endogenous or exogenous stem cells with the capacity to differentiate into committed cardiomyocytes and repopulate the lost myocardium. As a result of these efforts, dozens of stem cell types have been reported to have cardiac potential. These include pluripotent embryonic stem cells, as well various adult stem cells resident in compartments including bone marrow, peripheral tissues, and the heart itself. Some of these cardiogenic progenitors have been reported to contribute replacement muscle through endogenous reparative processes or via cell transplantation in preclinical cardiac injury models. However, considerable disagreement exists regarding the efficiency and even the reality of cardiac differentiation by many of these stem cell types, making these issues a continuing source of controversy in the field.

As for stem cell injections, these are random and may distort the described architecture of the heart. Furthermore, assuming that cells are injected into the area of infarction and scar, they will be exposed to a hostile collagenous environment devoid of vascularization. In fact, up to 75% of the injected cells, no matter what their origin, die within 48 h of injection. It should therefore be a prominent goal of cardiac scientists to enhance viability and engraftment, an issue that has attracted increased attention lately. For instance, growth factors have been added to stem cell injections to improve the surrounding vascularization and donor cell viability. However, it seems that the eventual site of administration of these cells could be better decided on the basis of the knowledge of the new helical model of myocardial fiber distribution.

As more clinical trials utilizing stem cells emerge, it is imperative to establish the mechanisms by which stem cells confer benefit in cardiac diseases. Molecular imaging provides accurate noninvasive information about myocardial perfusion and contractile function and viability, which enables the assessment of the clinical benefits of stem cell therapy. In addition, molecular imaging has provided more specific tracers targeting cellular and subcellular biologic events, which are expected to shed more light upon the mechanisms of cell therapy. Sodium iodide symporter gene transfer has been employed for the sequential detection and quantification of reporter gene expression in the transplanted heart [23]. In future, gene therapy may develop in combination with stem/progenitor cell transplantation therapy [24].

Heart transplantation remains as the treatment of choice for patients with end-stage heart failure without other medical or surgical options. During heart transplantation, the allograft becomes completely denervated. Lack of autonomic nerve supply is associated with major physiologic limitations. The inability to perceive pain does not allow symptomatic recognition of accelerated allograft vasculopathy, and heart transplant patients often develop acute ischemic events or left ventricular dysfuction or die suddenly. In addition, denervation of the sinus node does not allow either adequate acceleration of heart rate during stress or efficient increase in cardiac output. Reinnervation in transplanted hearts is functionally important, as patients with reinnervation have longer exercise times and an enhanced contractile response to exercise compared with patients with denervated transplanted hearts. Sympathetic reinnervation, measured by the regional distribution and intensity of myocardial MIBG uptake, increases with time after transplantation [25, 26]. Early vasculopathy may inhibit the process of sympathetic reinnervation of the transplanted heart. Fluorodeoxyglucose (FDG) positron emission tomography (PET) studies in heart transplant patients show marked enhancement of FDG uptake in the heart at 1 month of transplantation, which decreases over time, suggesting persistent ischemia/reperfusion damage after surgery. Ammonia PET studies after adenosine infusion provide a very sensitive means for the detection and evaluation of gtaft vasculopathy. Early detection of graft vasculopathy is important because it may respond to more aggressive immunosuppression [27]. In addition, cardiac-efferent adrenergic signals play an important role in modulating myocardial blood flow during activation of the sympathetic nervous system. The relationship between reinnervation status and graft vasculopathy deserves further investigation and may help to characterize subsets of transplant patients with different clinical outcomes. The restoration of sympathetic innervation is associated with improved response of the heart rate and contractile function to exercise, indicating the functional importance of reinnervation in transplanted hearts.

Conclusion

The appropriate interplay among the wide spectrum of imaging modalities will provide a powerful integration that transcends the single perspective provided by any one modality. The integrated global technology of anatomy, function, and molecular information will provide novel explanations to answer questions posed and unanswered by individual imaging tools. These integrated imaging technologies will bridge the newly discovered structural and functional phenomena with new integrated, targeted, and tailored therapies.

References

1. Torrent-Guasp F, Ballester M, Buckberg GD et al (2001) Spatial orientation of the ventricular muscle band: physiologic contribution and surgical implications. J Thorac Cardiovasc Surg 122: 389–392
2. Kocica ML, Corno AF, Carreras F et al (2006) The helical myocardial band: global, three dimensional, functional architecture of the ventricular myocardium. Eur J Cardiothorac Surg 9(Suppl 1): S21–40
3. Ballester M, Ferreira A, Carreras F (2008) The myocardial band. Heart Fail Clin 4: 261–272
4. Ballester M, Flotats A, Carrió I et al (2006) The sequence of regional ventricular motion. Eur J Cardiothorac Surg 29: S139-S144
5. Wyman BT, Hunter WC, Prinzen WP (1999) Mapping propagation of mechanical activation in the paced heart with MRI tagging. Am J Physiol 50: 682–688
6. Carreras F, Ballester M, Pujadas S et al (2006) Morphologic and functional evidences of the helical heart from non-invasive cardiac imaging. Eur J Cardiothorac Surg (Suppl 1): S50–55
7. Sengupta PP. Korinek J. Belohlavek M et al (2006) Left ventricular structure and function: basic science for cardiac imaging. J Am Coll Cardiol 48: 1988–2001
8. Sengupta PP, Krishnamoorthy VK, Kprinek J et al (2007) Left ventricular form and function revisited: applied translational science to cardiovascular ultrasound imaging. J Am Soc Echocardiogr 20: 539–551
9. Moore CC, Lugo-Oliveri CH, McVeigh ER et al (2000) Three-dimensional systolic strain patterns in the normal human left ventricle: characterization with tagged MR imaging. Radiology 2: 97–108
10. Notomi Y, Setser RM, Shiota T et al (2005) Assessment of left ventricular torsion deformation by Doppler tissue imaging : validation study with tagged MRI. Circulation 111(9): 1141–1147
11. Helle-Valle T, Crosby J, Edvardsen T et al (2005) New non-invasive method for assessment of left ventricular rotation : speckle tracking echocardiography. J Am Coll Cardiol 45: 2034–2041
12. Sallin EA (1969) Fiber orientation and ejection fraction in the human ventricle. Biophys J 9: 954–964
13. Amudsen BH, Hele-Valle T, Edvardsen T et al (2006) Noninvasive myocardial strain measurement by speckle tracking echocardiography. J Am Coll Cardiol 47: 798–793
14. Schmid P, Jaermann T, Boesinger P et al (2005) Ventricular myocardial architecture as visualised in postmortem swine hearts using magnetic resonance diffusion tensor imaging. Eur J Cardiothorac Surg 27: 468–474
15. Foster E, Lease KE (2006) New untwist on diastole: what goes around comes back. Circulation 113: 2477–2479

16. Vannan MA, Pedrizzetti G, Li P et al (2005) Effect of cardiac resynchronization therapy on longitudinal and circumferential left ventricular mechanics by velocity vector imaging. Echocardiography 22: 826–830
17. Torrent-Guasp F, Ballester M, Caralps JM et al (2002) Spatial orientation of the ventricular muscle band and approach to partial ventriculotomy in heart failure. In: Narula J, Virmani R, Ballester M, Carrió I (eds) Heart failure: pathogenesis and treatment. Martin Dunitz, London, pp 685–693
18. Wyman BY, Hunter WC, Prinzen WP et al (1999) Mapping propagation of mechanical activation in the paced heart with MRI tagging. Am J Physiol 276: H881–H891
19. Liakopoulos OJ, Tomioka H, Buckberg GD et al (2006) Sequential deformation and physiologic considerations in unipolar right or left ventricular pacing. Eur J Cardiothorac Surg 29: S198–S206
20. Burri H, Sunthorn H, Somsen A et al (2008) Improvement in cardiac sympathetic nerve activity in responders to resynchronization therapy. Europace 10: 374–378
21. Nishioka SA, Martinelli Filho M (2007) Cardiac sympathetic activity pre and post resynchronization therapy evaluated by MIBG myocardial scintigraphy. J Nucl Cardiol 14: 852–859
22. Kipshidze NN, Serruys PW (eds) (2004) Handbook of cardiovascular cell transplantation. Martin Dunitz, London
23. Hung Q, Frangioni J, Hajjar J (2008) Imaging in cardiac cell-based therapy: in vivo tracking of the biological fate of therapeutic cells. Nature Cardiovascular Medicine 5 (Suppl 2): S96–S102
24. Inubushi M, Tamaki N (2007) Radionuclide reporter gene imaging for cardiac gene therapy. Eur J Nucl Med Mol Imaging 34: S27–S33
25. Estorch M, Camprecios M, Flotats A et al (1999) Sympathetic reinnervation of cardiac allografts evaluated by 123I-MIBG imaging. J Nucl Med 40: 911–916
26. Bengel F, Ueberfuhr P, Schafer D et al (2006) Effect of diabetes mellitus on sympathetic neuronal regeneration studied in the model of transplant reinnervation. J Nucl Med 47: 1413–1419
27. Lamich R, Ballester M, Marti V et al (1998) Efficacy of augmented immunosuppressive therapy for early vasculopathy in heart transplantation. J Am Coll Cardiol 2: 413–419

Molecular Imaging of Angiogenesis

Ambros J. Beer, Hans-Jürgen Wester, and Markus Schwaiger

Summary

Molecular imaging attempts to visualize biologic processes in order to improve diagnosis and to direct therapy. Angiogenesis is an important process in oncogenesis, inflammation, and wound healing. Most recently, new drugs have been introduced to stop tumor growth in patients with cancer.

Angiogenesis is a complex process closely linked to tissue hypoxia and proliferation. Besides growth factors such as vascular endothelial growth factor (VEGF), the expression of integrins has been selected as a potential target for imaging. The integrin $\alpha v\beta 3$ is overexpressed on activated endothelial cells in the process of angiogenesis. Cyclic arginine-glycine-aspartate (RGD) peptides specific for the integrin $\alpha v\beta 3$, such as the positron emission tomography (PET) tracer [^{18}F] galacto-RGD, have been successfully used for imaging angiogenesis. However, tracer uptake is also observed in tumor cells and inflammatory cells. Future studies have to address the specificity of the tracer approach. Using multimodal probes and combining functional and molecular imaging studies might help to further define the prognostic and diagnostic value of this exciting new imaging approach.

Key words Angiogenesis · Molecular Imaging · PET · RGD · $\alpha v\beta 3$

Introduction

Angiogenesis is a fundamental process for physiological as well as pathological processes. As a physiological process, it is required for development, wound repair, reproduction, and response to ischemia. However, it is also associated with pathological conditions such as arthritis, psoriasis, retinopathies, and cancer [1]. Judah Folkman in 1971 first articulated the importance of angiogenesis for tumor growth

Department of Nuclear Medicine, Technische Universität München, Klinikum rechts der Isar; Ismaningerstr. 22, 81675 Munich, Germany

N. Tamaki, Y. Kuge (eds.)
Molecular Imaging for Integrated Medical Therapy and Drug Development.
DOI: 10.1007/978-4-431-98074-2_11, ©Springer 2010

[2]. Recently, encouraging results have been achieved with the vascular endothelial growth factor (VEGF) antibody Avastin (bevacizumab, Roche, Basel, Switzerland), in combination with standard cytotoxic chemotherapy, first in metastasized colorectal cancer, and subsequently in breast cancer and non-small cell lung cancer [3, 4]. Consequently, there will be a growing demand for imaging modalities for the assessment of the optimum doses of new antiangiogenic agents and for response evaluation of antiangiogenic therapies. Up to now, clinical trials with conventional cytotoxic chemotherapeutic agents have mainly used morphological imaging to provide indices of therapeutic response, mostly computed tomography (CT) or magnetic resonance imaging (MRI). However, as antiangiogenic agents rather lead to a stop of tumor progression than to tumor shrinkage, the approach of measuring tumor response by a reduction of tumor size might take months or years to assess. Therefore, there is great interest in reliable biomarkers of early tumor response to noncytotoxic drugs [5]. Imaging techniques could potentially be used as such biomarkers as early indicators of effectiveness at a functional or molecular level. On a functional level, changes in hemodynamic parameters such as blood flow, blood volume, or vessel permeability may be promising biomarkers for response evaluation. Current clinical trials employ techniques such as dynamic contrast-enhanced MRI (DCE-MRI), and less often, ultrasound, positron emission tomography (PET; especially with [^{15}O] water), and dynamic contrast-enhanced CT [6]. On a molecular level, targeting specific molecular markers of angiogenesis, such as the VEGF pathway or cell surface markers such as the integrin $\alpha v\beta 3$, might also be used for response assessment of antiangiogenic therapies.

Functional Imaging of Tumor Vasculature

Magnetic Resonance Imaging and Computed Tomography

Dynamic contrast-enhanced CT (DCE-CT) and DCE-MRI are noninvasive methods of investigating microvascular structure and function by tracking the pharmacokinetics of injected low-molecular-weight contrast agents (LMWCA) as they pass through the tumor vasculature. For both techniques the same principles apply when LMWCA are used. An important difference and advantage of CT is, however, that it is a quantitative technique, because the measured Hounsfield Units (HU) are linearly correlated with the concentration of the contrast material in the tissue [7].

Changes in vascular permeability and in extracellular, extravascular, and vascular volumes, and changes in blood flow can be measured by these techniques. MRI has the advantage over CT that it does not involve ionizing radiation, and that MRI contrast agents generally have a better toxicity profile than iodine contrast agents. In T1-weighted DCE-MRI, an intravenous bolus of gadolinium contrast agent enters tumor arterioles, passes through the capillary bed, and then drains via tumor veins. The signal enhancement is dependent on physiological and physical factors, including tissue perfusion, capillary surface area and permeability, and the volume of the extracellular extravascular space (EES). However, signal enhancement will also be affected by the native T1-relaxation time of each tissue, by the contrast agent dose, and by the chosen sequence parameters, which makes quantitation of the results more difficult than with DCE-CT [8, 9]. In general, anatomical information to localize the tumor is acquired initially. Next, sequences that allow calculation of baseline tissue T1-values before contrast agent administration are performed to enable subsequent analysis. Finally, dynamic data are acquired every few seconds in T1-weighted images over a period of around 5–10 min. Several analysis methods can then be applied to the acquired data. Parameters that describe the shape of the contrast agent concentration-time curve represent a combination of blood flow, blood volume, vessel permeability, and EES volume. One such parameter, the initial area under the contrast agent concentration-time curve (IAUC), is easy to calculate, reasonably reproducible, and is routinely used as a biomarker in drug trials. However, the IAUC has a complicated relationship with underlying tumor physiology and represents a composite of physiological processes, including blood flow, capillary permeability, and blood volume. A more quantitative analysis can be achieved by applying pharmacokinetic models to contrast agent concentration data to enable estimates of physiological characteristics such as flow and capillary endothelial permeability [10]. These parameters are the volume transfer constant, K^{trans} (min^{-1}), which describes the rate of flux of contrast agent into the EES within a given volume; the volume of the EES per unit volume of tissue, υ_e; and the rate constant for the back flux from the EES to the vasculature, κ_{ep} (min^{-1}). These terms are related to each other by the equation: $\kappa_{ep} = K^{trans} / \upsilon_e$. The relationship of the different parameters K^{trans}, κ_{ep}, and υ_e can be expressed by the equation:

$$\frac{dC_t}{dt} = K^{trans} * \left(C_p - \frac{C_t}{V_e} \right) \tag{1}$$

or

$$\frac{dC_t}{dt} = K^{trans} * C_p - \kappa_{ep} * C_t \tag{2}$$

where C_t = the tracer concentration in tissue, C_p = the tracer concentration in plasma, t = time (in seconds).

In a 1999 consensus publication, this set of terms was recommended by an international group of investigators developing DCE-MRI methodologies. However, the exact physiological meaning of these parameters is complex and is not related to a single process such as blood flow or blood volume only. The interpretation of K^{trans} varies depending on the relationship between blood flow and the capillary permeability-surface area product (PS). Concerning the reproducibility of these hemodynamic parameters, the maximum enhancement and the area under the curve (AUC) appear to be the most reproducible among the semiquantitative parameters. The slope or rate of enhancement appears to be the least reproducible, which may reflect the dependence of these measurements on the rate of injection of the contrast agent bolus, as well as variations in cardiac input. Among the quantitative parameters, υ_e seems to have the least variability, with interpatient variability being much greater than the intrapatient variability. However, K^{trans} and κ_{ep} also have sufficient reproducibility to be useful for measuring changes over time [11, 12]. Evidence of drug efficacy has been demonstrated with DCE-MRI in several trials of antiangiogenic drugs [13, 14]. DCE-MRI seems promising as an imaging biomarker in clinical trials; however, data concerning its performance, especially for response assessment, are not uniform and seem to depend strongly on the therapy protocol and tumor type.

Radiotracer Techniques

A major advantage of the nuclear medicine techniques using radiotracers is that they are truly quantitative and that the tissue concentration, C_t, can be measured noninvasively.

Fick described the central relationship between blood flow and tissue clearance of circulating tracers first in 1870, as:

$$C_t = P * \int (C_i - C_e)\,dt \tag{3}$$

where C_t = tissue concentration ($mol*ml_{tissue}^{-1}$); C_i = influx concentration ($mol*ml_{carrier}^{-1}$), C_e = eflux concentration ($mol*ml_{carrier}^{-1}$); P = Perfusion, ($ml_{carrier} * min^{-1} * ml_{tissue}^{-1}$).

In literal terms, the amount of tracer cleared by the tissue over time t is the product of perfusion P and tracer extraction. Perfusion is calculated by rearranging term (3) so that

$$P = \frac{C_t}{\int (C_i - C_e)\,dt} \tag{4}$$

Most PET-perfusion measurements nowadays are performed using [^{15}O] water. [^{15}O]H_2O satisfies all the requirements for a perfusion tracer in Fick's model because it is biologically and metabolically inert, and freely diffusible into and out of tissue water. Thus "tissue water" can be modelled as a single compartment including both tissue and its draining fluids (lymphatics and veins). The [^{15}O]-dynamic water method described by Lammertsma and Jones [15] is currently used most often for perfusion studies [16]. The tracer is administered by inhalation or by peripheral venous bolus injection. Continuous arterial data are obtained either by image-based arterial input functions (a large vessel such as the aorta or the left ventricle) or by peripheral sampling to a well counter device. The change in tissue concentration over time is modelled as:

$$\frac{dC_t(t)}{dt} = P * C_a(t) - (P/V_D + \lambda) * C_t(t) \tag{5}$$

where V_D = "volume of distribution", the "proportion of the region of interest in which the radioactive water is distributed" ($ml_{blood}/ml_{tissue} = 1/\rho$), $C_t(t)$ = instantaneous tissue concentration of $[^{15}O]H_2O$ at time t (Bq/ml_{tissue}), $C_a(t)$ = corrected instantaneous arterial concentration of $[^{15}O]H_2O$ at time t (Bq/mlt_{issue}).

The mathematics for solving P and V_D from the dynamic curves depend on convolution of the arterial and tissue datasets. The expression for tissue concentration at each time t is given by the convolution integral:

$$C_t(t) = \int P * C_a(T) * e^{-(P/V_D+\lambda)*(t-T)} dT \quad (6)$$

or

$$C_t(t) = P * C_a(t) \otimes e^{-(P/V_D+\lambda)*t} \quad (7)$$

where ⊗ is the operation of convolution. $C_t(t)$ describes a biphasic curve with an initial peak followed by a longer tail of decay. P and V_D can be determined from this curve using nonlinear least-squares fitting. The validity and reproducibility of this method was initially assessed for the brain and myocardium, but subsequently also for tumors of pancreas, brain, breast, and liver [17, 18]. In locally advanced breast cancer first results with dynamic $[^{15}O]H_2O$ PET are promising, as blood flow decreased in the responder group after chemotherapy, whereas it increased in the nonresponder group [19].

Imaging of Molecular Markers of Tumor Vasculature

While techniques such as DCE MRI and $[^{15}O]H_2O$ PET are widely used for the assessment of hemodynamic parameters, the interpretation of the results with regard to their physiological meaning often remains difficult. Therefore, more specific markers of angiogenic activity in tumors are necessary for the pretherapeutic assessment of angiogenesis and response evaluation during therapy. One approach is to identify molecular markers of angiogenesis such as receptors, enzymes, or extracellular matrix proteins and to use specific ligands to these targets conjugated with imaging probes for PET, single photon emission computed tomography (SPECT), MRI, optical imaging, or ultrasound. One of the most promising targets in this respect is the integrin αvβ3, which is one of the few markers of angiogenesis to date that has been successfully imaged in patients by using PET and SPECT techniques [20, 21].

Integrins are heterodimeric transmembrane glycoproteins consisting of different α- and β-subunits which play an important roles in cell-cell and cell-matrix-interactions. Especially well examined is the integrin αvβ3 and the role it plays in angiogenesis and tumor metastasis by facilitating endothelial and tumor cell migration. It has been found that several extracellular matrix (ECM) proteins such as vitronectin, fibrinogen, and fibronectin interact with integrins via the amino acid sequence arginine-glycine-aspartic acid or RGD in the single letter code [22].

Kessler and co-workers developed the pentapeptide cyclo(-Arg-Gly-Asp-DPhe-Val-), which shows high affinity and selectivity for αvβ3 [22]. For the first evaluation of this approach, Haubner et al. synthesised radioiodinated RGD peptides which showed comparable affinity and selectivity to the lead structure [23]. Several strategies to improve the pharmacokinetics of radiohalogenated peptides have been developed. The glycosylation approach is based on the introduction of sugar derivatives which are conjugated to the ε-amino function of a corresponding lysine in the peptide sequence. By conjugating the RGD-containing cyclic pentapeptide cyclo (-Arg-Gly-Asp-DPhe-Val-) with glucose- or galactose-based sugar amino acids, [*I]gluco-RGD and [^{18}F]galacto-RGD have been developed for PET and SPECT imaging. Both compounds demonstrated improved pharmacokinetics with predominantly renal tracer elimination and increased uptake and retention in a murine tumor model compared with the first-generation peptides [23]. Extensive preclinical evaluations concerning monomeric compounds were carried out using [*I]gluco-RGD and [^{18}F]galacto-RGD [24, 25]. Initial in vivo evaluation was carried out using the human melanoma M21 model, which is well characterized concerning αvβ3 expression [26]. Using this model, [^{18}F]galacto-RGD and [^{125}I]gluco-RGD uptake in the tumor at 120 min post-injection was 1.5% and 1.8% injection dose (ID)/g, respectively. Blocking experiments injecting 6 mg c(RGDfV) per kg mouse 10 min prior to tracer injection reduced tumor accumulation to approximately 15% of control for [^{125}I]gluco-RGD and to approximately 35% of control for [^{18}F]galacto-RGD, which demonstrated receptor-specific accumulation. Furthermore, imaging studies with mice bearing melanoma tumors with increasing amounts of αvβ3-positive cells (produced by mixing M21 and M21-L cells) showed that there was a correlation between integrin expression and tracer accumulation [25]. These data demonstrate that noninvasive determination of αvβ3 expression and quantification with radiolabelled RGD peptides is feasible with static emission scans. Moreover, results were obtained in an animal PET study with increasing amounts of c(RGDfV) that indicated that the dose-dependent blocking of tracer uptake in the receptor-positive tumor could be monitored.

Up to now, the only approach of imaging αvβ3 expression which has made the transition into the clinic is the radiotracer approach. [^{18}F]Galacto-RGD was the first PET tracer applied in patients and could successfully image αvβ3 expression in human tumors with good tumor/background ratios (Fig. 1). In all patients, rapid, predominantly renal tracer elimination was observed, resulting in low background activity in most regions of the body [27]. High inter- and intraindividual variance in tracer accumulation in tumor lesions was noted, suggesting great diversity of αvβ3 expression. These findings emphasize the potential value of noninvasive techniques for the appropriate selection of patients entering clinical trials with αvβ3-targeted therapies. Further biodistribution and dosimetry studies confirmed the rapid clearance of [^{18}F]galacto-RGD from the blood pool and its primarily renal excretion. Background activity in lung and muscle tissue was low and the calculated effective dose found was approximately 19 μSv/MBq, which is very similar to an [^{18}F] fluorodeoxyglucose (FDG) scan [28]. We also investigated whether [^{18}F] galacto-RGD uptake correlated with αvβ3 expression. Nineteen patients with solid

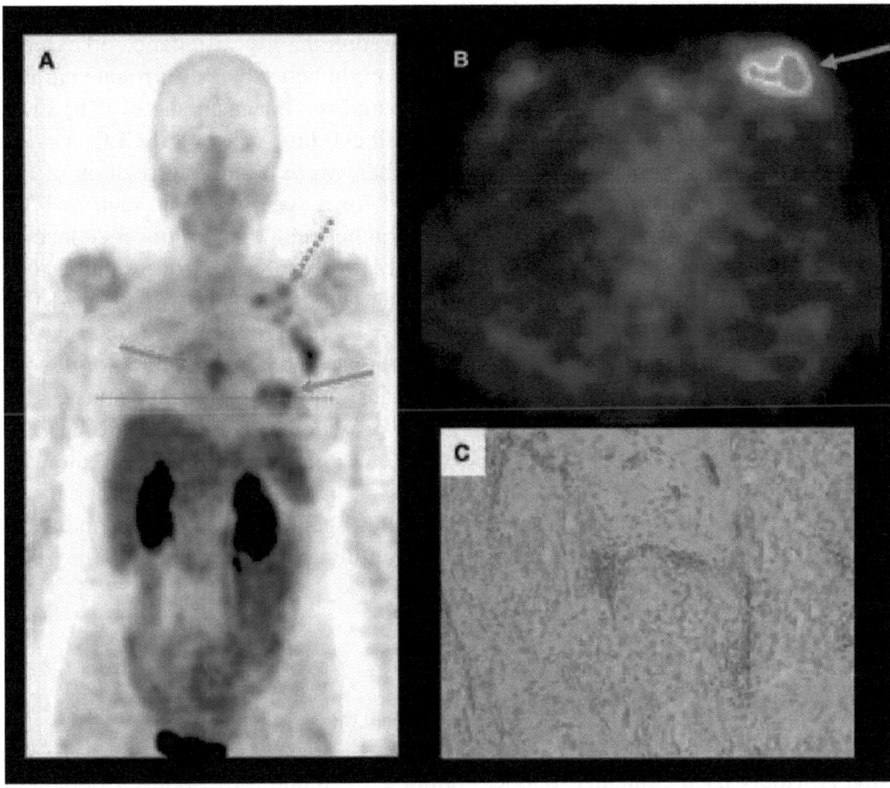

Fig. 1A–C. [^{18}F]Galacto-RGD positron emission tomography (PET) of a male patient with invasive ductal breast cancer on the left side (*closed-tip arrows with continuous shaft in* **A** *and* **B**). The maximum intensity projection (MIP; **A**) and the axial section (**B**) show intense uptake in the tumor. Note the physiological biodistribution of the tracer with predominantly renal elimination (kidneys; *dotted line and arrow with dotted-line shaft in* **A**), and to a lesser extent hepatobiliary elimination (gallbladder; *open-tip arrow in* **A**). Moderate tracer uptake is notable in the liver, spleen, and intestine, whereas there is only low tracer uptake in the lungs, mediastinum, and musculoskeletal system. **C** Immunohistochemistry of αvβ3 expression shows intense staining predominantly of the neovasculature.

tumors (musculoskeletal system, n = 10; melanoma, n = 4; head and neck cancer, n = 2; glioblastoma, n = 2; breast cancer, n = 1) were examined with PET using [^{18}F]galacto-RGD before surgical removal of the lesions [29]. Standardized uptake values (SUVs) and tumor/blood ratios were found to correlate significantly with the intensity of immunohistochemical staining, as well as with the microvessel density. Moreover, immunohistochemistry confirmed the lack of αvβ3 expression in normal tissue and in the two tumors without tracer uptake. We are now systematically examining different tumor entities with respect to their αvβ3 expression patterns as shown by [^{18}F]galacto-RGD PET. In squamous cell carcinoma of the head and neck (SCCHN) we could demonstrate good tumor/background ratios with

[^{18}F]galacto-RGD PET, but again also there was a widely varying intensity of tracer uptake. Immunohistochemistry demonstrated predominantly vascular αvβ3 expression; thus, in SCCHN, [^{18}F]galacto-RGD PET might be used as a surrogate parameter of angiogenesis [30]. We have also compared the tracer uptake of [^{18}F] FDG and [^{18}F]galacto-RGD in patients with non-small cell lung cancer (NSCLC, $n = 10$) and various other tumors ($n = 8$), because if there were a close correlation of the two tracers, there would probably be no need for a specific tracer such as [^{18}F] galacto-RGD. The results showed no correlation between the two tracers concerning all lesions ($r = 0.157$). For the subgroup of [^{18}F] FDG-avid lesions and lesions in patients with NSCLC, there was a slight trend towards a higher [^{18}F]galacto-RGD uptake in the more [^{18}F] FDG-avid lesions ($r = 0.337$). However, the correlation coefficient was very low. Our results suggests that αvβ3 expression and glucose metabolism are not closely correlated in tumor lesions and that, consequently, [^{18}F] FDG cannot provide information similar to that provided by [^{18}F]galacto-RGD [31]. Recently, the SPECT tracer [^{99m}Tc] NC100692 was introduced by GE Healthcare for imaging αvβ3 expression in humans and was first evaluated in breast cancer. Nineteen of 22 tumors could be detected with this agent, which was safe and well-tolerated by the patients. It is therefore expected that commercial agents for αvβ3 imaging will soon be available.

Magnetic Resonance Imaging

MRI is widely used clinically to assess tumor growth and for response evaluation. Anatomical information can be co-registered with functional and molecular information within a single imaging method. A further advantage compared to radiotracer techniques is that MRI does not use ionizing radiation and generally is more widely available than PET. MRI also offers good depth penetration and its resolution is usually higher than that of clinical PET scanners, although this depends on the exact protocol applied. A major disadvantage of MRI compared to radiotracer techniques is its lower sensitivity for the detection of targeted agents. Therefore, targeted molecular imaging agents for MRI have not entered clinical trials yet, except for the fibrin-specific contrast agent EP2140R [32]. The first MRI approach for molecular imaging of angiogenesis was the imaging of αvβ3 expression. By using Gd^{3+}-containing paramagnetic liposomes and the αvβ3 specific antibody LM609 as a ligand, MRI of squamous cell carcinomas in a rabbit model was successfully achieved [33]. However, Gd^{3+} for enhancing the T1 contrast can only be reliably detected at millimolar levels. Superparamagnetic iron oxide (SPIO) nanoparticles can be detected at a much lower concentration because of the high susceptibility induced by this particle, which leads to a decrease of the signal in T2 and especially T2* weighted sequences ("negative contrast") [34]. In a recent study, αvβ3 integrin-targeted ultrasmall SPIO (USPIO) nanoparticles were used for the noninvasive differentiation of tumors with high and lower area fractions of αvβ3-positive tumor vessels [35]. After RGD-USPIO injection, T2*-weighted MRI identified the heterogeneous distribution of $\alpha v\beta 3$-positive tumor vessels by an irregular signal intensity decrease, whereas the signal intensity decreased more homogeneously in the control tumor with predominantly small and uniformly distributed vessels.

Conclusion

A multitude of imaging techniques is available for the assessment of tissue vasculature on structural, functional, and molecular levels. All these methods have been successfully used preclinically and will, it is hoped, aid in antiangiogenic drug development in animal studies. At present, only the imaging of functional hemodyamic parameters such as K^{trans}, blood flow, and blood volume is being used in the clinical arena for the evaluation of antiangiogenic and cytotoxic chemotherapies. However, the physiological meaning of the results is often hard to interpret. Concerning the imaging of molecular parameters of angiogenesis, only a few radiotracers have been used in humans up to now, and their role in the assessment of antiangiogenic therapies is still unsettled. Although $\alpha v\beta 3$ is by far the most extensively studied angiogenic factor for imaging, future trials still have to show which target structure is optimal for the assessment of angiogenic activity. Concerning the optimum imaging technique, the radiotracer approach will probably be the first to be used on a wider scale in patients in the intermediate term, due to its high sensitivity and the low amounts of tracer which have to be used. In the long term, MRI might be an attractive alternative, due to its lack of ionizing radiation and high spatial resolution. However, it is likely that no one single parameter, target structure, or imaging technique will be used for the assessment of angiogenesis in the future, but that a combination of parameters will be used, which will allow for the evaluation of the angiogenic cascade in its full complexity. In summary, assessment of the different aspects of angiogenesis at the functional and molecular levels will, it is hoped, become a reality in the not too distant future and will be implemented in therapy planning and response evaluation as part of the concept of "personalized medicine".

References

1. Folkman J (1995) Angiogenesis in cancer, vascular, rheumatoid and other disease. Nat Med 1: 27–31
2. Folkman J (1971) Tumor angiogenesis: therapeutic implications. N Engl J Med 285: 1182–1186
3. Kerbel RS (2006) Antiangiogenic therapy: a universal chemosensitization strategy for cancer? Science 312: 1171–1175
4. Hurwitz H, Fehrenbacher L, Novotny W et al (2004) Bevacizumab plus irinotecan, fluorouracil, and leucovorin for metastatic colorectal cancer. N Engl J Med 350: 2335–2342

5. Tortora G, Melisi D, Ciardiello F (2004) Angiogenesis: a target for cancer therapy. Curr Pharm Des 10: 11–26
6. Galbraith SM (2003) Antivascular cancer treatments: imaging biomarkers in pharmaceutical drug development. Br J Radiol 76: 83–86
7. Barrett T, Brechbiel M, Bernardo M et al (2007) MRI of tumor angiogenesis. J Magn Reson Imaging 26: 235–249
8. Tofts PS, Brix G, Buckley DL et al (1999) Estimating kinetic parameters from dynamic contrast-enhanced T(1)-weighted MRI of a diffusable tracer: standardized quantities and symbols. J Magn Reson Imaging 10: 223–232
9. Koh TS, Tan CK, Cheong LH et al (2006) Cerebral perfusion mapping using a robust and efficient method for deconvolution analysis of dynamic contrast-enhanced images. Neuroimage 32: 643–653
10. Leach MO, Boggis CR, Dixon AK et al (2005) Screening with magnetic resonance imaging and mammography of a UK population at high familial risk of breast cancer: a prospective multicentre cohort study (MARIBS). Lancet 365: 1769–1778
11. Galbraith SM, Lodge MA, Taylor NJ et al (2002) Reproducibility of dynamic contrast-enhanced MRI in human muscle and tumours: comparison of quantitative and semi-quantitative analysis. NMR Biomed 15: 132–142
12. Padhani AR, Hayes C, Landau S et al (2002) Reproducibility of quantitative dynamic MRI of normal human tissues. NMR Biomed 15: 143–153
13. Conrad C, Friedman H, Reardon D et al (2004) A phase I/II trial of single-agent PTK 787/ZK 222584 (PTK/ZK), a novel, oral angiogenesis inhibitor, in patients with recurrent glioblastoma multiforme (GBM). J Clin Oncol 22: 1512
14. Morgan B, Thomas AL, Drevs J et al (2003) Dynamic contrast-enhanced magnetic resonance imaging as a biomarker for the pharmacological response of PTK787/ZK 222584, an inhibitor of the vascular endothelial growth factor receptor tyrosine kinases, in patients with advanced colorectal cancer and liver metastases: results from two phase I studies. J Clin Oncol 21: 3955–3964
15. Lammertsma AA, Jones T (1992) Low oxygen extraction fraction in tumours measured with the oxygen-15 steady state technique: effect of tissue heterogeneity. Br J Radiol 65: 697–700
16. Wilson CB, Lammertsma AA, McKenzie CG et al (1992) Measurements of blood flow and exchanging water space in breast tumors using positron emission tomography: a rapid and noninvasive dynamic method. Cancer Res 52: 1592–1597
17. Anderson H, Price P (2002) Clinical measurement of blood flow in tumours using positron emission tomography: a review. Nucl Med Commun 23: 131–138
18. Tseng J, Dunnwald LK, Schubert EK et al (2004) 18F-FDG kinetics in locally advanced breast cancer: correlation with tumor blood flow and changes in response to neoadjuvant chemotherapy. J Nucl Med 45: 1829–1837
19. Haubner R, Weber WA, Beer AJ et al (2005) Non-invasive visualization of the activated $\alpha v\beta 3$ integrin in cancer patients by positron emission tomography and [^{18}F]Galacto-RGD. PLoS Medicine 2, e70
20. Bach-Gansmo T, Danielsson R, Saracco A et al (2006) Integrin receptor imaging of breast cancer: a proof-of-concept study to evaluate 99mTc-NC100692. J Nucl Med 47: 1434–1439
21. Ruoslahti E, Pierschbacher MD (1987) New perspectives in cell adhesion: RGD and integrins. Science 238: 491–497
22. Haubner R, Finsinger D, Kessler H (1997) Stereoisomeric peptide libraries and peptidomimetics for designing selective inhibitors of the $\alpha v\beta 3$ integrin for a new cancer therapy. Angew Chem Int Ed Engl 36: 1374–1389
23. Haubner R, Wester HJ, Reuning U et al. (1999). Radiolabeled $\alpha v\beta 3$ integrin antagonists: a new class of tracers for tumor targeting. J Nucl Med 40: 1061–1071
24. Haubner R, Wester HJ, Weber WA et al (2001) Noninvasive imaging of $\alpha v\beta 3$ integrin expression using 18F-labeled RGD-containing glycopeptide and positron emission tomography. Cancer Res 61: 1781–1785

25. Haubner R, Wester HJ, Burkhart F et al (2001) Glycosylated RGD-containing peptides: tracer for tumor targeting and angiogenesis imaging with improved biokinetics. J Nucl Med 42: 326–336
26. Haubner R, Kuhnast B, Mang C et al (2004) [^{18}F]Galacto-RGD: synthesis, radiolabeling, metabolic stability, and radiation dose estimates. Bioconjug Chem 15: 61–69.
27. Felding-Habermann B, Mueller BM, Romerdahl CA et al (1992) Involvement of integrin αv gene expression in human melanoma tumorigenicity. J Clin Invest 89: 2018–2022
28. Beer AJ, Haubner R, Goebel M et al (2005) Biodistribution and pharmacokinetics of the αvβ3 selective tracer 18F Galacto-RGD in cancer patients. J Nucl Med 46: 1333–1341
29. Beer AJ, Haubner R, Wolf I et al (2006) PET-based human dosimetry of ^{18}F-galacto-RGD, a new radiotracer for imaging alpha v beta3 expression. J Nucl Med 47: 763–769
30. Beer AJ, Haubner R, Sarbia M et al (2006). Positron emission tomography using [^{18}F]Galacto-RGD identifies the level of integrin αvβ3 expression in man. Clin Cancer Res 12: 3942–3949
31. Beer AJ, Grosu AL, Carlsen J (2007) [^{18}F]Galacto-RGD PET for imaging of αvβ3 expression on neovasculature in patients with squamous cell carcinoma of the head and Neck. Clin Cancer Res 13: 6610–6616
32. Beer AJ, Lorenzen S, Metz S et al (2008). Comparison of integrin αvβ3 expression and glucose metabolism in primary and metastatic lesions in cancer patients: a PET study using [^{18}F]Galacto-RGD and [^{18}F]FDG. J Nucl Med 49: 22–29
33. Spuentrup E, Botnar RM (2006) Coronary magnetic resonance imaging: visualization of vessel lumen and the vessel wall and molecular imaging of arteriothrombosis. Eur Radiol 16: 1–14
34. Sipkins DA, Cheresh DA, Kazemi MR et al (1998) Detection of tumor angiogenesis in vivo by αvβ3-targeted magnetic resonance imaging. Nat Med 4: 623–626
35. Thorek DL, Chen AK, Czupryna, J et al (2006) Superparamagnetic iron oxide nanoparticle probes for molecular imaging. Ann Biomed Eng 34: 23–38

Spontaneous In Vivo Regeneration of the Articular Cartilage Using a Novel Double-Network Hydrogel

Nobuto Kitamura[1], Kazunori Yasuda[1], Jian Ping Gong[2], and Yoshihito Osada[3]

Summary

It has been a commonly established concept that articular (hyaline) cartilage tissue cannot spontaneously regenerate in vivo. We have challenged this concept by using a plug created with an originally developed double-network (DN) gel—consisting of poly-(2-acrylamido-2-methylpropanesulfonic acid; PAMPS) and poly-(N,N′-dimethyl acrylamide; PDMAAm)—into a large osteochondral defect in a rabbit patellofemoral joint. The defect was made so that it intentionally remained with a depth of 1 to 3 mm after surgery. At 4 weeks, cartilage tissue rich in proteoglycan and type 2 collagen had regenerated in the defect. In cells of the regenerated tissue, type 2 collagen, Aggrican, and SRY (sex determining region Y)-box 9 (SOX9) mRNAs were highly expressed. On the other hand, cartilage regeneration was rarely found in a defect without any treatment. We conclude that spontaneous regeneration of the articular cartilage can be induced in vivo in an osteochondral defect by implanting a PAMPS/PDMAAm DN gel plug into the bottom of the cavity, leaving the defect itself vacant. This finding suggests a significant modification of the above commonly established concept. Also, the findings of this study have prompted an innovative strategy in the field of joint surgery, to repair an osteochondral defect using an advanced high-function gel. Concerning the mechanism of the spontaneous cartilage regeneration, we speculate that the bioactive DN gel may have the potential to induce the differentiation of mesenchymal stem cells contained in the blood clot into chondrocytes.

Key words Hyaline cartilage · Double-network hydrogel · Patellofemoral joint

[1]Department of Sports Medicine and Joint Surgery, Hokkaido University Graduate School of Medicine, Kita-15 Nishi-7, Kita-ku, Sapporo 060-8638, Japan
[2]Division of Biological Sciences, Graduate School of Science, Hokkaido University, Kita-10 Nishi-8, Kita-ku, Sapporo 060-0810, Japan
[3]Molecular and Informative Life Science Unit, RIKEN Wako Institute, 2-1 Hirosawa, Wako, Saitama 351-0198, Japan

N. Tamaki, Y. Kuge (eds.)
Molecular Imaging for Integrated Medical Therapy and Drug Development.
DOI: 10.1007/978-4-431-98074-2_12, ©Springer 2010

Introduction

The articular (hyaline) cartilage is frequently injured due to trauma, pathologic alterations, and age-related degeneration. Such cartilage injury is a significant and increasing problem for young athletes as well as aged people. It has been a commonly established concept that the articular cartilage tissue cannot spontaneously regenerate in vivo [1,2]. Therefore, the implantation of a tissue-engineered cartilage-like tissue or a cell-seeded scaffold material has been the prevalent strategy to treat articular cartilage injury. A number of studies have been conducted applying various kinds of cells cultured in various types of scaffolds to create an implantable tissue-engineered cartilage tissue [3–10]. However, it has been pointed out that this strategy has many problems for current clinical application [11–16]. Thus, functional repair of articular osteochondral defects remains a major challenge for tissue regeneration medicine.

Is articular cartilage tissue really unable to spontaneously regenerate in vivo? We raise a challenge to answer this question. We have paid attention to the clinical fact that fibrous cartilage tissue can spontaneously regenerate in an osteochondral defect. Namely, it is well known that, clinically, an osteochondral defect is spontaneously filled with a blood clot containing mesenchymal stem cells derived from the bone marrow, and that the blood clot has a high potential to induce cartilage-like cells [17]. Therefore, we have considered that the mesenchymal stem cells contained in the blood clot may differentiate into chondrocytes if a certain material with an ideal quality is implanted just beneath the blood clot, because several studies have reported that the material quality of the tissue located just beneath the transplanted tissue significantly affects the degree of hyaline cartilage regeneration in autologous chondrocyte transplantation [18–20].

We anticipated that a double-network (DN) gel, composed of poly-(2-acrylamido-2-methylpropanesulfonic acid; PAMPS) and poly-(N,N′-dimethyl acrylamide; PDMAAm) (PAMPS/PDMAAm) [21], may have the potential to induce the differentiation of mesenchymal stem cells contained in the blood clot into chondrocytes, because this DN gel is a stable [22] but bioactive hydrogel [23] having a high degree of stiffness (elastic modulus of 0.20 MPa) and mechanical strength (compressive fracture strength of 3.1 MPa) [24]. The DN gel is completely different in concept from the common interpenetrated polymer gel or fiber-reinforced hydrogel. The DN gel consists of two independently cross-linked polymer networks, which are physically entangled with each other. There is no other interaction between the two networks and no micro-phase separation between the two components.

Thus, we have developed a novel method to induce spontaneous hyaline cartilage regeneration in vivo for a large osteochondral defect by implanting a plug made from the PAMPS/PDMAAm DN gel in the bottom of the cavity, leaving the defect itself vacant [25]. Here, we introduce this novel method.

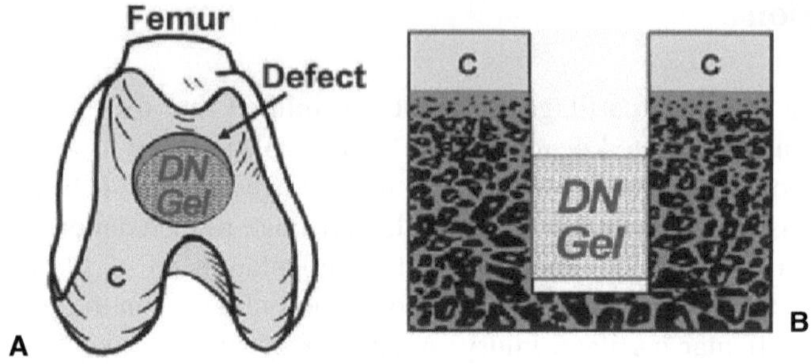

Fig. 1A,B. How to induce cartilage regeneration. **A** We implanted a double-network (*DN*) gel plug into an osteochondral defect created in the femoral groove of the patellofemoral joint. **B** A schematic cross-section of the osteochondral defect into which the plug was implanted. Note that a defect a few millimeters in depth from the cartilage surface remained after surgery. *C*, cartilage

Method to Induce Cartilage Regeneration

The PAMPS/PDMAAm DN gel was synthesized using a previously reported two-step sequential polymerization method [21]. Animal experimentation was performed using mature female Japanese white rabbits, according to the rules and regulations of the Animal Care and Use Committee, Hokkaido University School of Medicine. The operation for each animal was performed under intravenous anesthesia (pentobarbital, 25 mg/kg) with sterile conditions.

From the PAMPS/PDMAAm DN gel, we created a cylindrical plug having a 4.5-mm diameter and a 9-mm length, and implanted it into an osteochondral defect created in the femoral groove of the right patellofemoral joint of the rabbit, so that a defect having a certain degree of depth remained after surgery (Fig. 1). The actual defect depth was precisely measured in the histological sections after the animal was killed. In the left knee, an osteochondral defect having the same degree of depth was created and this remained without any treatment to obtain the nontreated control. The incised joint capsule and the skin wound were closed in layers with 3-0 nylon sutures, and an antiseptic spray dressing was applied. Postoperatively, each animal was allowed unrestricted activity in a cage (310 × 550 ×320 mm) without any joint immobilization.

Effects of Gel Implantation in the Defect

We conducted an in vivo experimental study using a total of 60 rabbits. In the right knee, we implanted a cylindrical plug made from the PAMPS/PDMAAm DN gel, polyvinyl alcohol (PVA) gel, or an ultra-high-molecular-weight polyethylene (UHMWPE) into the defect so that a defect having a 1.5-, 2.5-, or 3.5-mm depth

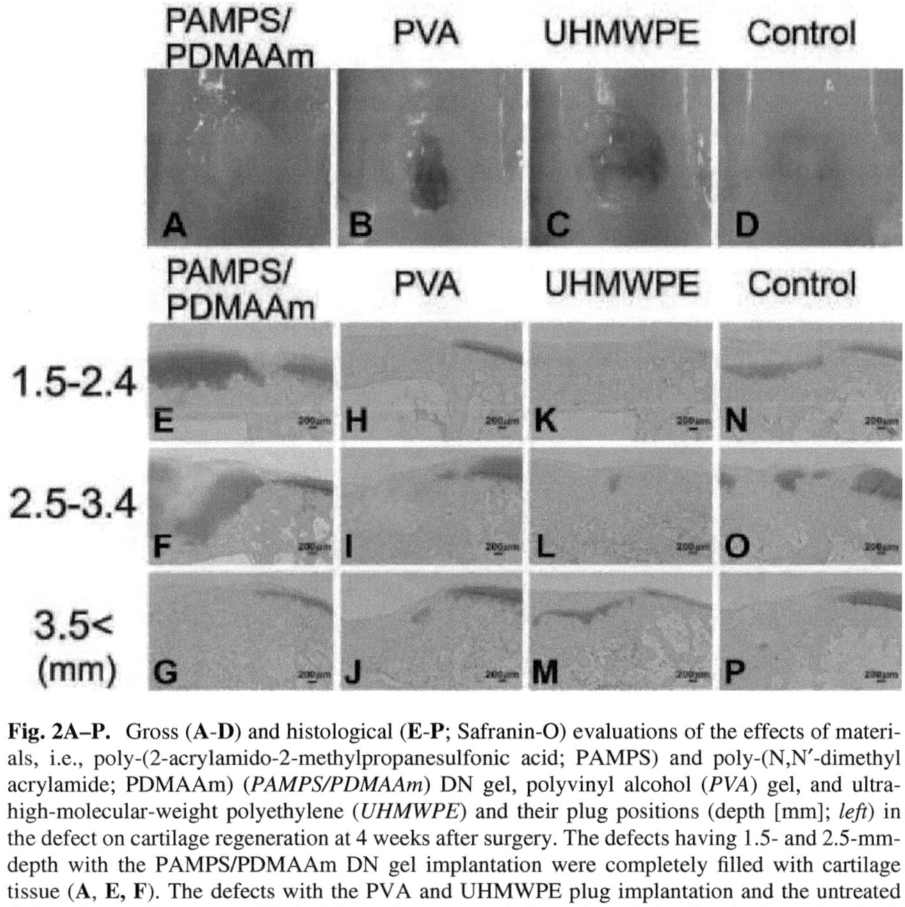

Fig. 2A–P. Gross (**A-D**) and histological (**E-P**; Safranin-O) evaluations of the effects of materials, i.e., poly-(2-acrylamido-2-methylpropanesulfonic acid; PAMPS) and poly-(N,N′-dimethyl acrylamide; PDMAAm) (*PAMPS/PDMAAm*) DN gel, polyvinyl alcohol (*PVA*) gel, and ultra-high-molecular-weight polyethylene (*UHMWPE*) and their plug positions (depth [mm]; *left*) in the defect on cartilage regeneration at 4 weeks after surgery. The defects having 1.5- and 2.5-mm-depth with the PAMPS/PDMAAm DN gel implantation were completely filled with cartilage tissue (**A**, **E**, **F**). The defects with the PVA and UHMWPE plug implantation and the untreated defects (*control*) were not filled with cartilage tissue, independent of the plug implantation depth (**H-P**)

remained after surgery. The latter two materials were chosen as representative of biomaterials that have been experimentally applied to create artificial cartilage. In each left knee, an osteochondral defect having 1.5-, 2.5-, or 3.5-mm depth was created without any treatment. Postoperatively, each animal was allowed unrestricted activity. Each animal was killed at 4 weeks after surgery.

In gross observation of the joint surface repair, the defects having 1.5- and 2.5-mm depth treated with the PAMPS/PDMAAm DN gel implantation were completely filled with a white opaque tissue (Fig. 2A). When the PVA and UHMWPE plugs were implanted, all defects were insufficiently filled with white and reddish opaque, patchy tissues, independent of the plug implantation depth (Fig. 2B, C). The untreated (control) defect surface showed white and reddish opaque, patchy, stiff tissues, independent of the depth (Fig. 2D). In histological observations (Fig. 2E-P), when the PVA and UHMWPE plugs were implanted, all defects were insufficiently filled with a fibrous tissue, independent of the plug implantation depth (Fig. 2H-M). For statistical comparisons, the gross appearance and the histological findings were evaluated by the scoring system reported by Wayne et al. [26]. The gross appearance score of the 1.5-mm-deep defect with the DN gel bottom was significantly greater than that for the other defects ($P < 0.0001$, except for the 3.5-mm-deep defect where $P = 0.0015$; Fig. 3A). The score of the defects with a PVA or UHMWPE bottom was not significantly greater than the score of the control defect; rather, the score was less than that of the control. The histological score of the 1.5-mm-deep defect with the DN gel bottom was significantly greater than those of the other defects ($P < 0.0001$ except for the 1.5- and 3.5-mm deep defects, where $P = 0.0006$; Fig. 3B). The score of the defects with the PVA or UHMWPE bottom was not significantly greater than that of the control defect; rather, the score was less than that of the control.

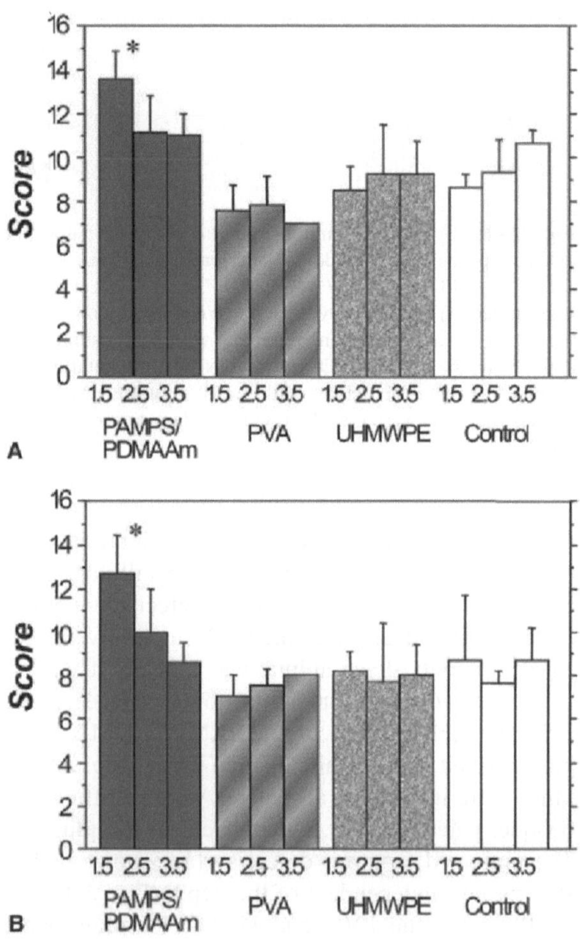

Fig. 3. A The gross appearance score was significantly greater in the 1.5-mm-deep defect with the bottom made from the DN gel than in the other defects (*$P < 0.0001$, except for the 3.5-mm-deep defect, where $P = 0.0015$). **B** The histological score was significantly greater in the 1.5-mm-deep defect with the bottom made from the DN gel than in the other defects (*$P < 0.0001$, except for the 1.5- and 3.5-mm-deep defects, where $P = 0.0006$). The gross appearance score and the histological score of the defects with the bottom made from the PVA or UHMWPE plugs were not significantly greater, but were, rather, less than the scores for the control defect

These results demonstrated that the PAMPS/PDMAAm DN gel implanted at the bottom of the osteochondral defect successfully induced spontaneous cartilage regeneration in the defect, and that this effect was affected by the depth of the defect that was intentionally created by the gel plug implantation.

In Vivo Evaluation of Regenerated Tissues

We implanted the same DN gel plug as that used above into an osteochondral defect created in the femoral groove of the right patellofemoral joint, so that a defect having 1.5-mm depth remained after surgery. In the left knee, an osteochondral defect having 1.5-mm depth was created and remained without any treatment to obtain the nontreated control. Five rabbits were killed by pentobarbital injection at 1, 2, 3, and 4 weeks after surgery, respectively.

Low-magnification histology showed that the untreated (control) defect was filled with fibrous and bone tissues even at 4 weeks, while a small amount of proteoglycan-rich tissue was occasionally and irregularly seen among these tissues (Fig. 4A-D-1). Type 2 collagen expression was not found in the tissue regenerated in the untreated defect, except for a limited amount in the peripheral portion (Fig. 4D-2). On the other hand, the treated defect was filled with a blood clot at 1 week. In the defect filled with a blood clot, symmetrical triangular (in cross-section) zones rich in cells were observed in a localized zone close to the bony wall and the implanted gel (Fig. 4E). In the triangular zones, a cartilage-like tissue rich in proteoglycan appeared at 2 weeks (Fig. 4F), increasing at 3 weeks (Fig. 4G). At 4 weeks, the defect was filled by a sufficient volume of the proteoglycan-rich tissue with regenerated subchondral bone tissue (Fig. 4H-1). Immunohistochemical observations showed that type-2 collagen was abundantly expressed in the proteoglycan-rich tissue regenerated in the treated defect. Specifically, at 4 weeks, the type-2 collagen expression was richer in a zone close to the joint surface than in a deep zone (Fig. 4H-2), resembling normal articular cartilage.

In high-magnification histology at 1 week, we found a number of elliptical and spindle-shaped cells in the above-described triangular zones in the defect. The elliptical cells were observed in the core portion of the triangular zones and the portion close to the bony wall, and the spindle-shaped cells resembling stem cells were predominantly seen in the marginal (transitional) portion of the triangular zone (Fig. 4a). At 2 weeks, the core portion of the triangular zone consisted of a number of large elliptical or round cells rich in cytoplasm, which were scattered separately in a proteoglycan-rich matrix. At the marginal (transitional) portion of the triangular zone, the size of these cells was smaller and the matrix surrounding these cells showed less intense staining with Safranin-O than the other portion (Fig. 4b). Type-2 collagen was expressed within and surrounding the large elliptical or round cells and this finding showed that the hyaline cartilage tissue had regenerated. At 4 weeks, fairly large round cells rich in cytoplasm were scattered singly or as isogenous groups in the proteoglycan-rich matrix (Fig. 4c-1). In these cells, type-2

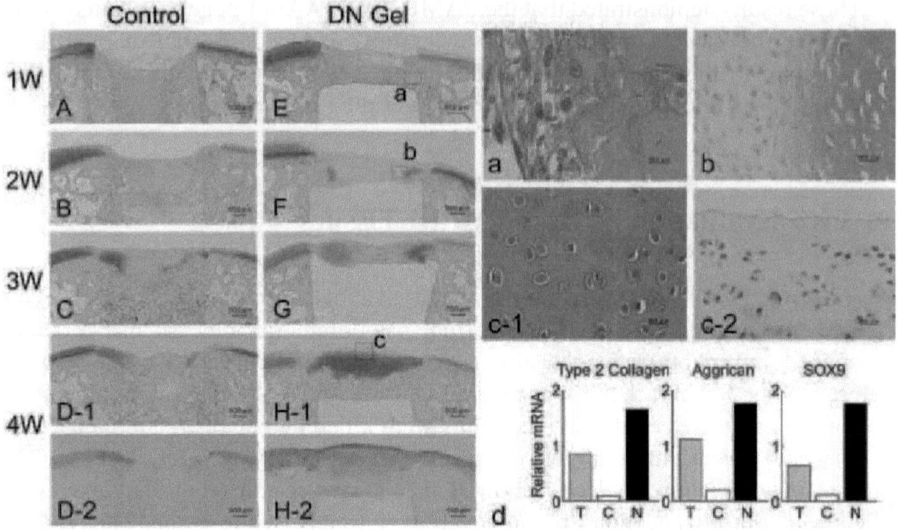

Fig. 4. A-H-2 In vivo evaluation of regenerated tissues; low-magnification histology. The untreated (control) defect was filled with fibrous and bone tissues (**A-D1**; Safranin-O, D2; anti-type-2 collagen). Type-2 collagen expression was not found in the tissue regenerated in the untreated defect (**D-2**; anti-type-2 collagen). The treated defect was filled with a blood clot at 1 week (**E**; Safranin-O). In the defect, symmetrical triangular zones rich in cells were observed in a localized zone close to the bony wall and the implanted gel. At 2 weeks, a cartilage-like tissue rich in proteoglycan appeared in the triangular zones (**F**; Safranin-O), increasing at 3 weeks (**G**; Safranin-O). At 4 weeks, the defect was filled by a sufficient volume of the proteoglycan-rich tissue with regenerated subchondral bone tissue (**H-1**; Safranin-O). Type-2 collagen expression was richer in the zone close to the joint surface than in a deep zone at 4 weeks (**H-2**; anti-type 2-collagen). **a-c** High-magnification histology of the treated defect (high magnification of insets *a*, *b*, and *c*, in **E**, **F**, and **H-1**). At 1 week, elliptical cells were observed in both the core portion of the triangular zones and the portion close to the bony wall, and spindle-shaped cells resembling stem cells were predominantly seen in the marginal (transitional) portion of the triangular zone (**a**; HE). At 2 weeks, a number of large elliptical or round cells rich in cytoplasm were separately scattered in the proteoglycan-rich matrix in the core portion of the triangular zone (**b**; Safranin-O). At 4 weeks, fairly large round cells rich in cytoplasm were scattered singly or as isogenous groups in the proteoglycan-rich matrix (**c-1**; Safranin-O). The most superficial part was devoid of cells, resembling the normal lamina splendens (**c-2**; Safranin-O). **d** Real-time polymerase chain reaction analyses performed at 4 weeks (*each bar* shows average of results from three rabbits) showed that the mRNAs of type-2 collagen, Aggrican, and SOX9 were obviously expressed in the cells regenerated in the defect, while such expression was seldom seen in the cells regenerated in the untreated defect. *T*; DN gel treated, *C*; untreated control defect, *N*; normal cartilage.

collagen was richly expressed. In the superficial layer of this tissue, the cells were rather small and sparse, while some cells were aligned as cell columns parallel to the surface. In addition, the most superficial part was devoid of cells, resembling the lamina splendens in the normal articular cartilage (Fig. 4c-2). Thus, the regenerated proteoglycan-rich tissue showed a four-layer structure similar to the normal articular cartilage structure, although the orientation of the chondrocyte columns in the radial layer was rather poor and the calcified layer was irregular.

We also performed real-time polymerase chain reaction (PCR) analyses of the expression of type-2 collagen, Aggrican, and SOX9 mRNAs in the large round cells observed at 4 weeks. These mRNAs were obviously expressed (with expression comparable to that in normal chondrocytes) while such expression was seldom seen in the tissues regenerated in the untreated defect (Fig. 4d).

A series of these evaluations demonstrated that the implantation of PAMPS/PDMAAm DN gel on the bottom of an osteochondral defect could induce spontaneous hyaline cartilage regeneration in vivo. We think that the migrated undifferentiated cells, including mesenchymal stem cells from the bone marrow differentiated into chondrocytes in the in vivo biochemical and biomechanical environment created by the PAMPS/PDMAAm DN gel.

Discussion

Our studies showed that spontaneous hyaline cartilage regeneration could be induced in vivo within a large osteochondral defect in the rabbit by means of implanting a cylindrical PAMPS/PDMAAm DN gel plug into the bottom of the defect so that a 1- to 3-mm-deep vacant space was intentionally left in the defect. This finding has significantly modified the commonly established concept that articular cartilage tissue cannot spontaneously regenerate in vivo. Also, this finding has prompted a novel strategy for the repair of an osteochondral defect.

At the present time, the most common strategy used to repair an osteochondral defect is the implantation of various tissue-engineered cartilage tissues. Concerning this strategy, however, various practical problems have been reported [11–16]. For example, cell sources for culture are limited; donor site morbidity is inevitable; it is difficult for tissue regenerated in vitro to remain stable in the defect; the patient has to undergo two surgical procedures, which include cell harvesting and implantation; and it takes a few months until weight-bearing is allowed. In addition, it is difficult to constantly regenerate a sufficient amount of hyaline cartilage having a normal function in an animal serum-free medium, while animal serum contained in the medium for cell culture has a risk of serious disease transmission. Economically, the total cost of the treatment per patient is estimated to be prohibitively expensive. Our novel strategy of the induction of spontaneous cartilage regeneration may have the potential to solve almost all of the above-described problems that are inherent with the common strategy, leading to an innovation in the clinical field of articular cartilage repair.

Recently, in experimental trials, some investigators have tried to completely fill osteochondral defects with acellular polymer scaffolds with signaling molecules [27–29]. However, these experimental trials have some limitations to be improved for clinical application at this time [25]. The above-described novel spontaneous regeneration strategy developed by us is completely different from these strategies. This novel strategy should be studied in greater detail in the near future as a realistic research focus.

Conclusion

We have developed a novel method to induce spontaneous hyaline cartilage regeneration in vivo for a large osteochondral defect by implanting a plug made from PAMPS/PDMAAm DN gel in the bottom of the cavity, leaving the defect itself vacant. At 4 weeks, cartilage tissue rich in proteoglycan and type-2 collagen had regenerated in the defect. In the cells of the regenerated tissue, type-2 collagen, Aggrican, and SOX9 mRNAs were highly expressed. On the other hand, cartilage regeneration was rarely found in defects without any treatment or in defects treated with PVA gel or UHMWPE plugs in the same manner. The findings of the present study have demonstrated that spontaneous regeneration of the articular cartilage can be induced in vivo in an osteochondral defect. Cartilage regeneration by using this PAMPS/PDMAAm DN gel has given a significant modification to the commonly established concept that articular cartilage tissue cannot spontaneously regenerate in vivo.

References

1. Mandelbaum B, Browne JE, Fu F et al (2007) Treatment outcomes of autologous chondrocyte implantation for full-thickness articular cartilage defects of the trochlea. Am J Sports Med 35: 915–921
2. Buckwalter JA, Mankin HJ (1998) Articular cartilage repair and transplantation. Arthritis Rheum 41: 1331–1342
3. Brittberg M, Lindahl A, Nilsson A et al (1994) Treatment of deep cartilage defects in the knee with autologous chondrocyte transplantation. N Engl J Med 331: 889–895
4. Peterson L, Minas T, Brittberg M et al (2003) Treatment of osteochondritis dissecans of the knee with autologous chondrocyte transplantation: results at 2 to 10 years. J Bone Joint Surg Am 85: 17–24
5. Hangody L, Füles P (2003) Autologous osteochondral mosaicplasty for the treatment of full-thickness defects of weight-bearing joints: 10 years of experimental and clinical experience. J Bone Joint Surg Am 85: 25–32
6. Horas U, Pelinkovic D, Herr G et al (2003) Autologous chondrocyte implantation and osteochondral cylinder transplantation in cartilage repair of the knee joint. A prospective, comparative trial. J Bone Joint Surg Am 85: 185–192
7. Knutsen G, Engebretsen L, Ludvigsen TC et al (2004) Autologous chondrocyte implantation compared with microfracture in the knee. A randomized trial. J Bone Joint Surg Am 86: 455–464
8. Browne JE, Anderson AF, Arciero R et al (2005) Clinical outcome of autologous chondrocyte implantation at 5 years in US subjects. Clin Orthop Relat Res 436: 237–245
9. Henderson I, Francisco R, Oakes B et al (2005) Autologous chondrocyte implantation for treatment of focal chondral defects of the knee—a clinical, arthroscopic, MRI and histologic evaluation at 2 years. Knee 12: 209–216
10. Ochi M, Adachi N, Nobuto H et al (2004) Articular cartilage repair using tissue engineering technique–novel approach with minimally invasive procedure. Artif Organs 28: 28–32
11. Smith GD, Knutsen G, Richardson JB (2005) A clinical review of cartilage repair techniques. J Bone Joint Surg Br 87: 445–449
12. Feczkó P, Hangody L, Varga J et al (2003) Experimental results of donor site filling for autologous osteochondral mosaicplasty. Arthroscopy 19: 755–761

13. Driesang IM, Hunziker EB (2000) Delamination rates of tissue flaps used in articular cartilage repair. J Orthop Res 18: 909–911
14. Micheli LJ, Browne JE, Erggelet C et al (2001) Autologous chondrocyte implantation of the knee: multicenter experience and minimum 3-year follow-up. Clin J Sport Med 11: 223–228
15. Redman SN, Oldfield SF, Archer CW (2005) Current strategies for articular cartilage repair. Eur Cell Mater 9: 23–32
16. Buckwalter JA (2002) Articular cartilage injuries. Clin Orthop Relat Res 402: 21–37
17. Steadman JR, Rodkey WG, Rodrigo JJ (2001) Microfracture: surgical technique and rehabilitation to treat chondral defects. Clin Orthop Relat Res 391: 362–369
18. Ahsan T, Sah RL (1999) Biomechanics of integrative cartilage repair. Osteoarthritis Cartilage 7: 29–40
19. Qiu YS, Shahgaldi BF, Revell WJ et al (2003) Observations of subchondral plate advancement during osteochondral repair: a histomorphometric and mechanical study in the rabbit femoral condyle. Osteoarthritis Cartilage 11: 810–820
20. Shapiro F, Koide S, Glimcher MJ (1993) Cell origin and differentiation in the repair of full-thickness defects of articular cartilage. J Bone Joint Surg Am 75: 532–553
21. Gong JP, Katsuyama Y, Kurokawa T et al (2003) Double-network hydrogels with extremely high mechanical strength. Adv Mater 15: 1155–1158
22. Azuma C, Yasuda K, Tanabe Y et al (2007) Biodegradation of high-toughness double network hydrogels as potential materials for artificial cartilage. J Biomed Mater Res A 81: 373–380
23. Tanabe Y, Yasuda K, Azuma C, et al (2008) Biological responses of novel high-toughness double network hydrogels in muscle and the subcutaneous tissues. J Mater Sci Mater Med 19: 1379–1387
24. Yasuda K, Gong JP, Katsuyama Y et al (2005) Biomechanical properties of high-toughness double network hydrogels. Biomaterials 26: 4468–4475
25. Yasuda K, Kitamura N, Gong JP et al (2009) A novel double-network hydrogel induces spontaneous articular cartilage regeneration in vivo in a large osteochondral defect. Macromol Biosci 9: 307–316
26. Wayne JS, McDowell CL, Shields KJ et al (2005) In vivo response of polylactic acid-alginate scaffolds and bone marrow-derived cells for cartilage tissue engineering. Tissue Eng 11: 953–963
27. Holland TA, Bodde EW, Baggett LS et al (2005) Osteochondral repair in the rabbit model utilizing bilayered, degradable oligo(poly(ethylene glycol) fumarate) hydrogel scaffolds. J Biomed Mater Res A 75: 156–167
28. Gotterbarm T, Richter W, Jung M et al (2006) An in vivo study of a growth-factor enhanced, cell free, two-layered collagen-tricalcium phosphate in deep osteochondral defects. Biomaterials 27: 3387–3395
29. Fukuda A, Kato K, Hasegawa M et al (2005) Enhanced repair of large osteochondral defects using a combination of artificial cartilage and basic fibroblast growth factor. Biomaterials 26: 4301–4308

Bone Marrow Stromal Cell Transplantation for Central Nervous System Disorders: Perspectives for Translational Research and Clinical Application

Satoshi Kuroda[1], Yuji Kuge[2], Nagara Tamaki[3], and Yoshinobu Iwasaki[1]

Summary

Objectives: There is increasing evidence that transplanted bone marrow stromal cells (BMSCs) significantly promote functional recovery after central nervous system (CNS) damage in animal models of various kinds of CNS disorders, including cerebral infarct, brain contusion, and spinal cord injury. However, there is a lack of information when considering the clinical application of BMSC transplantation for patients with neurological disorders. Therefore, we discuss here what we should clarify to establish cell transplantation therapy in the clinical situation, and we describe our recent work for this purpose.
Methods and Results: BMSCs have the ability to alter their gene expression profile and phenotype in response to the surrounding circumstances and to protect neurons by producing some neurotrophic factors. They also promote neurite extension and rebuild the neural circuits in the injured CNS. Using optical imaging techniques, transplanted BMSCs can be tracked noninvasively in living animals for at least 8 weeks after transplantation. Functional imaging such as positron emission tomography (PET) scans may have the potential to assess the beneficial effects of BMSC transplantation.
Conclusion: It is an urgent issue to develop clinical imaging techniques for tracking transplanted cells in the CNS and evaluating the therapeutic significance of BMSC transplantation in order to establish it as a definite therapeutic strategy in the clinical situation in the future.

Key words Bone marrow stromal cell · Transplantation · Cerebral stroke · Spinal cord injury · Translational study

[1]Department of Neurosurgery, Hokkaido University Graduate School of Medicine, North 15 West 7, Kita-ku, Sapporo 060-8638, Japan
[2]Department of Tracer Kinetics and Bioanalysis, Hokkaido University Graduate School of Medicine, North 15 West 7, Kita-ku, Sapporo 060-8638, Japan
[3]Department of Nuclear Medicine, Hokkaido University Graduate School of Medicine, North 15 West 7, Kita-ku, Sapporo 060-8638, Japan

N. Tamaki, Y. Kuge (eds.)
Molecular Imaging for Integrated Medical Therapy and Drug Development.
DOI: 10.1007/978-4-431-98074-2_13, ©Springer 2010

Introduction

Even now, central nervous system (CNS) disorders can easily cause longstanding disability. Recent studies, however, have shown that cell transplantation may potentially promote the recovery of neurological function in various pathological conditions of the CNS, including cerebral infarct, spinal cord injury, and traumatic brain injury. A variety of cell types have been studied as cell sources of transplantation into animal models of CNS disorders, including embryonic stem (ES) cells, neural stem cells, and bone marrow stromal cells (BMSCs). Of these, BMSCs may have the greatest therapeutic potential, because they can be harvested from the patients themselves without posing ethical or immunological difficulties [1, 2]. There is increasing evidence that the transplanted BMSCs enhance functional recovery by differentiating into neural cells and/or by producing various cytokines or growth factors that can rescue the host neurons [3, 4]. Although the results are encouraging, a variety of questions or problems still remain to be solved [3, 5–18]. In this review, we present recent progress in basic and clinical research in the field of BMSC transplantation for CNS disorders; we also discuss and critically examine what we should clarify to introduce BMSC transplantation therapy in the clinical situation as a scientifically proven therapy.

Recent Progress in BMSC Transplantation

Although the exact mechanisms underlying the beneficial effects of BMSC transplantation have not been fully clarified yet, several hypotheses have been proposed to explain these effects.

First, BMSCs per se are believed to differentiate into neural cells in the host's brain ("transdifferentiation theory"). This theory is based on findings that BMSCs simulate neuronal morphology and express the proteins specific for neurons in vitro [19, 20] and in vivo [21, 22]. Although the transdifferentiation theory is quite attractive, several questions still remain. For example, how is the mesenchymal cell fate of BMSCs oriented to the neuronal lineage? Are the morphological changes and expression of neuronal phenotype in BMSCs identical to the morphology and phenotype that would be seen in the differentiation of functional neuronal cells? Actually, several studies have posed questions about the in vitro differentiation of BMSCs into neurons [23, 24]. In order to seek the answers, we chemically treated cultured BMSCs and found that they could potentially modify their gene expression profile in response to the surrounding environment [13]. Furthermore, a certain subpopulation of BMSCs, when co-cultured with neurons, morphologically simulated neurons and expressed neuron-specific proteins without evidence of cell fusion [3]. These findings strongly suggest that at least a certain subpopulation of BMSCs has the potential to alter their gene expression profile and to differentiate into neural cells in response to the surrounding environment.

Second, transplanted BMSCs have also been reported to fuse with host cells and to simulate differentiation into host cells ("cell fusion theory") [25]. Indeed, some BMSCS, when co-cultured with neurons, fuse with the neurons and acquire the phenotypes of both cells [3]. However, the roles and fates of the BMSCs fused with the neurons are still completely unknown.

Third, there is increasing evidence that BMSCs produce some neuroprotective or neurotrophic factors and support the survival of the host neural cells [26]. This "feeder theory" is quite natural, because BMSCs per se support the homing and proliferation of hematopoietic cells in the bone marrow by producing a variety of cytokines, such as stromal cell-derived factor-1α (SDF-1α) [27]. Indeed, the conditioned medium of BMSCs significantly promoted neurite outgrowth from the dorsal root ganglion [28]. Very recently, we have co-cultured BMSCs with neurons subjected to glutamate exposure, using a three-dimensional co-culture paradigm. We found that the co-cultured BMSCs released various kinds of soluble factors, including nerve growth factor (NGF), hepatocyte growth factor (HGF), and brain-derived neurotrophic factor (BDNF), and these co-cultured BMSCs ameliorated the glutamate-induced neuronal injury [3]. These observations strongly suggest that BMSCs consist of heterogeneous cell populations and that they protect and/or repair the damaged CNS through multiple mechanisms.

Since the reporting of several pioneering studies [21, 22], numerous studies have confirmed that BMSCs, when transplanted into animal models of CNS disorders, can survive, migrate into the lesion, express neural phenotypes, and enhance functional recovery. In the majority of these studies, the BMSCs are transplanted within 7 days after the insults, and the beneficial effects of BMSC transplantation can be observed approximately 4 weeks after transplantation [1, 2]. Recently, we evaluated whether or not transplanted BMSCs can proliferate in the CNS. For this purpose, we labeled green fluorescence protein (GFP)-expressing BMSCs with a superparamagnetic iron oxide (SPIO) agent and transplanted the cells into the ipsilateral striatum of mouse brain subjected to permanent focal ischemia, at 7 days after the insult. Fluorescence immunohistochemistry revealed that the GFP-positive cells were widely distributed around the infarct and that some of these cells also expressed microtubule-associated protein 2 (MAP2) and neuronal nuclear antigen (NeuN) 3 months after transplantation. However, only a small number of SPIO-positive cells could be detected on Turnbull blue staining. The proportion of SPIO-positive to GFP-positive cells was approximately 2.7%. This value was comparable to the finding that the proportion of SPIO-positive BMSCs gradually decreased from 93.6% at passage (P) 3 to 6.5% at P7 when the passages were repeated in vitro. These results strongly suggested that BMSCs repeat proliferation many times, migrate into the lesion, and that some of the cells express the neuronal phenotype in the host brain during 3 months after transplantation [15]. Previous studies have shown that transplanted BMSCs migrate towards the lesion, although the underlying mechanisms have not been clarified. Recent studies have shown that some chemokines, such as monocyte chemoattractant protein-1 (MCP-1) and SDF-1α are expressed around the damaged CNS tissue and play an important role in the migration of the transplanted cells [29, 30]. Recently, we investigated the role of CXCR4, a specific receptor for

SDF-1α, in the migration of BMSCs in the CNS. The BMSCs were isolated from wild-type (WT) and CXCR4-knockout (KO) mice, and were transplanted into ischemic mouse brain. The results showed that recovery of motor function in the WT BMSC-transplanted mice was more pronounced than that in the CXCR4-KO-transplanted mice. SDF-1α was extensively expressed in the peri-infarct area. In the WT BMSC-transplanted mice, the transplanted cells were extensively distributed in the ipsilateral hemisphere, although these findings were not observed in the CXCR4-KO BMSC-transplanted mice. The results suggest that the SDF-1α/CXCR4 system may play a critical role in the migration of transplanted BMSCs and contribute to the recovery of neurological function [11]. Likewise, Son et al. [31] also reported that the SDF-1/CXCR4 and HGF/c-Met axes were involved in the recruitment of BMSCs to damaged tissue. It may be quite valuable to elucidate the temporal profile of the expression of these chemokines around damaged CNS tissue to determine the optimal timing of BMSC transplantation.

BMSCs may also promote axon regeneration by secreting neuroprotective and/or neurotrophic factors. Thus, Hofstetter et al. [32] transplanted BMSCs into an injured spinal cord and found that the engrafted BMSCs were tightly associated with longitudinally arranged immature astrocytes and formed bundles bridging the epicenter of the injury. Interestingly, BMSCs dramatically promoted neurite extension toward their cluster when topically applied onto an organotypic spinal cord slice [10]. A very recent study in our laboratory demonstrated that transplanted BMSCs not only acquired neural cell phenotypes but that they were also integrated into the neural circuits of the host around the injured spinal cord, promoting the recovery of neurological function [5].

Knowledge of the mechanisms underlying functional recovery after BMSC transplantation is largely based on histological findings. Alternatively, recent autoradiographic studies have demonstrated that transplanted BMSCs express the protein for the γ-aminobutyric acid (GABA) receptor and improve the binding potential for ^{125}I-iomazenil around CNS lesions [12, 16]. Mori et al. [33] transplanted BMSCs into a rat cold-injury model and found, using an autoradiography technique, that the transplanted BMSCs improved glucose metabolism in response to sensory stimuli.

The majority of previous animal studies on BMSC transplantation have focused on the effects of such transplantation on motor function [1, 2]. In fact, BMSCs promote the recovery of motor function, when transplanted into animal models of CNS disorders such as cerebral infarct and spinal cord injury, within 7 days after the insults. However, BMSC transplantation may also support the recovery of cognitive function after traumatic brain injury (TBI). Thus, mild to severe TBI often induces widespread damage to axons and white matter through shearing forces, called diffuse axonal injury (DAI). Much of the DAI in humans occurs in a scattered, multifocal distribution without focal cerebral contusion. Nowadays, DAI is well known to play a major role in the development of cognitive dysfunction, emotional difficulties, and behavioral disturbances in patients following TBI [17]. Maruichi et al. [18] have shown that, when transplanted into rat brain subjected to DAI, BMSCs are widely distributed in the injured neocortex and significantly improve cognitive function, measured by the Morris water-maze test. Further experiments would clarify the beneficial effects of cell transplantation therapy on cognitive dysfunction caused by various kinds of CNS disorders.

Preliminary Clinical Trials of BMSC Transplantation for CNS Disorders

Based on the above observations obtained from animal experiments, some preliminary clinical testing has already been started [34–39]. Thus, Bang et al. [34] intravenously injected autologous BMSCs into 5 patients with severe neurological deficits due to ischemic stroke at 5 to 9 weeks after the onset, and they concluded that autologous BMSC infusion was a feasible and safe therapy that may improve functional recovery. On the other hand, Sykova et al. [37] injected unmanipulated bone marrow cells, through an intra-arterial or intravenous route, in 20 patients with spinal cord injury in the acute or chronic stage. Yoon et al. [38] isolated autologous bone marrow mononuclear cells from 35 patients with spinal cord injury, and directly injected them into the injured spinal cord in the acute or chronic stage. They found an improvement of neurological function in patients who were treated in the acute or subacute stage [38]. Saito et al. [36] intrathecally infused autologous BMSCs into a 35-year-old patient with spinal cord injury 2 weeks after the onset. Of note, Zhang et al. [39] expanded autologous BMSCs in culture, and directly injected them into the brain during surgery in 7 patients with traumatic brain injury. Lee et al. [35] transplanted BMSCs into 11 patients with multiple system atrophy through three consecutive intra-arterial and intravenous routes, and found significant improvement in their neurological scores. These studies indicate that the transplantation of bone marrow-derived cells may cause no serious adverse effects. However, further studies will be essential to scientifically confirm the therapeutic effects of BMSC transplantation in CNS disorders. In fact, some investigators take a cautious attitude even if favorable results are reported, because a blinded placebo-controlled study is very difficult to employ in these clinical trials of cell transplantation therapy [40, 41]. Considering better strategies to establish BMSC transplantation as a scientifically proven therapy in the clinical situation, we should learn the lessons from the long (>50 years) history of the development of neuroprotective drugs.

Lessons from Preclinical Studies of Neuroprotective Drugs

It is well known that despite much animal research concerning the pathophysiology of focal brain injury, little of this work has translated into effective treatment modalities for stroke in humans [42]. As recently pointed out by Savitz and Fisher

[43], a large number of neuroprotective drugs have demonstrated varying degrees of effectiveness in preclinical models of ischemic stroke. Of these, a total of 15 agents have advanced to phase III clinical trials to date. However, none of them was proven to improve the outcome of patients with ischemic stroke. Savitz and Fisher [43] raised some major problems that account for the failures of these phase III clinical trials. First, inadequate preclinical testing may account for the failures. For example, in certain studies, in rat models of focal cerebral ischemia, some agents were administered just after the insult, and their effects on infarct volume were examined only at 24 h. Clinical trials were started on the basis of these results, although no data on long-term neurological outcome or on testing in aged, diseased animals or a second species beyond rats have been published. Publication bias may be responsible, in part, for the failure of clinical trials, because any scientific journals hesitate to accept negative results in animal studies of neuroprotective drugs. Second, Savitz and Fisher [43] have pointed out that inadequate clinical testing may explain why some phase III trials of neuroprotective drugs were not beneficial. Thus, many experimental studies have previously found the neuroprotective effects when the agents were administered before or just after the insult of cerebral ischemia. In the majority of phase III clinical trials, however, the agents were administered 6 to 8 h or even longer after stroke onset. Therefore, a significant dissociation in the therapeutic time window exists between animal experiments and phase III clinical trials. In addition to time-window considerations, the stroke subtype of patients enrolled in phase III clinical trials varies widely. Thus, the majority of experimental studies have used transient or permanent middle cerebral artery (MCA) occlusion in rodents, which simulates embolic infarct in humans. However, previous clinical trials have enrolled patients with multiple stroke types such as lacunar infarct and subcortical white matter infarct. Some neuroprotective drugs may have no impact on white matter ischemic injury.

Based on these historical considerations, the first Stroke Therapy Academic Industry Roundtable (STAIR) meeting was organized, and their recommendations for standards regarding preclinical neuroprotective and restorative drug development were published in 1999 [42]. As shown in Table 1, the recommendations

Table 1. STAIR recommendations for preclinical stroke drug development [48]

1. Evaluate the candidate drug in permanent and temporary occlusion models and in both rodent and gyrencephalic species.
2. Evaluate an adequate dose-response effect over a reasonable time window.
3. Appropriate physiological monitoring and blinding should be performed.
4. Histological and functional outcome measures should be assessed with prolonged survival to ensure that early treatment effects are not lost.
5. If feasible, treatment effects should be confirmed in both sexes and in aged animals.
6. Treatment effects should be replicated in several laboratories, including both industry and academic locations.
7. Data, both positive and negative, should be published.

STAIR, Stroke Therapy Academic Industry Roundtable

propose that the effects of neuroprotective drugs should be assessed by analyzing both histological and functional outcomes over an extended period, using appropriate animal stroke models. Precise evaluations of an adequate dose-response effect over a reasonable time window are also recommended.

Subsequently, a phase III clinical trial has been adopted to evaluate the beneficial effects of NXY-059 in acute ischemic stroke. NXY-059 is a newly developed spin trap agent that aggressively scavenges reactive oxygen species, and it is proven to ameliorate tissue damage due to focal cerebral ischemia in both rodents and primates. The agent has a clinically relevant therapeutic window, because it significantly reduces infarct volume and improves neurological function even when administered 4 to 6 h after the insult [44, 45]. Many investigators believed that NXY-059 was the first neuroprotective agent that fulfilled many of the STAIR recommendations for preclinical testing. In the first phase III clinical trial, named the Stroke-Acute-Ischaemic-NXY-Treatment (SAINT-I) trial, 1722 patients were enrolled up to 6 h after stroke onset, and a modified Rankin scale at 90 days was chosen to measure their functional outcome. The results showed that NXY-059 significantly improved patients' outcome [46]. The second, SAINT-II, trial enrolled 3195 patients, but could not reproduce the beneficial effects of NXY-059 for them [47]. Consequently, NXY-059 was then withdrawn from further development.

As pointed out by some investigators [43, 48], the failure of NXY-059 in the SAINT-II trial has cast a pall on the field of the development of neuroprotective drugs for acute ischemic stroke. These investigators [43, 48] have proposed that it is essential to bridge the still-existing gap between preclinical studies and clinical investigations in order to achieve clinical application of neuroprotective drugs. We note that most animal studies measure infarct volume as a fundamental indicator to assess the neuroprotective effects of the agents being tested. However, previous phase III clinical trials used disability and neurological deficit scales such as a modified Rankin scale and the Barthel index. Even the phase III clinical trials of NXY-059 did not determine the location and size of the cerebral infarct. In their review article, therefore, Savitz and Fisher [43] have suggested that future clinical trials should include a biologically relevant end-point. For example, it may be important to directly visualize the activity of neuroprotective agents on damaged tissue in clinical trials. Further, stroke type was not considered when enrolling the patients in the SAINT trials, although the neuroprotective effects had been examined in MCA occlusion models in the majority of animal studies. Therefore, Savitz and Fisher [43] have prop osed that future clinical trials should enroll only patients with an MCA infarct, but not those with a small-vessel infarct.

Responding to the STAIR call, neurologists and neurosurgeons in the United States organized the first Stem Cell Therapeutics as an Emerging Paradigm in Stroke (STEPS) meeting in 2007, and recommended translational criteria for designing laboratory studies on cell therapy for stroke (see Table 2). They concluded that the efficacy of cell transplantation therapy should be tested in multiple models of focal stroke, in two species, in both sexes, and in multiple laboratories prior to clinical translation [49].

Table 2. STEPS recommendations for preclinical cell therapy studies [7]

- The use of appropriate species and clinically relevant type of stroke models
- The standardization of outcome measures and treatment protocols
- The need for imaging of cell tracking and host response
- The requirement for safety indices
- A call for demonstrating mechanisms of action underlying restorative therapies in ischemic stroke

STEPS, Stem Cell Therapeutics as an Emerging Paradigm in Stroke

Questions to Be Answered Before the Clinical Application of BMSC Transplantation

There are several questions to be answered prior to the clinical application of BMSC transplantation for CNS disorders.

First, it is still undetermined when the BMSCs should be transplanted into the damaged CNS to achieve maximal therapeutic effects. As described above, the BMSCs are transplanted within 7 days after the insults in the majority of animal studies. Bakshi et al. [50] reported that transplantation within 14 days of spinal cord injury provided significantly greater grafting efficiency than more delayed delivery, when the BMSCs were intrathecally injected. However, there are few studies that show the effects of BMSC transplantation in the chronic stage of CNS disorders, except for spinal cord injury [51–53]. Despite these observations in animal experiments, the bone marrow-derived cells were usually transplanted several weeks (or even several months) after the onset of the insult in previous clinical trials [34–39]. Therefore, a considerable gap in treatment protocols exists between animal experiments and clinical trials, and this gap may correspond to the "inadequate preclinical testing" in the development of neuroprotective drugs mentioned above. The difficulty in rapidly expanding the yield of autologous BMSCs may explain the "delayed" therapy protocols in clinical trials. Unfortunately, there is still no optimal culture protocol that enables us to expand the yield of autologous BMSC to provide enough cells for transplantation therapy within 7 days after the onset of CNS disorders such as cerebral infarct. Indeed, in a recent clinical study, about 30 days were required to obtain 1×10^8 cells of autologous BMSCs with a conventional culture technique [34]. In order to solve the problem of the dissociation between animal experiments and clinical testing, we recently cultured mouse BMSCs with granulocyte colony-stimulating factor (G-CSF); almost all of the BMSCs were immunologically positive for the G-CSF receptor. G-CSF significantly enhanced their proliferation by modulating their cell cycle, and it also upregulated their production of NGF, HGF, and SDF-1α [54]. Such pharmacological "activation" of cultured BMSCs may contribute to the successful clinical application of BMSC transplantation therapy for CNS disorders in the near future.

Second, BMSCs have been transplanted directly, intravenously or intrathecally, in the majority of previous animal experiments. Although direct injection permits the efficient delivery of the donor cells to the damaged tissue, it has a potential risk of causing additional CNS injury. The injection of cells afloat in the medium may also result in limited cell retention and transplant survival. Intravenous or intrathecal transplantation is attractive because of its noninvasive, safe technique for the host CNS, but it has been reported to result in less pronounced cell migration and functional recovery than direct cell transplantation [51]. The blood-brain barrier should also be permeable to the intravenously administered cells to enable them to migrate into the brain [55]. Therefore, an optimal transplantation technique should be developed to provide maximally safe and efficient results, when applying BMSC transplantation in the clinical setting. Alternatively, the intra-arterial injection of BMSCs may be valuable to noninvasively deliver them to the damaged CNS [56]. More interestingly, the emerging field of tissue engineering may also provide promising alternatives. Tissue engineering approaches are designed to repair lost or damaged tissue through the use of cellular transplantation and biomaterial scaffolding. Nowadays, degradable biomaterials have been accepted as a valuable "scaffold" on which to fix and stabilize transplanted cells in organs such as bone, cartilage, heart, and skin. Until recently, however, there have been only a small number of studies that showed effective scaffolds for cell transplantation for CNS disorders [57]. Therefore, we recently assessed whether a fibrin matrix could act as an injectable, valuable scaffold in BMSC transplantation for injured CNS tissue. The results showed that the fibrin matrix markedly improved the survival and migration of BMSCs transplanted into the hemisected spinal cord or injured neocortex of rats [6, 58]. Thermoreversible gelation polymer (TGP) hydrogel may also be a candidate as a scaffold to provide a suitable environment for donor cells (see the chapter by Osanai et al., this volume). Such a tissue-engineering strategy could be one of the therapeutic options for CNS regeneration in patients with an injured CNS.

Third, it is essential to develop techniques to track the fate of the transplanted cells in the host CNS continuously and noninvasively in order to guide further advances in neurotransplantation research and its future clinical applications. A cell-tracking technique would also be important as a "biologically relevant endpoint" in the clinical situation (see above). Recent studies have suggested that magnetic resonance (MR) imaging, nuclear imaging, and optical imaging could be suitable candidates for cell-tracking. In previous animal experiments, the donor cells were identified on MR imaging by labeling with an SPIO agent [59–61]. MR imaging can image intact, opaque organisms in three dimensions with good spatial resolution, but it requires long imaging times and consequently slows data acquisition because of the low sensitivity. More importantly, the magnetic nanoparticles that label the donor cells cannot label all the cells during their proliferation. Nuclear imaging can detect transplanted cells by labeling them with radioactive tracers. Correa et al. [62] recently labeled autologous bone marrow mononuclear cells with ^{99m}Tc-hexamethylpropylene (HMPAO) and injected them into a patient with ischemic stroke through a balloon catheter. The transplanted cells were visualized on single photon emission tomography (SPECT). Nuclear imaging can detect the target with high sensitivity, but it is difficult to monitor the donor cells for periods of several weeks because of the relatively short half-life of clinically available tracers. On the other hand, optical imaging techniques may also serve future technology to visualize BMSCs engrafted in the damaged CNS. We isolated BMSCs from GFP transgenic mice, and transplanted the GFP-expressing BMSCs into mouse brain subjected to cerebral infarct. The results showed that, with an intravital fluorescence imaging technique, the BMSCs engrafted in the ipsilateral neocortex could be serially visualized through the skull [9]. Similarly, GFP-expressing BMSCs, when transplanted into the injured spinal cord in mice, could be identified through the dura mater [14]. However, it is difficult to detect the fluorescence emitted from GFP through the skin because of its relatively short wavelength. Very recently, therefore, we examined a near-infrared (NIR) fluorescence tracer with a much longer wavelength and showed that it could be a valuable tracer for the noninvasive tracking of donor cells transplanted into the brain (see the chapter by Sugiyama et al., this volume).

Conclusion

In this review, we have discussed recent progress in basic and clinical research in the field of BMSC transplantation for CNS disorders. Furthermore, based on the history of the development of neuroprotective drugs, we have critically examined the questions to be clarified in order to introduce BMSC transplantation therapy into the clinical situation as a scientifically proven therapy.

Acknowledgments This study was supported by a Grant-in-Aid from the Ministry of Education, Science and Culture of Japan. The authors thank all of their colleagues for their experimental work and discussions.

References

1. Bliss T, Guzman R, Daadi M et al (2007) Cell transplantation therapy for stroke. Stroke 38: 817–826
2. Parr AM, Tator CH, Keating A (2007) Bone marrow-derived mesenchymal stromal cells for the repair of central nervous system injury. Bone Marrow Transplant 40: 609–619.
3. Hokari M, Kuroda S, Shichinohe H et al (2008) Bone marrow stromal cells protect and repair damaged neurons through multiple mechanisms. J Neurosci Res 86: 1024–1035.
4. Prockop DJ, Gregory CA, Spees JL (2003) One strategy for cell and gene therapy: harnessing the power of adult stem cells to repair tissues. Proc Natl Acad Sci USA 100 (Suppl 1): 11917–11923.
5. Chiba Y, Kuroda S, Maruichi K et al (2009) Transplanted bone marrow stromal cells promote axonal regeneration and improve motor function in a rat spinal cord injury model. Neurosurgery 64: 991–999; discussion 999–1000.
6. Itosaka H, Kuroda S, Shichinohe H et al (2008) Fibrin matrix provides a suitable scaffold for bone marrow stromal cells transplanted into injured spinal cord: A novel material for CNS tissue engineering. Neuropathology 29: 248–257.

7. Lee JB, Kuroda S, Shichinohe H et al (2003) Migration and differentiation of nuclear fluorescence-labeled bone marrow stromal cells after transplantation into cerebral infarct and spinal cord injury in mice. Neuropathology 23: 169–180.
8. Lee JB, Kuroda S, Shichinohe H et al (2004) A pre-clinical assessment model of rat autogeneic bone marrow stromal cell transplantation into the central nervous system. Brain Res Brain Res Protoc 14: 37–44.
9. Shichinohe H, Kuroda S, Lee JB et al (2004) In vivo tracking of bone marrow stromal cells transplanted into mice cerebral infarct by fluorescence optical imaging. Brain Res Brain Res Protoc 13: 166–175.
10. Shichinohe H, Kuroda S, Tsuji S et al (2008) Bone marrow stromal cells promote neurite extension in organotypic spinal cord slice: significance for cell transplantation therapy. Neurorehabil Neural Repair 22: 447–457.
11. Shichinohe H, Kuroda S, Yano S et al (2007) Role of SDF-1/CXCR4 system in survival and migration of bone marrow stromal cells after transplantation into mice cerebral infarct. Brain Res 1183: 138–147.
12. Shichinohe H, Kuroda S, Yano S et al (2006) Improved expression of gamma-aminobutyric acid receptor in mice with cerebral infarct and transplanted bone marrow stromal cells: an autoradiographic and histologic analysis. J Nucl Med 47: 486–491.
13. Yamaguchi S, Kuroda S, Kobayashi H et al (2006) The effects of neuronal induction on gene expression profile in bone marrow stromal cells (BMSC)–a preliminary study using microarray analysis. Brain Res 1087: 15–27.
14. Yano S, Kuroda S, Lee JB et al (2005) In vivo fluorescence tracking of bone marrow stromal cells transplanted into a pneumatic injury model of rat spinal cord. J Neurotrauma 22: 907–918.
15. Yano S, Kuroda S, Shichinohe H et al (2005) Do bone marrow stromal cells proliferate after transplantation into mice cerebral infarct? A double labeling study. Brain Res 1065: 60–67.
16. Yano S, Kuroda S, Shichinohe H et al (2006) Bone marrow stromal cell transplantation preserves gammaaminobutyric acid receptor function in the injured spinal cord. J Neurotrauma 23: 1682–1692.
17. Maruichi K, Kuroda S, Chiba Y et al (2009) Graded model of diffuse axonal injury for studying head injury-induced cognitive dysfunction in rats. Neuropathology 29: 132–139.
18. Maruichi K, Kuroda S, Chiba Y et al (2009) Transplanted bone marrow stromal cells improves cognitive dysfunction due to diffuse axonal injury in rats. Neuropathology 2009 Jan 2 [E pub ahead of print].
19. Sanchez-Ramos J, Song S, Cardozo-Pelaez F et al (2000) Adult bone marrow stromal cells differentiate into neural cells in vitro. Exp Neurol 164: 247–256.
20. Woodbury D, Schwarz EJ, Prockop DJ et al (2000) Adult rat and human bone marrow stromal cells differentiate into neurons. J Neurosci Res 61: 364–370.
21. Azizi SA, Stokes D, Augelli BJ et al (1998) Engraftment and migration of human bone marrow stromal cells implanted in the brains of albino rats—similarities to astrocyte grafts. Proc Natl Acad Sci USA 95: 3908–3913
22. Kopen GC, Prockop DJ, Phinney DG (1999) Marrow stromal cells migrate throughout forebrain and cerebellum, and they differentiate into astrocytes after injection into neonatal mouse brains. Proc Natl Acad Sci USA 96: 10711–10716.
23. Lu P, Blesch A, Tuszynski MH (2004) Induction of bone marrow stromal cells to neurons: differentiation, transdifferentiation, or artifact? J Neurosci Res 77: 174–191.
24. Neuhuber B, Timothy Himes B, Shumsky JS et al (2005) Axon growth and recovery of function supported by human bone marrow stromal cells in the injured spinal cord exhibit donor variations. Brain Res 1035: 73–85.
25. Terada N, Hamazaki T, Oka M et al (2002) Bone marrow cells adopt the phenotype of other cells by spontaneous cell fusion. Nature 416: 542–545.
26. Zhong C, Qin Z, Zhong CJ et al (2003) Neuroprotective effects of bone marrow stromal cells on rat organotypic hippocampal slice culture model of cerebral ischemia. Neurosci Lett 342: 93–96.

27. Kortesidis A, Zannettino A, Isenmann S et al (2005) Stromal-derived factor-1 promotes the growth, survival, and development of human bone marrow stromal stem cells. Blood 105: 3793–3801.
28. Neuhuber B, Gallo G, Howard L et al (2004) Reevaluation of in vitro differentiation protocols for bone marrow stromal cells: disruption of actin cytoskeleton induces rapid morphological changes and mimics neuronal phenotype. J Neurosci Res 77: 192–204.
29. Wang L, Li Y, Chen J et al (2002) Ischemic cerebral tissue and MCP-1 enhance rat bone marrow stromal cell migration in interface culture. Exp Hematol 30: 831–836.
30. Askari AT, Unzek S, Popovic ZB et al (2003) Effect of stromal-cell-derived factor 1 on stem-cell homing and tissue regeneration in ischaemic cardiomyopathy. Lancet 362: 697–703
31. Son BR, Marquez-Curtis LA, Kucia M et al (2006) Migration of bone marrow and cord blood mesenchymal stem cells in vitro is regulated by stromal-derived factor-1-CXCR4 and hepatocyte growth factor-c-met axes and involves matrix metalloproteinases. Stem Cells 24: 1254–1264.
32. Hofstetter CP, Schwarz EJ, Hess D et al (2002) Marrow stromal cells form guiding strands in the injured spinal cord and promote recovery. Proc Natl Acad Sci USA 99: 2199–2204.
33. Mori K, Iwata J, Miyazaki M et al (2005) Functional recovery of neuronal activity in rat whisker-barrel cortex sensory pathway from freezing injury after transplantation of adult bone marrow stromal cells. J Cereb Blood Flow Metab 25: 887–898.
34. Bang OY, Lee JS, Lee PH et al (2005) Autologous mesenchymal stem cell transplantation in stroke patients. Ann Neurol 57: 874–882
35. Lee PH, Kim JW, Bang OY et al (2008) Autologous mesenchymal stem cell therapy delays the progression of neurological deficits in patients with multiple system atrophy. Clin Pharmacol Ther 83: 723–730.
36. Saito F, Nakatani T, Iwase M et al (2008) Spinal cord injury treatment with intrathecal autologous bone marrow stromal cell transplantation: the first clinical trial case report. J Trauma 64: 53–59.
37. Sykova E, Homola A, Mazanec R et al (2006) Autologous bone marrow transplantation in patients with subacute and chronic spinal cord injury. Cell Transplant 15: 675–687.
38. Yoon SH, Shim YS, Park YH et al (2007) Complete spinal cord injury treatment using autologous bone marrow cell transplantation and bone marrow stimulation with granulocyte macrophage-colony stimulating factor: Phase I/II clinical trial. Stem Cells 25: 2066–2073.
39. Zhang ZX, Guan LX, Zhang K et al (2008) A combined procedure to deliver autologous mesenchymal stromal cells to patients with traumatic brain injury. Cytotherapy 10: 134–139.
40. Bakay RA (2005) Neural transplantation. J Neurosurg 103: 6–8; discussion 8.
41. Quinn N, Barker RA, Wenning GK (2008) Are trials of intravascular infusions of autologous mesenchymal stem cells in patients with multiple system atrophy currently justified, and are they effective? Clin Pharmacol Ther 83: 663–665.
42. Stroke Therapy Academic Industry Roundtable (1999) Recommendations for standards regarding preclinical neuroprotective and restorative drug development. Stroke 30: 2752–2758.
43. Savitz SI, Fisher M (2007) Future of neuroprotection for acute stroke: in the aftermath of the SAINT trials. Ann Neurol 61: 396–402.
44. Kuroda S, Tsuchidate R, Smith M-L et al (1999) Neuroprotective effects of a novel nitrone, NXY-059, after transient focal cerebral ischemia in the rat. J Cereb Blood Flow Metab 19: 778–787.
45. Marshall JWB, Duffin KJ, Green AR et al (2001) NXY-059, a free radical-trapping agent, substantially lessons the functional disability resulting from cerebral ischemia in a primate species. Stroke 32: 190–198.
46. Lees KR, Zivin JA, Ashwood T et al (2006) NXY-059 for acute ischemic stroke. N Engl J Med 354: 588–600.

47. Shuaib A, Lees KR, Lyden P et al (2007) NXY-059 for the treatment of acute ischemic stroke. N Engl J Med 357: 562–571.
48. Feuerstein GZ, Zaleska MM, Krams M et al (2008) Missing steps in the STAIR case: a translational medicine perspective on the development of NXY-059 for treatment of acute ischemic stroke. J Cereb Blood Flow Metab 28: 217–219.
49. Borlongan CV, Chopp M, Steinberg GK et al (2008) Potential of stem/progenitor cells in treating stroke: the missing steps in translating cell therapy from laboratory to clinic. Regen Med 3: 249–250
50. Bakshi A, Barshinger AL, Swanger SA et al (2006) Lumbar puncture delivery of bone marrow stromal cells in spinal cord contusion: a novel method for minimally invasive cell transplantation. J Neurotrauma 23: 55–65
51. Vaquero J, Zurita M, Oya S et al (2006) Cell therapy using bone marrow stromal cells in chronic paraplegic rats: systemic or local administration? Neurosci Lett 398: 129–134.
52. Zurita M, Vaquero J (2004) Functional recovery in chronic paraplegia after bone marrow stromal cells transplantation. Neuroreport 15: 1105–1108.
53. Zurita M, Vaquero J (2006) Bone marrow stromal cells can achieve cure of chronic paraplegic rats: functional and morphological outcome 1 year after transplantation. Neurosci Lett 402: 51–56.
54. Hokari M, Kuroda S, Chiba Y et al (2009) Synergistic effects of granulocyte-colony stimulating factor on bone marrow stromal cell transplantation for mice cerebral infarct. Cytokine 46: 260–266.
55. Lu D, Mahmood A, Wang L et al (2001) Adult bone marrow stromal cells administered intravenously to rats after traumatic brain injury migrate into brain and improve neurological outcome. Neuroreport 12: 559–563.
56. Shen LH, Li Y, Chen J et al (2006) Intracarotid transplantation of bone marrow stromal cells increases axon-myelin remodeling after stroke. Neuroscience 137: 393–399.
57. Lu D, Mahmood A, Qu C et al (2007) Collagen scaffolds populated with human marrow stromal cells reduce lesion volume and improve functional outcome after traumatic brain injury. Neurosurgery 61: 596–602; discussion 602–593.
58. Yasuda H, Kuroda S, Shichinohe H et al (2009) Effect of biodegradable fibrin scaffold on survival, migration, and differentiation of transplanted bone marrow stromal cells after cortical injury in rats. J Neurosurg 2009 Mar 6 [Epub ahead of print].
59. Hoehn M, Kustermann E, Blunk J et al (2002) Monitoring of implanted stem cell migration in vivo: a highly resolved in vivo magnetic resonance imaging investigation of experimental stroke in rat. Proc Natl Acad Sci USA 99: 16267–16272.
60. Jendelova P, Herynek V, DeCroos J et al (2003) Imaging the fate of implanted bone marrow stromal cells labeled with superparamagnetic nanoparticles. Magn Reson Med 50: 767–776.
61. Modo M, Mellodew K, Cash D et al (2004) Mapping transplanted stem cell migration after a stroke: a serial, in vivo magnetic resonance imaging study. Neuroimage 21: 311–317.
62. Correa PL, Mesquita CT, Felix RM et al (2007) Assessment of intra-arterial injected autologous bone marrow mononuclear cell distribution by radioactive labeling in acute ischemic stroke. Clin Nucl Med 32: 839–841.

Noninvasive Optical Tracking of Bone Marrow Stromal Cells Transplanted into Rat Cerebral Infarct

Taku Sugiyama[1], Satoshi Kuroda[1], Toshiya Osanai[1], Katsuhiko Maruichi[1], Yasuhiro Chiba[1], Hideo Shichinohe[1], Yuji Kuge[2], Nagara Tamaki[3], and Yoshinobu Iwasaki[1]

Summary

Background and Purpose: It is quite important to track donor cells transplanted into the central nervous system (CNS) in order to evaluate the therapeutic effects of cell transplantation therapy. Therefore, in this preliminary study we aimed to evaluate whether fluorescent quantum dots (QDs) would be useful in noninvasive cell tracking in cerebral infarcts of rats.
Methods: Cultured rat bone marrow stromal cells (BMSCs) were labeled with Qtracker800 (QD 800; Invitrogen, Hayward, CA). They were stereotactically transplanted into the ipsilateral striatum of rats subjected to permanent middle cerebral artery occlusion 7 days after the insult. Using an in vivo optical imaging technique, the emitted fluorescence was serially monitored for 3 weeks after transplantation. Then histological analysis was performed.
Results: Optical imaging revealed that the intensity of the fluorescence emitted from the transplanted cells through the scalp and skull gradually increased in the ipsilateral peri-infarct area for 3 weeks after transplantation. Histological evaluation supported the findings on the optical images.
Conclusion: In vivo optical imaging using near-infrared (NIR) fluorescent QDs may be valuable to monitor the behavior of donor cells transplanted into the brain.

Key words Optical imaging · Cell tracking · Bone marrow stromal cell · Quantum dot

[1]Department of Neurosurgery, Hokkaido University Graduate School of Medicine, North 15 West 7, Kita-ku, Sapporo 060-8638, Japan
[2]Department of Tracer Kinetics and Bioanalysis, Hokkaido University Graduate School of Medicine, North 15 West 7, Kita-ku, Sapporo 060-8638, Japan
[3]Department of Nuclear Medicine, Hokkaido University Graduate School of Medicine, North 15 West 7, Kita-ku, Sapporo 060-8638, Japan

N. Tamaki, Y. Kuge (eds.)
Molecular Imaging for Integrated Medical Therapy and Drug Development.
DOI: 10.1007/978-4-431-98074-2_14, ©Springer 2010

Introduction

Cell transplantation therapy has been expected to promote the recovery of neurological function after central nervous system (CNS) disorders, including cerebral infarcts. When considering the clinical application of cell transplantation therapy, it is quite important to establish a noninvasive technique to track the transplanted cells in order to longitudinally monitor the behavior of these cells.

Optical imaging (OI) has some advantages over other types of imaging for tracking transplanted cells, including low cost, rapid acquisition, no radiation toxicity, and relatively high sensitivity, although its spatial resolution and penetration depth are limited because of light scattering and absorption. Previously, we succeeded in visualizing green fluorescence protein (GFP)-labeled bone marrow stromal cells (BMSCs) transplanted into the brain in mice. However, because of its short wavelength, the green fluorescence, could be detected only through the skull [1].

Therefore, in this preliminary study we aimed to evaluate whether a near-infrared (NIR) fluorescence tracer with a much longer wavelength than GFP would be useful in the "noninvasive" tracking of donor cells transplanted into the brain. For this purpose, we labeled BMSCs with NIR-emitting quantum dots (QDs); these QDs have recently been investigated for use as fluorescent biological probes, because of their nanometer dimensions, attractive optical characteristics, high resistance to photo-bleaching, and their strong fluorescence [2, 3].

Materials and Methods

Isolation and Labeling of BMSCs

The BMSCs were isolated under sterile conditions from 6- to 10-week-old Sprague-Dawley (SD) rats, as described previously [1, 4, 5]. The cultured BMSCs were labeled with Qtracker800 (QD800; Invitrogen, Hayward, CA) after three passages, according to the manufacturer's protocol.

Permanent Middle Cerebral Artery (MCA) Occlusion Model and Transplantation of QD-labeled BMSCs.

Adult SD rats were anesthetized. To induce a focal cerebral infarct, the right middle cerebral artery (MCA) was exposed through a 5 × 5-mm craniotomy and was ligated with a 10-0 nylon thread. Then the bilateral common carotid arteries were occluded with surgical microclips for 1 h [6].

The QD-labeled BMSCs were transplanted into the ipsilateral striatum of the rats 7 days after the onset of MCA occlusion. Under aseptic conditions, on a clean bench, the animals were fixed to a stereotactic apparatus and the cranium was exposed through a midline skin incision. A burr hole was made 3 mm to the right of the bregma, and a Hamilton syringe was inserted into the brain parenchyma 6 mm from the surface of the dura mater. A total of 20 μl of cell suspension

(1.0×10^6 cells) was introduced into the striatum using an automatic microinjection pump [1, 4, 5].

In Vivo Optical Imaging and Histological Analysis

Using an IVIS Imaging System 200 (Xenogen, Alameda, CA, USA), the labeled cells were serially tracked in the living animals for 3 weeks after transplantation. Subsequently, the animals were killed and ex vivo images were also obtained.

The brain slices were immersed in 4% paraformaldehyde for 2 days and embedded in paraffin. Then 5-μm-thick coronal sections were prepared for histological analysis.

Results

BMSCs Can Be Labeled with QD800

Observations under an inverted fluorescence microscope revealed that the majority of BMSCs were labeled with QD800 15 h after the start of incubation with QD800 (Fig. 1A). Fluorescence signals emitted from the QD800-labeled BMSC suspension could also be detected with the IVIS Imaging System 200 (Fig. 1B).

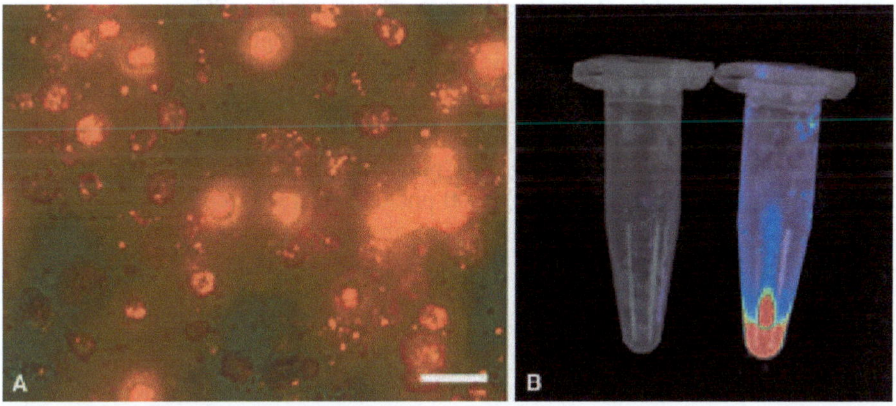

Fig. 1. **A** Photomicrograph of rat bone marrow stromal cells (BMSCs) after incubation with Qtracker800 cell-labeling kit (QD800; Invitrogen, Hayward, CA). *Scale bar,* 50 μm. The photomicrograph clearly shows that the majority of BMSCs were labeled with QD800 15 h after the start of incubation with QD800. **B** Fluorescence images of nonlabeled (*left*) and QD800-labeled BMSC suspensions (*right*) showing strong fluorescence signal emitted from the QD800-labeled BMSC suspension

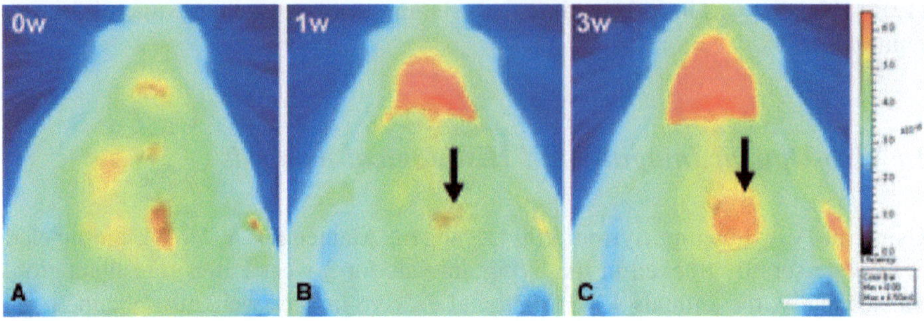

Fig. 2. **A–C** Serial in vivo fluorescence optical images after QD800-labeled BMSC transplantation in mice subjected to permanent middle cerebral artery (MCA) occlusion (also see text). Near-infrared (NIR) fluorescence emitted from QD800-labeled BMSCs could not be detected through the scalp immediately after transplantation into the right striatum (**A**). The NIR fluorescence, however, could be visualized in the right parietal region 1 week after transplantation (**B**; *arrow*). The intensity was significantly increased by 3 weeks after transplantation (**C**; *arrow*). *Scale bar*, 10 mm

Optical Imaging Can Noninvasively Visualize Transplanted BMSCs

As shown in Fig. 2, the NIR fluorescence emitted from the QD800-labeled BMSCs could not be detected through the scalp immediately after transplantation into the right striatum (Fig. 2A). The NIR fluorescence, however, could be visualized in the right parietal region 1 week after transplantation (Fig. 2B). The intensity was significantly increased by 3 weeks after transplantation (Fig. 2C).

To confirm the origin of the fluorescence on in vivo optical imaging (Fig. 3A), the animals were killed 3 weeks after BMSC transplantation. NIR fluorescence could be identified when the scalp was removed (Fig. 3B) and also when the cranium was removed (Fig. 3C). Ex vivo optical imaging demonstrated that the NIR fluorescence was emitted from the peri-infarct area (Fig. 3D). Histological analysis revealed that many of the QD800-positive cells were engrafted in the peri-infarct neocortex, supporting the findings on in vivo optical imaging.

Discussion

To date, several noninvasive cell-tracking techniques have been proposed, including magnetic resonance imaging (MRI) [7–9] and nuclear imaging [10]. Although MRI has the highest spatial resolution, it has lower sensitivity and requires a long imaging time. Although nuclear imaging has the highest sensitivity, which permits cell quantification, the radioactive tracers used have potential radiation toxicity and are less stable than the tracers used for other imaging modalities.

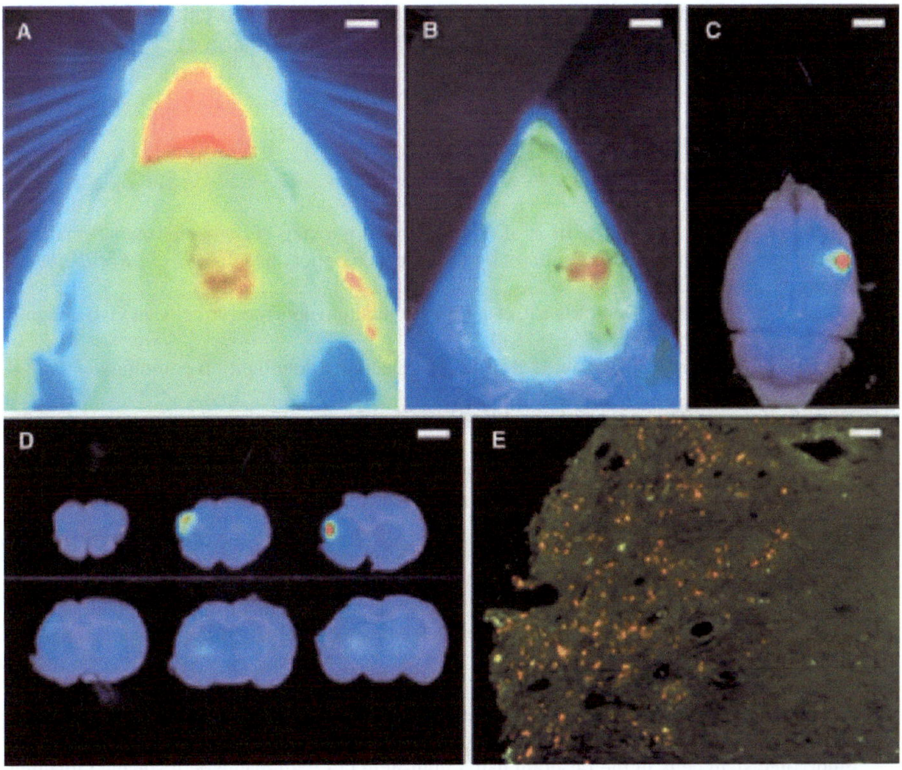

Fig. 3. A–E Representative fluorescence optical images in the living animal (**A**), after the removal of the scalp (**B**) and after the removal of the skull (**C**) clearly show that the NIR fluorescence is emitted from the infarct brain. **D** Fluorescence optical image of 2-mm-thick coronal brain slice at 3 weeks after transplantation demonstrates that the NIR fluorescence is emitted from the peri-infarct neocortex. **E** Photomicrograph under fluorescence microscopy reveals that many of the QD800-positive cells are engrafted in the peri-infarct neocortex. **A–D** *Scale bars,* 500 μm; **E** *scale bar,* 50 μm

The present study clearly showed that intravital optical imaging could visualize the QD800-labeled BMSCs through the intact scalp and skull at 1 and 3 weeks after transplantation into the striatum of rat brain subjected to cerebral infarct. The QD800-labeled BMSCs could not be identified just after transplantation; it appeared that they required at least 1 week to migrate toward the peri-infarct neocortex (Fig. 2). These findings are quite different from our previous observations that GFP-labeled BMSCs engrafted in the neocortex in mice could be detected only through the very thin skull [1].

The present findings strongly suggest that in vivo optical imaging may be useful to noninvasively visualize donor cells transplanted into the brain, through the scalp and skull, using a fluorescence tracer with a long wavelength. Continued accumulation of these observations would make it possible to establish cell transplantation as a novel therapeutic concept in the clinical situation.

References

1. Shichinohe H, Kuroda S, Lee JB et al (2004) In vivo tracking of bone marrow stromal cells transplanted into mice cerebral infarct by fluorescence optical imaging. *Brain Res Brain Res Protoc*, 13: 166–175.
2. Michalet X, Pinaud FF, Bentolila LA et al (2005) Quantum dots for live cells, in vivo imaging, and diagnostics. *Science*, 307: 538–544.
3. Tholouli E, Sweeney E, Barrow E et al (2008) Quantum dots light up pathology. *J Pathol*, 216: 275–285.
4. Yano S, Kuroda S, Shichinohe H et al (2005) Do bone marrow stromal cells proliferate after transplantation into mice cerebral infarct? A double labeling study. *Brain Res*, 1065: 60–67.
5. Shichinohe H, Kuroda S, Yano S et al (2007) Role of SDF-1/CXCR4 system in survival and migration of bone marrow stromal cells after transplantation into mice cerebral infarct. *Brain Res*, 1183: 138–147.
6. Chen ST, Hsu CY, Hogan EL et al (1986) A model of focal ischemic stroke in the rat: reproducible extensive cortical infarction. Stroke 17: 738–743
7. Jendelova P, Herynek V, DeCroos J et al (2003) Imaging the fate of implanted bone marrow stromal cells labeled with superparamagnetic nanoparticles. *Magn Reson Med*, 50: 767–776.
8. Modo M, Mellodew K, Cash D et al (2004) Mapping transplanted stem cell migration after a stroke: a serial, in vivo magnetic resonance imaging study. *Neuroimage*, 21: 311–317.
9. Stroh A, Faber C, Neuberger T et al (2005) In vivo detection limits of magnetically labeled embryonic stem cells in the rat brain using high-field (17.6 T) magnetic resonance imaging. *Neuroimage*, 24: 635–645.
10. Correa PL, Mesquita CT, Felix RM et al (2007) Assessment of intra-arterial injected autologous bone marrow mononuclear cell distribution by radioactive labeling in acute ischemic stroke. *Clin Nucl Med*, 32: 839–841.

Thermoreversible Gelation Polymer (TGP) Hydrogel as a Degradable Scaffold for Bone Marrow Stromal Cell Transplantation

Toshiya Osanai, Satoshi Kuroda, Hiroshi Yasuda, Yasuhiro Chiba, Katsuhiko Maruichi, Masaaki Hokari, Taku Sugiyama, Hideo Shichinohe, and Yoshinobu Iwasaki

Summary

Recent studies have indicated that bone marrow stromal cells (BMSCs) have the potential to improve neurologic function when transplanted into animal cerebral infarct models. However, it is still undetermined how the BMSCs should be transplanted to safely obtain the most efficient therapeutic benefits. Therefore, this study was carried out aiming to assess whether a thermoreversible gelation polymer (TGP) hydrogel would act as a noninvasive, valuable scaffold in BMSC transplantation for brain infarct. Mice were subjected to permanent middle cerebral artery occlusion. Vehicle, BMSC suspension, or a BMSC-TGP construct was transplanted onto the ipsilateral intact neocortex at 7 days after the insult. The TGP hydrogel completely disappeared and provoked no inflammation in the host brain. In the BMSC-TGP construct-treated mice, many transplanted cells were widely engrafted in the ipsilateral cerebrum, including the dorsal neocortex adjacent to the cerebral infarct. The number of these transplanted cells was significantly larger than that in the BMSC-treated mice; the majority of the cells were positive for both neuronal nuclear antigen (NeuN) and microtubule-associated protein 2 (MAP2) and morphologically simulated neurons. These findings suggest that the surgical transplantation of tissue-engineered BMSCs onto the intact neocortex enhances the engraftment of donor cells around a cerebral infarct. TGP hydrogel, because of its unique biochemical properties, could be a promising candidate as a valuable scaffold in BMSC transplantation for central nervous system disorders.

Key words Bone marrow stromal cell · Tissue engineering · Transplantation · Thermoreversible gelation polymer

Department of Neurosurgery, Hokkaido University Graduate School of Medicine, North 15 West 7, Kita-ku, Sapporo 060-8638, Japan

N. Tamaki, Y. Kuge (eds.)
Molecular Imaging for Integrated Medical Therapy and Drug Development.
DOI: 10.1007/978-4-431-98074-2_15, ©Springer 2010

Introduction

There is increasing evidence that bone marrow stromal cells (BMSCs) may be one of candidates as the donor cells for cell transplantation therapy for central nervous system (CNS) disorders. They can be harvested from the patients themselves without ethical or immunological problems [1–7]. However, it is undetermined through which routes the BMSCs should be transplanted into the CNS in order to obtain therapeutic effects most safely and efficiently. The recently emerging field of tissue engineering may provide promising alternative routes for cell transplantation. Accordingly, this study was carried out aiming to assess whether a thermoreversible gelation polymer (TGP) hydrogel (Mebiol, Mebiol Inc., Hiratsuka, Japan) acted as a noninvasive, valuable scaffold in BMSC transplantation for brain infarct.

Materials and Methods

Preparation of BMSCs

BMSCs were isolated under sterile conditions from 6-week-old Balb/c mice (CLEA Japan, Tokyo, Japan) or from transgenic mice expressing green fluorescence protein (GFP; Jackson Laboratory, Bar Harbor, ME, USA), as described previously [1–7]. The cells were utilized for subsequent experiments after three passages.

Permanent Middle Cerebral Artery (MCA) Occlusion Model and BMSC Transplantation

Eight-week-old male Balb/c mice were subjected to permanent middle cerebral artery (MCA) occlusion, as described before [3, 5–7]. The BMSCs harvested from the Balb/c mice were labeled with PKH-26 (2×10^{-6} M; Sigma, St. Louis, MO, USA) according to the manufacturer's protocol. The animals were anesthetized 7 days after the insult. The cranium was exposed through a midline skin incision, and the surface of the intact neocortex was exposed through a small craniectomy (3 × 5 mm) made in the right parietal region (Fig. 1A), using a dental drill under surgical microscope. The dura mater was carefully removed so as not to damage the brain surface. The animals were assigned to three groups. In group A ($n = 6$), 50 μl of phosphate-buffered saline [phosphate-buffered saline (PBS); vehicle] was poured onto the brain surface, and the wound was closed. In group B ($n = 6$), 50 μl of the PKH-26-labeled BMSC suspension (5×10^5 cells) was poured onto the brain surface, and the wound was closed. In group C ($n = 6$), the PKH-26-labeled BMSCs (4×10^6 cells) were embedded into 1.0 ml of TGP-culture medium mixture at 4°C, and 0.5 ml of the BMSC-TGP construct was placed into a flexible cell culture

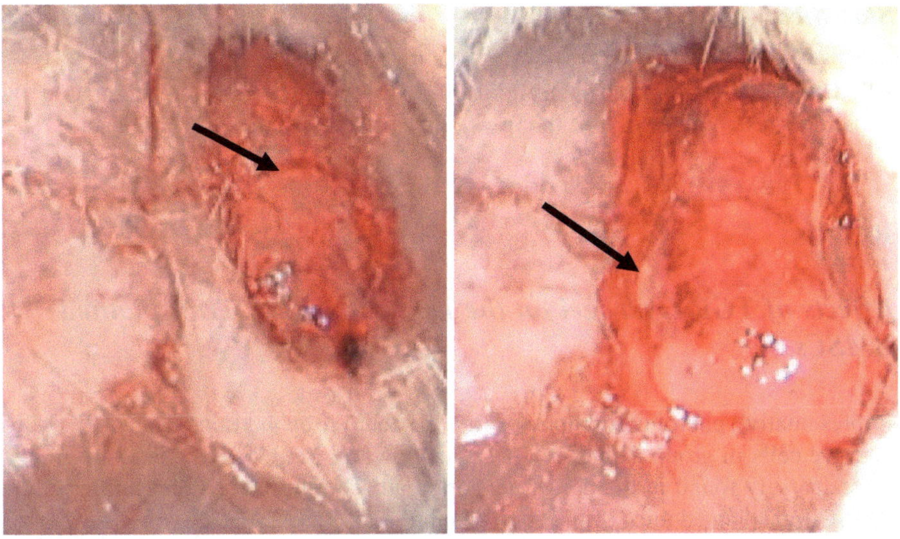

A B

Fig. 1. A At 7 days after the onset of permanent middle cerebral artery occlusion, a small craniectomy (3 × 5 mm; *arrow*) was made in the ipsilateral parietal region to expose the surface of the intact neocortex. **B** The bone marrow stromal cell (BMSC)-thermoreversible gelation polymer (TGP) construct (*arrow*) was solidified at 37°C and transplanted onto the surface of the intact neocortex in group C (*arrow*)

chamber (flexiPERM Disc; growth area, 1.8 cm^2; Greiner Bio-One, San Francisco, CA, USA). The BMSC-TGP construct was solidified at 37°C for 1 h, yielding a 0.5-mm-thick gel in each chamber. Subsequently, one-fourth of the construct, including 5×10^5 BMSCs, was mounted onto the surface of the intact neocortex (Fig. 1B).

Histological Analysis

At 8 weeks after BMSC transplantation, the animals were deeply anesthetized with 4.0% isoflurane in N_2O/O_2 (70: 30) and transcardially perfused with 100 ml of heparinized saline, followed by 100 ml of 4% paraformaldehyde. The brain was removed, immersed in 4% paraformaldehyde for 2 days, and embedded in paraffin. Then, 4-μm-thick coronal sections were prepared for subsequent analysis. In order to semiquantitatively analyze the effect of TGP on BMSC engraftment, two regions of interest (ROI; 1000 × 1000 μm) were symmetrically placed on the ipsi- and contralateral neocortex in coronal sections including the striatum, and the number of PKH-26-positive cells in each ROI was counted under a fluorescence microscope.

Double-fluorescence immunohistochemistry was employed to assess the phenotype of the engrafted cells, as described elsewhere [1–7].

Statistical Analysis

All data were expressed as means ± SD. Continuous data were compared by unpaired *t*-test. Values of $P < 0.05$ were considered statistically significant.

Results

Effects of TGP Hydrogel on the Engraftment of Transplanted BMSCs

Representative photomicrographs of the distribution of the transplanted PKH-26-positive cells in group B and group C are shown in Fig. 2. PKH-26-positive cells were sparsely found in the dorsal neocortex adjacent to the cerebral infarct in the BMSC-transplanted mice (Fig. 2A). The number of PKH-26-positive cells was 15.3 ± 4.1 in the 1-mm^2 ROI ($n = 6$). On the other hand, many PKH-26-positive cells were seen in the dorsal neocortex adjacent to the cerebral infarct in the BMSC-TGP construct-transplanted mice (Fig. 2B). The number of these cells was 46.2 ± 5.6 in the 1-mm^2-ROI ($n = 6$), being significantly higher than that in the BMSC-transplanted mice ($P < 0.01$).

A 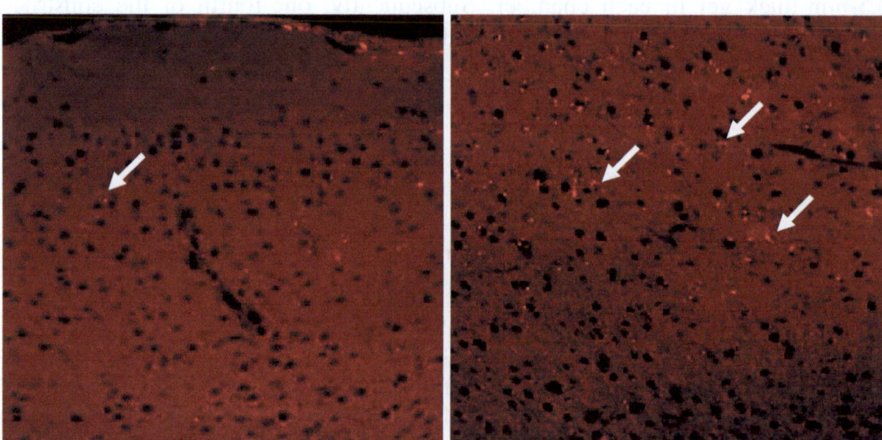 B

Fig. 2A,B. Photomicrographs under fluorescence microscopy. **A** In BMSC-transplanted animals (group B), PKH-26-positive cells were sparsely found in the dorsal neocortex adjacent to the cerebral infarct (*arrow*). **B** In the BMSC-TGP construct-transplanted animals (group C), many PKH-26-positive cells were seen in the dorsal neocortex adjacent to the cerebral infarct (*arrows*). **A** and **B** × 100

PKH-26-positive cells were also found in the hippocampus, corpus callosum, and contralateral neocortex in the BMSC-TGP construct-transplanted mice, but they were not found in these regions in the BMSC-transplanted mice.

Phenotype of Transplanted BMSCs in Peri-Infarct Area

In group B, the engrafted PKH-26-positive cells in the dorsal neocortex adjacent to the cerebral infarct expressed both NeuN and MAP2, although the number of these cells was small, as described above. In group C, however, many of the PKH-26-positive cells in the dorsal neocortex adjacent to the cerebral infarct simulated neurons morphologically and expressed both NeuN and MAP2 (Fig. 3).

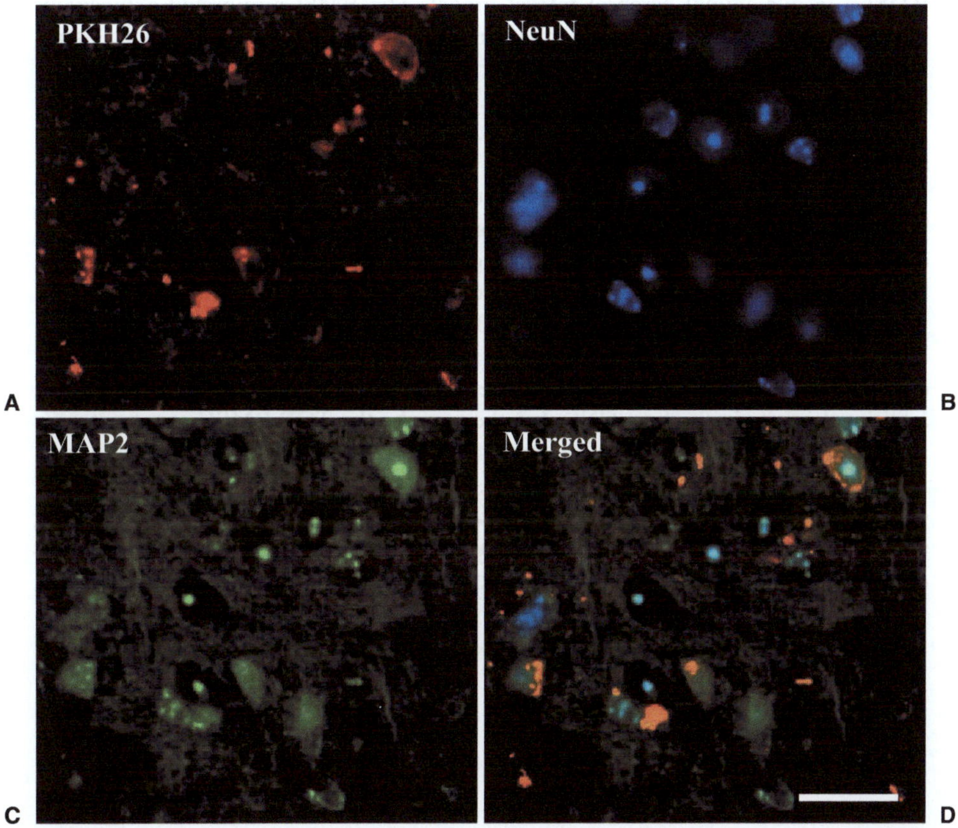

Fig. 3A–D. Fluorescence immunohistochemistry for neuronal nuclear antigen (*NeuN*) and microtubule-associated protein 2 (*MAP2*) in the ipsilateral peri-infarct neocortex of the BMSC-TGP construct-transplanted animals (group C). Photomicrographs show that many of the PKH-26-positive cells (*red;* **A** and **D**) also express NeuN (*blue;* **B** and **D**) and MAP2 (*green*; **C** and **D**) and morphologically simulate neurons. *Scale bar*, 50 μm

Discussion

Very recently, we evaluated the utility of a fibrin matrix as a biodegradable scaffold in cell transplantation for a hemisected spinal cord in rats. The results showed that the animals treated with the BMSC-fibrin matrix construct showed significantly more pronounced recovery of neurologic function than vehicle- or BMSC-treated animals. Histological analysis revealed that the fibrin scaffold markedly improved the survival and migration of the transplanted cells [2].

However, these techniques of tissue engineering for CNS disorders are based on the direct injection of donor cells and biomaterial into the injured CNS tissue. Such techniques may provoke additional CNS injury. In the present study, therefore, we aimed to develop a noninvasive tissue-engineering technique for CNS disorders; we found that the implantation of the BMSC-TGP construct onto the intact neocortical surface promoted the survival, migration, and differentiation of BMSCs in our animal cerebral infarct model. The BMSCs migrated extensively into the peri-infarct neocortex, hippocampus, corpus callosum, and contralateral neocortex, and the majority of them morphologically simulated neurons and expressed neuronal markers.

This is the first report that denotes the utility of a noninvasive tissue-engineering procedure for CNS disorders. This study clearly showed that BMSCs could penetrate through the pia mater of the intact neocortex and migrate towards the damaged brain within 4 weeks. The engrafted cells morphologically simulated neurons and expressed NeuN and MAP2, markers specific for neurons. However, their engraftment was much less pronounced when the cell suspension was only decanted onto the surface of brain. Therefore, the TGP-BMSC construct transplantation technique may be safer and less invasive than a direct injection of cell suspension or tissue-engineered graft into the injured CNS, because it has no potential risk of inducing additional injury in the host CNS. In other words, the TGP-BMSC construct may act as a noninvasive "biofilm" or "poultice" which includes the donor cells and regenerates the damaged tissue of the host CNS.

In conclusion, this study confirms that surgical transplantation of tissue-engineered BMSCs onto the intact neocortex enhances the engraftment of donor cells around a cerebral infarct. Furthermore, TGP hydrogel, because of its unique biochemical properties, could be a promising candidate as a valuable scaffold in BMSC transplantation for CNS disorders.

References

1. Hokari M, Kuroda S, Shichinohe H et al (2008) Bone marrow stromal cells protect and repair damaged neurons through multiple mechanisms. J Neurosci Res 86: 1024–1035
2. Itosaka H, Kuroda S, Shichinohe H et al (2009) Fibrin matrix provides a suitable scaffold for bone marrow stromal cells transplanted into injured spinal cord: a novel material for CNS tissue engineering. Neuropathology 29: 248–257

3. Shichinohe H, Kuroda S, Lee JB et al (2004) In vivo tracking of bone marrow stromal cells transplanted into mice cerebral infarct by fluorescence optical imaging. Brain Res Brain Res Protoc 13: 166–175
4. Shichinohe H, Kuroda S, Tsuji S et al (2008) Bone marrow stromal cells promote neurite extension in organotypic spinal cord slice: significance for cell transplantation therapy. Neurorehabil Neural Repair 22: 447–457
5. Shichinohe H, Kuroda S, Yano S et al (2007) Role of SDF-1/CXCR4 system in survival and migration of bone marrow stromal cells after transplantation into mice cerebral infarct. Brain Res 1183: 138–147
6. Shichinohe H, Kuroda S, Yano S et al (2006) Improved expression of gamma-aminobutyric acid receptor in mice with cerebral infarct and transplanted bone marrow stromal cells: an autoradiographic and histologic analysis. J Nucl Med 47: 486–491
7. Yano S, Kuroda S, Lee JB et al (2005) In vivo fluorescence tracking of bone marrow stromal cells transplanted into a pneumatic injury model of rat spinal cord. J Neurotrauma 22: 907–918

Beneficial Effects of Bone Marrow Stromal Cell Transplantation on Axonal Regeneration in Injured Spinal Cord

Yasuhiro Chiba, Satoshi Kuroda, Katsuhiko Maruichi, Toshiya Osanai, Masaaki Hokari, Shunsuke Yano, Hideo Shichinohe, Kazutoshi Hida, and Yoshinobu Iwasaki

Summary

Recent studies have indicated that bone marrow stromal cells (BMSCs) have the potential to improve neurological function when transplanted into animal models of spinal cord injury (SCI). However, it is still unclear how the transplanted BMSCs promote functional recovery after SCI. In this study, therefore, we evaluated how transplanted BMSCs restore the function of the dorsal corticospinal tract (dCST) in the injured spinal cord. Rats were subjected to incomplete SCI, using a pneumatic impact device. Then BMSC suspension or vehicle was transplanted into the rostral site of the SCI at 7 days after the injury. Fluoro-ruby (FR; Molecular Probes), a fluorescent axonal tracer, was injected into the dorsal funiculus of the rostral site of the SCI 63 days after the injury. BMSC transplantation significantly enhanced functional recovery of the hind limbs. The number of FR-labeled fibers in the dCST at the caudal site of the SCI was significantly higher in the BMSC-transplanted animals than in the vehicle-transplanted animals. Some of the engrafted BMSCs were positive for FR, neuronal nuclear antigen (NeuN), and microtubule-associated protein 2 (MAP2) in the gray matter. The findings suggest that the transplanted BMSCs acquire neural cell phenotypes around the injury site and contribute to rebuilding neural circuits, including those in the CST, promoting functional recovery of the hind limbs.

Key words Bone marrow stromal cell · Axon · Spinal cord injury · Differentiation · Transplantation

Introduction

There is increasing evidence that bone marrow stromal cells (BMSCs) can survive, migrate toward injured tissue, express proteins specific for neural cells, and improve neurological function, when transplanted into animal models of central nervous

Department of Neurosurgery, Hokkaido University Graduate School of Medicine, North 15 West 7, Kita-ku, Sapporo 060-8638, Japan

N. Tamaki, Y. Kuge (eds.)
Molecular Imaging for Integrated Medical Therapy and Drug Development.
DOI: 10.1007/978-4-431-98074-2_16, ©Springer 2010

system (CNS) disorders, including spinal cord injury (SCI) [1, 2]. Although several cell sources have been proposed for such transplantation, BMSCs may be the most likely source, because they can be harvested from the patients themselves without ethical or immunological problems. Recent studies have shown that cultured BMSCs morphologically simulate neurons and express proteins specific for neural cells under both in vitro and in vivo conditions [3, 4]. Especially, BMSCs acquire a neuronal phenotype without cell fusion, when cocultured with neurons [5]. More interestingly, transplanted BMSCs express the gamma-aminobutyric acid (GABA) receptor and improve its function in the marginal region of injured spinal cord [6]. These findings strongly suggest that transplanted BMSCs may have the potential to differentiate into neural cells and replace injured CNS tissues. Alternatively, the BMSCs may support the survival of the host CNS tissue by secreting some neuro-protective or neurotrophic factors, including vascular endothelial growth factor (VEGF), fibroblast growth factor-2 (FGF-2), hepatocyte growth factor (HGF), nerve growth factor (NGF), and brain-derived neurotrophic factor (BDNF). Hokari et al. [5] recently reported that BMSCs ameliorated glutamate-induced neuronal injury by producing these factors.

In spite of this recent progress, however, the precise mechanisms through which transplanted BMSCs promote functional recovery after SCI are still unclear. In this study, therefore, we aimed to evaluate how transplanted BMSCs restore the lost function of the corticospinal tracts in the injured spinal cord, using axon-tracing and immunohistochemical techniques. Detailed information on this study has been published elsewhere [7].

Materials and Methods

Isolation of BMSCs from Green Fluorescent Protein (GFP)-Transgenic Mice

BMSCs were isolated from 4- to 8-week-old transgenic mice expressing (GFP; the Jackson Laboratory, Bar Harbor, ME, USA), as described previously [1, 2, 5, 6, 8–10].

Rat Spinal Cord Injury and BMSC Transplantation

Adult female Sprague-Dawley rats were subjected to SCI induced with a pneumatic impact device at the level of T11, as described previously [2, 6]. The animals were then housed until transplantation surgery had been done. Neurological symptoms were assessed with the Basso-Beattie-Bresnahan (BBB) Locomotor Rating Scale before and after the injury insult.

BMSC suspension ($n = 7$) or vehicle ($n = 7$) was transplanted into the spinal cord at the level of T9 at 7 days after the induction of SCI [2, 6]. A 7-μl suspension of BMSCs (1×10^4 cells/μl) or phosphate-buffered saline (PBS) was injected into the spinal cord, using an automatic microinjection pump.

Anterograde Labeling of Dorsal Corticospinal Tract

To trace the dorsal corticospinal tract (dCST), the animals were anesthetized 63 days after SCI. A needle of a 10-μl Hamilton syringe was positioned 10 mm rostral to the center of the injury and was inserted into the spinal cord 1.5 mm from the surface of the dura mater. Then, a total amount of 2 μl of Fluoro-ruby (FR; MW 10 000, D-1817; Molecular Probes, Eugene, OR, USA), a fluorescent axonal tracer, was injected into the spinal cord 0.5 mm lateral to the midline, on both sides, over a period of 3 min, using an automatic microinjection pump.

Histological Analysis

Five days after the FR injection, the animals were deeply anesthetized and were transcardially perfused with 50 ml of physiological saline, followed by 150 ml of 4% paraformaldehyde. Transverse sections of the spinal cord at 10 mm caudal to the injury site were observed under a fluorescence microscope (BX51; Olympus, Tokyo, Japan). Because the dCST in rats is known to exist within the ventral one-third of the dorsal funiculus, the number of FR-labeled axons within the ventral one-third of the dorsal funiculus in each section was counted bilaterally.

Double-fluorescence immunohistochemistry was employed to identify the distribution and differentiation of the transplanted BMSCs and their spatial relationship with FR-labeled axons, as reported previously [2, 6, 8, 9, 11].

Results

BMSC Transplantation Improves Hind Limb Function

The vehicle-transplanted animals did not show any significant improvement of hind limb function throughout the experiments. On the other hand, the BMSC-transplanted animals exhibited significant recovery of hind limb function. Thus, BBB scores were significantly higher in the BMSC-transplanted animals than in the vehicle-transplanted animals at 14, 35, 42, and 49 days ($P < 0.05$) and at 56 and 63 days after SCI ($P < 0.01$).

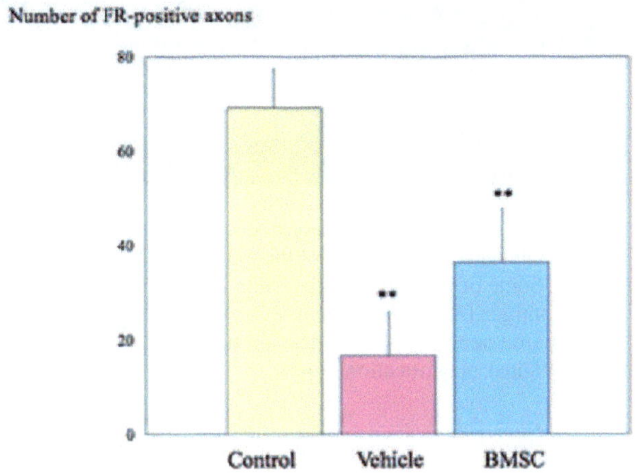

Fig. 1. Bar graphs show the numbers of Fluoro-ruby (*FR*)-positive axons in the ventral one-third of the dorsal funiculus in the control group (*yellow bar*), vehicle-transplanted group (*red bar*), and bone marrow stromal cell (*BMSC*)-transplanted group (*blue bar*). *Error bars* show SD. $^{**}P < 0.01$

BMSC Transplantation Improves Axonal Continuity in the dCST

As shown in Fig. 1, the number of FR-positive axons was 69.0 ± 8.1 in the sham-operated, uninjured group (control; $n = 4$). The number was 16.6 ± 9.1 in the vehicle-transplanted group ($n = 7$), being significantly lower than that in the control group ($P < 0.01$). On the other hand, the number of FR-positive axons in the BMSC-transplanted group ($n = 7$) was 36.4 ± 11.3, being significantly higher than that in the vehicle-transplanted group ($P < 0.01$).

Transplanted BMSCs Are Integrated into Neural Circuits

The transplanted GFP-expressing BMSCs migrated extensively towards the lesion and were found in the rostral and partially in the caudal regions of the injury. In the gray matter, most of the GFP-positive cells were large and had an oval shape. Some of them were also positive for FR (Fig. 2), but none of them expressed glial fibrillary acidic protein (GFAP). On the other hand, the GFP-positive cells in the white matter and glial scar were rather small, and morphologically simulated astrocytes. Most of them were also positive for GFAP, but negative for FR.

Many of the GFP-positive cells in the gray mater were doubly positive for neuronal markers such as NeuN and MAP2. More interestingly, they were also positive for FR that was exogenously injected into the dorsal funiculus at 10 mm rostral to the center of injury. The findings strongly suggested that a certain subpopulation of BMSCs had not only acquired a neuronal phenotype but had also built synaptic connections with the FR-injected axons of the host dCST within 8 weeks after transplantation. The other GFP-positive cells morphologically simulated astrocytes and were also positive for GFAP in the white matter and glial scar. Many of cells

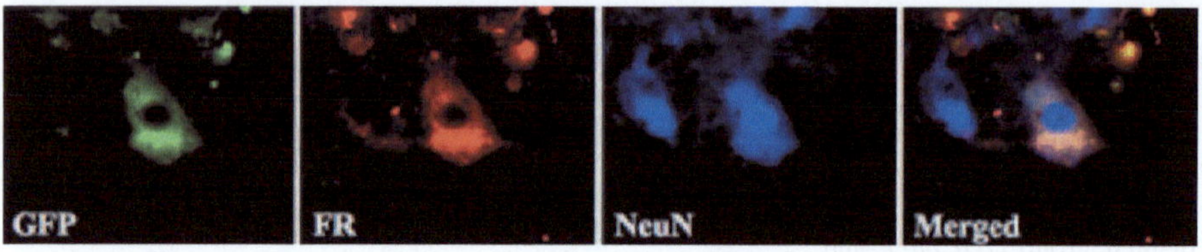

Fig. 2. Photomicrographs of fluorescence immunohistochemistry around the injury site show the morphology and differentiation of the transplanted green fluorescence protein (*GFP*)-positive cells (*green*). Some of the GFP-positive cells were also positive for Fluoro-ruby (*FR; red*) or neuronal nuclear antigen (*NeuN; blue*), suggesting that the transplanted bone marrow stromal cells (BMSCs) had acquired neuronal phenotypes and had built synaptic connections with the host's neural circuits. Original magnification, ×400

doubly positive for GFP and GFAP surrounded the FR-positive axons of the dCST in the white matter and glial scar (data not shown).

Discussion

In the present study, BMSCs significantly enhanced the functional recovery of hind limbs when stereotactically transplanted into the injured spinal cord. This finding agrees well with previous studies. However, the quantification of FR-positive axons revealed that BMSC transplantation significantly improved the axonal connections of the dCST beyond the lesion. Histological analysis also demonstrated that the transplanted BMSCs had migrated into the injury sites and were densely distributed around the core of the injury site. Some of them had acquired a neuronal phenotype and were incorporated into the host's neural circuits in the gray matter. Other transplanted BMSCs also expressed an astrocytic phenotype and supported the FR-positive axons of the dCST in the white matter. Based on these findings, we can assume that the transplanted cells not only differentiated into neurons but that they also supported the regeneration of damaged axons in the dCST. This study, using an axonal tracer and immunohistochemistry, is unique in clarifying the underlying mechanisms through which BMSC transplantation promotes functional recovery after spinal cord injury (SCI).

This study reveals that transplanted BMSCs may ameliorate the neurological dysfunction caused by SCI just after transplantation by secreting some neurotrophic factors. Subsequently, the transplanted BMSCs may acquire neuronal and astrocytic phenotypes around the injury site and contribute to rebuilding the neural circuits, including the corticospinal tract (CST), finally promoting functional recovery of the hind limbs. These findings can shed some light on the underlying mechanisms through which transplanted BMSCs restore the neurological function lost after SCI; these findings may lead to the establishment of scientifically verified strategies for cell transplantation therapy in patients with SCI.

References

1. Lee JB, Kuroda S, Shichinohe H et al (2003) Migration and differentiation of nuclear fluorescence-labeled bone marrow stromal cells after transplantation into cerebral infarct and spinal cord injury in mice. Neuropathology 23: 169–180
2. Yano S, Kuroda S, Lee JB et al (2005) In vivo fluorescence tracking of bone marrow stromal cells transplanted into a pneumatic injury model of rat spinal cord. J Neurotrauma 22: 907–918
3. Azizi SA, Stokes D, Augelli BJ et al (1998) Engraftment and migration of human bone marrow stromal cells implanted in the brains of albino rats–similarities to astrocyte grafts. Proc Natl Acad Sci U S A 95: 3908–3913
4. Kopen GC, Prockop DJ, Phinney DG (1999) Marrow stromal cells migrate throughout forebrain and cerebellum, and they differentiate into astrocytes after injection into neonatal mouse brains. Proc Natl Acad Sci U S A 96: 10 711–10 716
5. Hokari M, Kuroda S, Shichinohe H et al (2008) Bone marrow stromal cells protect and repair damaged neurons through multiple mechanisms. J Neurosci Res 86: 1024–1035
6. Yano S, Kuroda S, Shichinohe H et al (2006) Bone marrow stromal cell transplantation preserves gammaaminobutyric acid receptor function in the injured spinal cord. J Neurotrauma 23: 1682–1692
7. Chiba Y, Kuroda S, Maruichi K et al (2009) Transplanted bone marrow stromal cells promote axonal regeneration and improve motor function in a rat spinal cord injury model. Neurosurgery 64: 991–1000
8. Shichinohe H, Kuroda S, Lee JB et al (2004) In vivo tracking of bone marrow stromal cells transplanted into mice cerebral infarct by fluorescence optical imaging. Brain Res Brain Res Protoc 13: 166–175
9. Shichinohe H, Kuroda S, Yano S et al (2006) Improved expression of gamma-aminobutyric acid receptor in mice with cerebral infarct and transplanted bone marrow stromal cells: an autoradiographic and histologic analysis. J Nucl Med 47: 486–491
10. Yano S, Kuroda S, Shichinohe H et al (2005) Do bone marrow stromal cells proliferate after transplantation into mice cerebral infarct? A double labeling study. Brain Res 1065: 60–67
11. Shichinohe H, Kuroda S, Yano S et al (2007) Role of SDF-1/CXCR4 system in survival and migration of bone marrow stromal cells after transplantation into mice cerebral infarct. Brain Res 1183: 138–147

Transplanted Bone Marrow Stromal Cells Improve Cognitive Function after Diffuse Axonal Injury in Rats

Katsuhiko Maruichi, Satoshi Kuroda, Yasuhiro Chiba, Toshiya Osanai, Taku Sugiyama, Masaaki Hokari, Kazutoshi Hida, and Yoshinobu Iwasaki

Summary

Background and Purpose: Diffuse axonal injury (DAI) often leads to persistent cognitive dysfunction in spite of the lack of gross lesions on magnetic resonance (MR) imaging. Therefore, in this study, we aimed to develop a highly controlled and reproducible model of DAI that simulates post-traumatic cognitive dysfunction in humans; we aimed to investigate whether the intracerebral transplantation of bone marrow stromal cells (BMSCs) improved cognitive function in this model.
Methods: Female 7- to 9-week-old Sprague-Dawley (SD) rats were subjected to impact acceleration head injury, using a pneumatic high-velocity impactor. BMSCs were harvested from 6- to 8-week-old transgenic mice that express green fluorescent protein and were cultured. BMSC suspension (4.0×10^5 cells) or vehicle was stereotactically transplanted into the right striatum at 10 days postinjury. Cognitive function analysis was carried out at 1, 2, and 4 weeks postinjury, using the Morris water maze test. Histological analysis was performed at 2, 8, and 20 weeks postinjury, using double-fluorescence immunohistochemistry.
Results: The degree and duration of cognitive dysfunction were dependent on the force of impact. The transplanted BMSCs were widely distributed in the injured brain and gradually acquired the phenotypes of neurons and astrocytes over a period of 20 weeks. In addition, the transplanted BMSCs significantly improved the DAI-induced cognitive function as early as 2 weeks postinjury, although the process of neuronal differentiation had not been completed at this time point.
Conclusions: The findings suggest that engrafted BMSCs may exhibit an early beneficial effect on cognitive function by producing neuroprotective or neurotrophic factors. The direct transplantation of BMSCs may serve as a novel therapeutic strategy to enhance recovery from DAI-induced cognitive impairment.

Department of Neurosurgery, Hokkaido University Graduate School of Medicine, North 15 West 7, Kita-ku, Sapporo 060-8638, Japan

N. Tamaki, Y. Kuge (eds.)
Molecular Imaging for Integrated Medical Therapy and Drug Development.
DOI: 10.1007/978-4-431-98074-2_17, ©Springer 2010

Key words Diffuse axonal injury · Bone marrow stromal cell · Transplantation · Cognitive function · Differentiation · Neuroprotection

Introduction

Traumatic brain injury (TBI) often leads to persistent neurological sequelae and is a significant public health problem. Growing recent evidence indicates that TBI often induces widespread damage to axons and white matter through shearing forces, called diffuse axonal injury (DAI). Nowadays, it is widely recognized that DAI plays a major role in the development of cognitive dysfunction in patients following TBI, even when they have no definite abnormalities on conventional magnetic resonance (MR) imaging. Even now, however, therapeutic strategies to improve cognitive function in these patients are limited. There is increasing evidence that various kinds of stem cells can proliferate extensively, migrate into these lesions, and express neural cell markers, improving neurological function. Embryonic stem cells, neural stem cells, and bone marrow stromal cells (BMSCs) have been considered as candidate source cells for transplantation therapy. Of these sources, BMSCs may be the most suitable, because they can be harvested from the patients themselves without ethical or immunological problems. In fact, there are many studies that show the beneficial effects of transplanted BMSCs in animals subjected to TBI. However, there are only a few studies that prove the efficacy of BMSC transplantation therapy for alleviating post-traumatic cognitive dysfunction. In this study, therefore, we aimed to evaluate whether BMSCs could attenuate DAI-related cognitive dysfunction when transplanted into the injured brain. For this purpose, we produced a highly controlled and reproducible rat model of DAI that mimics DAI in humans, using a pneumatic high-velocity impactor.

Materials and Methods

Preparation of BMSCs

To harvest BMSCs, femurs were aseptically dissected from 6- to 8-week-old transgenic mice that express green fluorescent protein (GFP), as described previously [1–4]. The adherent cells were passaged three times.

Diffuse Axonal Injury Model in Rats

Female 7- to 9-week-old Sprague-Dawley (SD) rats were subjected to impact acceleration head injury, as previously described [5]. The dorsal surface of the skull was exposed through a midline incision, and a metal disc was fixed centrally between the bregma and the lambda, using dental cement. The driving pressure, duration, and depth of the impact determine the severity of the DAI produced. Based on our recent observations [5], in this study, these parameters were kept strictly constant, at 80 pounds per square inch (psi), 6 mm, and 40 ms, respectively. In this study, three groups of animals (total, $n = 27$) were studied: a sham-operated group ($n = 8$), a vehicle-transplanted group ($n = 6$), and a BMSC-transplanted group ($n = 13$). Following the impact, the metal disc was removed and the midline incision wound was closed.

Transplantation of BMSCs

BMSC suspension or phosphate-buffered saline (PBS) was transplanted into the right striatum 10 days after the TBI, as reported previously. A burr hole was made 1 mm posterior and 1.5 mm to the right of the bregma, using a small dental drill. Then the animals were mounted on a stereotactic fixation apparatus. The needle of a 10-μl Hamilton syringe was inserted into the brain parenchyma 6 mm from the surface of the dura mater, and 10 μl of cell suspension (4.0×10^5 cells; $n = 13$) or PBS ($n = 6$) was injected into the striatum over a period of 10 min, using an automatic microinjection pump. All animals were treated with 10 mg/kg of cyclosporine A subcutaneously every day for 4 weeks following transplantation.

Assessment of Cognitive Function

Cognitive testing (spatial learning) was assessed, using a hidden-platform version of the Morris water maze (MWM). The rats were trained to locate a hidden, submerged platform using constant extra-maze visual information. The latency to find the hidden platform within a 180-s maximal time was recorded, using a computerized video tracking system. Trials were conducted over 3 consecutive days at 1, 2, and 4 weeks postinjury.

Histology and Immunohistochemistry

To evaluate histological features, the brain was removed, embedded in paraffin, and cut coronally into 5-μm-thick sections. The deparaffinized sections were stained with hematoxylin and eosin (H & E). Double-fluorescence immunohistochemistry was employed to identify the distribution of GFP-positive cells and the differentiation of the BMSCs into neural cells (using primary antibodies against MAP2, GFAP, NeuN, nestin, or Tuj-1).

Statistical Analysis

All data were expressed as means ± SD. Continuous data between three groups were compared with one-factor analysis of variance (ANOVA) followed by Bonferroni's test. Differences with a *P* value of less than 0.05 were considered statistically significant.

Results

BMSC Transplantation Improves Post-Traumatic Cognitive Function

The effects of the pneumatic impactor-induced brain injury on cognitive function were quantitatively analyzed over 3 days at 1, 2, and 4 weeks postinjury in each group. Compared with the sham-operated animals, the vehicle-transplanted animals ($n = 6$) showed significantly prolonged latency throughout the experiment, except for the first day (Fig. 1) [6].

The results in the BMSC-transplanted animals ($n = 13$) were compared with those in the vehicle-transplanted animals. The latency at 15 days postinjury was

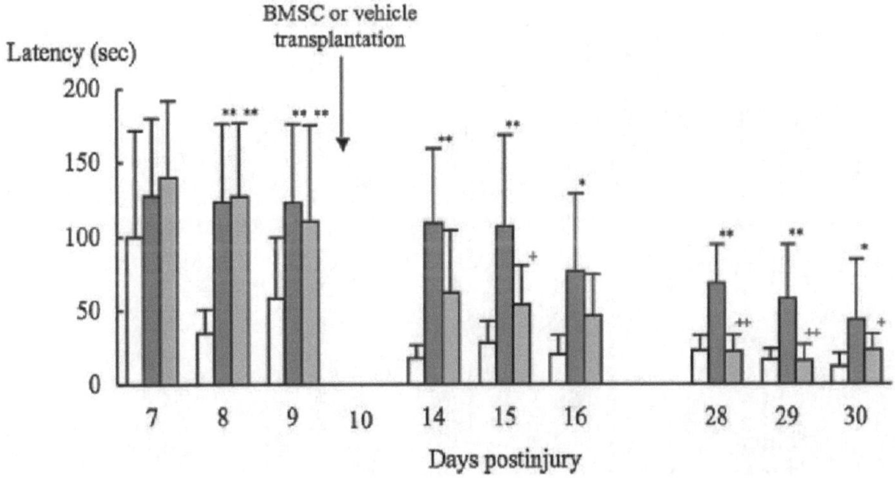

Fig. 1. Bar graph showing the latency to locate the platform on the Morris water-maze test at 1, 2, and 4 weeks postinjury. Vehicle or bone marrow stromal cell (*BMSC*) suspension was transplanted 10 days after the injury. *White bars* represent the data in the sham-operated animals. *Black and gray bars* represent the data in the vehicle- and BMSC-transplanted animals, respectively. BMSC transplantation significantly improved post-traumatic cognitive function at 15, 28, 29, and 30 days postinjury. $*P < 0.05$ and $**P < 0.01$ compared with the sham-operated group; $^{+}P < 0.05$ and $^{++}P < 0.01$ compared with the vehicle-transplanted group

significantly shorter than that in the vehicle-transplanted animals ($P < 0.05$). The latencies further decreased at 28, 29, and 30 days postinjury, and were significantly shorter than those in the vehicle-transplanted animals.

Transplanted BMSCs Are Widely Distributed and Gradually Acquire Neural Cell Phenotypes

The BMSC-transplanted animals were killed at 2 ($n = 2$), 8 ($n = 6$), and 20 weeks ($n = 5$) postinjury. As described above, they had exhibited significant improvement of cognitive function at 2 weeks postinjury, i.e., at 4 days after BMSC transplantation. Double-fluorescence immunohistochemistry revealed that GFP-positive BMSCs were widely distributed throughout the injured brain, especially in the neocortex on both sides, and these cells expressed nestin, Tuj-1, GFAP, and NeuN, but not MAP2 (Fig. 2) [6]. These findings indicated that the transplanted cells had started to acquire the phenotypes of neural cells, but their processes were not completed.

At 8 weeks postinjury, the number of GFP-positive cells in the injured brain had increased. Double-fluorescence immunohistochemistry revealed that a certain

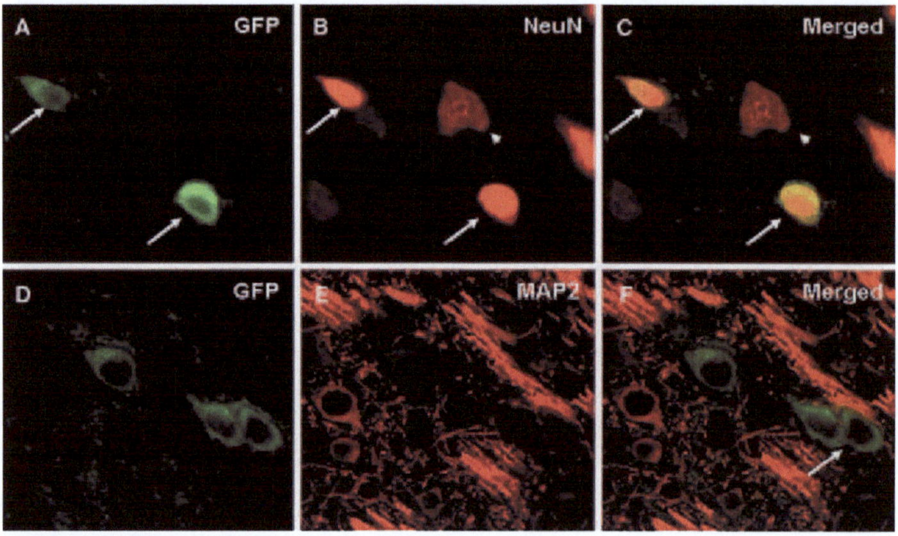

Fig. 2A–F. Photomicrographs of double-fluorescence immunohistochemistry using primary antibodies against green fluorescent protein (*GFP*; **A** and **D**), NeuN (**B**), and MAP2 (**E**) in the neocortex of BMSC-transplanted animals at 2 weeks postinjury. The merged images (**C** and **F**) show that the transplanted BMSCs are widely distributed in the neocortex and express NeuN (**A–C**; *arrows*), but not MAP2 (**D, F**). Note that the GFP-positive BMSCs are closely associated with the MAP-2 positive axons (**F**; *arrow*). The cells positive for NeuN, but negative for GFP are the host neurons (**B** and **C**; *arrowheads*). **A–F** × 400

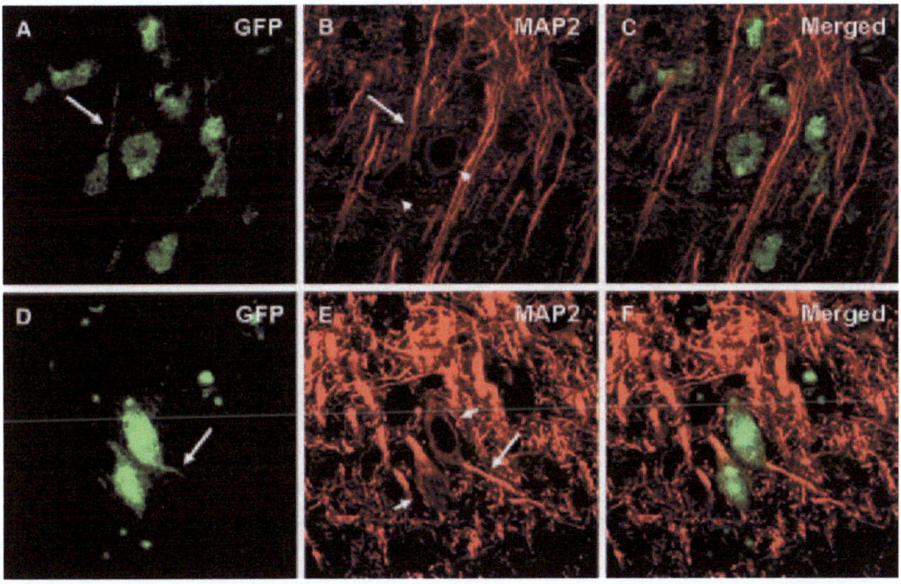

Fig. 3A–F. Photomicrographs of double-fluorescence immunohistochemistry using primary antibodies against GFP (**A** and **D**) and MAP2 (**B** and **E**) in the neocortex of BMSC-transplanted animals at 20 weeks postinjury. The merged images (**C** and **F**) show that the transplanted BMSCs are widely distributed in the neocortex. Many of the transplanted BMSCs morphologically simulate neurons, and express MAP2 in the cell body (*arrowheads*) and projections (*arrows*), suggesting that the engrafted BMSCs are integrated into the brain. **A–F** × 400

subpopulation of the GFP-positive BMSCs expressed GFAP, Tuj-1, NeuN, nestin, and also MAP2. The morphology of the cells that were doubly positive for GFP and GFAP was quite different from that of the host astrocytes positive only for GFAP (data not shown).

The GFP-positive cells were still engrafted throughout the injured brain at 20 weeks postinjury. Double-fluorescence immunohistochemistry revealed that the GFP-positive BMSCs were still distributed in the injured brain. Many of them morphologically simulated neurons, and expressed MAP2 in the cell body and projections (Fig. 3) [6]. These findings suggested that the differentiation processes in the engrafted BMSCs had further matured, and the engrafted BMSCs were integrated into the brain.

Discussion

The present study clearly showed that BMSCs significantly improved DAI-induced cognitive function when transplanted into the injured brain at 10 days postinjury, and the study also showed that the transplanted BMSCs could migrate widely into the brain, survive for at least 5 months, and gradually acquire the phenotypes of neural cells, especially neurons. Based on these observations, it seems that the intracerebral transplantation of BMSCs may serve as a novel therapeutic strategy for post-traumatic cognitive impairment.

Conclusion

BMSCs, when directly transplanted at 10 days postinjury, have the potential to migrate extensively into the injured brain and significantly improve DAI-related cognitive function at 2 and 4 weeks postinjury. It appears that the engrafted BMSCs may exhibit this early beneficial effect by producing neuroprotective or neurotrophic factors, because they had not completed the process of neuronal differentiation at 2 weeks postinjury. The long-term histological observations in the present study revealed that neuronal differentiation in transplanted BMSCs occurred both morphologically and biochemically for a period of 5 months postinjury; their role in the injured brain should be further studied.

References

1. Hokari M, Kuroda S, Shichinohe H et al (2008) Bone marrow stromal cells protect and repair damaged neurons through multiple mechanisms. J Neurosci Res 86: 1024–1035
2. Shichinohe H, Kuroda S, Yano S et al (2006) Improved expression of gamma-aminobutyric acid receptor in mice with cerebral infarct and transplanted bone marrow stromal cells: an autoradiographic and histologic analysis. J Nucl Med 47: 486–491
3. Yano S, Kuroda S, Lee JB et al (2005) In vivo fluorescence tracking of bone marrow stromal cells transplanted into a pneumatic injury model of rat spinal cord. J Neurotrauma 22: 907–918
4. Yano S, Kuroda S, Shichinohe H et al (2005) Do bone marrow stromal cells proliferate after transplantation into mice cerebral infarct?—A double labeling study. Brain Res 1065: 60–67
5. Maruichi K, Kuroda S, Chiba Y et al (2009) Graded model of diffuse axonal injury for studying head injury-induced cognitive dysfunction in rats. Neuropathology 29: 132–139
6. Maruichi K, Kuroda S, Chiba Y et al (2009) Transplanted bone marrow stromal cells improves cognitive dysfunction due to diffuse axonal injury in rats. Neuropathology [Epub ahead of print]

Part III
Molecular Imaging and Future Perspectives of Brain Research

Ultra-High Throughput Screening (uHTS) Chemical Genetics to Identify Novel Chronotherapeutics

Steve A. Kay and Tsuyoshi Hirota

Summary

The circadian clock controls daily oscillations of gene expression at the cellular level. We present the development of a high-throughput circadian functional assay system that consists of luminescent reporter cells, screening automation, and a data analysis pipeline. We applied this system to further dissect the molecular mechanisms underlying the mammalian circadian clock using a chemical biology approach. We initially analyzed the effect of 1280 pharmacologically active compounds with diverse structures on the circadian period length that is indicative of the core clock mechanism. Our screening paradigm identified many compounds previously known to change the circadian period or phase, demonstrating the validity of the assay system. Furthermore, we found that small molecule inhibitors of glycogen synthase kinase 3 (GSK-3) consistently caused a strong short-period phenotype in contrast to the well-known period lengthening by lithium, another presumed GSK-3 inhibitor. siRNA-mediated knock-down of GSK-3β also caused a short period, confirming the phenotype obtained with the small molecule inhibitors. These results indicate a novel role of GSK-3β in the period regulation of the mammalian clockwork and highlight the effectiveness of chemical biology in exploring unidentified mechanisms of the circadian clock. We have now continued screening a collection of more than 650 000 compounds for the ability to alter rhythmic attributes.

Key words Circadian clock · Chemical biology · High-throughput screening · Small molecule library · Glycogen synthase kinase 3

Division of Biological Sciences, University of California San Diego, 9500 Gilman Drive, La Jolla, CA 92093, USA

N. Tamaki, Y. Kuge (eds.)
Molecular Imaging for Integrated Medical Therapy and Drug Development.
DOI: 10.1007/978-4-431-98074-2_18, ©Springer 2010

Introduction

Genetic networks of regulated transcription and protein turnover lie at the core of circadian regulation in all organisms. Mutation in key nodes of the circadian networks causes changes in overt behavioral and physiological rhythms [1, 2]. For example, familial advanced sleep phase syndrome with early sleep times and early-morning awakening is attributed to missense mutations of human *PER2* and *CSNK1D* genes [3, 4]. The clock genes constitute the transcription/translation-based negative feedback loop of the core oscillator; CLOCK/BMAL1 heterodimers activate the transcription of *Per* and *Cry* genes, and PER and CRY proteins in turn inhibit their own transcription [5]. In addition to transcriptional regulation, post-translational modifications of clock proteins by phosphorylation, ubiquitination, and acetylation play essential roles in the oscillator mechanism [6, 7].

The molecular clock machinery resides at the cellular level, and each single cell shows circadian rhythmicity in a cell-autonomous manner [8–10]. At the organismal level, the cellular oscillators are organized in a hierarchy, in which the suprachiasmatic nucleus (SCN) constitutes the central circadian pacemaker [11]. In the SCN, the cellular clocks are synchronized to form a coherent oscillator through intracellular coupling [12], making the SCN clock more robust against genetic and environmental perturbations than peripheral oscillators [13]. Therefore, a cell-based assay system using cultured fibroblasts that lack intercellular coupling [9, 10] will provide a particularly responsive system to characterize the circadian clockwork through an unbiased, phenotype-driven screening [14, 15]. Perturbations may be revealed in such a cell-based approach that might otherwise be masked via coupling in the SCN and thus missed using behavioral genetic screens.

Although many clock genes forming the core oscillatory loop have been identified, evidence suggests the existence of more additional unknown clock components and modulators [16]. Chemical biology methods utilize small molecules as proof-of-concept probes for biological systems and can be effective in discovering novel biological mechanisms [14, 17]. The approach can complement the limitations of classical forward and reverse genetic screens associated with lethality, pleiotropy, and functional overlapping of closely related proteins. Chemical probes can be applied in a dose-dependent and reversible manner at multiple levels of biological organization. A set of compounds that potently affect the circadian clock function will lead to the identification of novel clock components and form the basis for therapeutic strategies directed towards circadian disorders.

In this study, we developed a robust cell-based circadian screening paradigm for the identification of compounds. To test the screening pipeline, we used a structurally diverse chemical library (LOPAC; Library of Pharmacologically Active Compounds) containing 1280 pharmacologically active compounds that span a broad range of biological pathways. Among them, we successfully identified 11 compounds causing reproducible period changes [18]. Therefore, the assay system is competent for large-scale compound screening to discover new chemical probes for dissecting novel circadian pathways.

Development of a High-Throughput Circadian Assay System

We utilized the circadian luciferase reporter *Bmal1-dLuc* [19] for monitoring circadian rhythms in cultured cells and developed a 384-well plate-based assay system to screen compound libraries. Among all cell lines tested, a human U2OS cell line showed prominent rhythmicity with high luminescence intensity. We established several clonal U2OS lines stably expressing the *Bmal1-dLuc* reporter and selected one clone with high amplitude and a low damping rate rhythm for further study.

For the luminescence recording, the cells were suspended in the culture medium and plated onto 384-well white solid-bottom plates by using a microplate dispenser. They were cultured for 2 days to reach confluence. Then, the explant medium containing hydroxyethylpiperazine ethanesulfonic acid (HEPES) buffer and luciferin was dispensed to each well, followed by the application of compounds by using 384-well head PinTool (GNF Systems, California). The plate was covered by an optically clear film and set to a luminescence monitoring system. This procedure is sufficient for the synchronization of the cellular rhythms accompanied by the induction of endogenous *PER1* and *PER2* genes. The luminescence was recorded every 2 h over 4 days by using a microplate reader (Infinite M200; Tecan, Männedorf, Switzerland) or a Genomics Institute of the Novartis Research Foundation (GNF) automated robotic system [20] equipped with a charge-coupled device (CCD) imager (ViewLux; Perkin Elmer).

In parallel, we have developed an automated curve fitting program CellulaRhythm for the analysis of large amounts of luminescence data to obtain rhythm parameters such as period length. Raw luminescence data were fitted to a damped cosine curve using nonlinear least squares to the following equation:

$$Luminescence = Baseline + mt + Amplitude\left(e^{-kt}\right)cos\left(\frac{2\pi(t - Phase)}{Period}\right)$$

where $m = Slope$, $k = Damping\ rate$, and $t = Time$.

By optimizing cell culture and luminescence measurement conditions, we obtained highly consistent rhythms from each well in a 384-well format. More than 97% of the wells were within the period range of mean ± 0.5 h, which is similar to and better than a recently reported system using a U2OS reporter line, 384-well plate assays, and kinetic luminescence measurements to perform siRNA-based perturbations to the circadian machinery [15]. Furthermore, treatment of the cells with D4476, a casein kinase I (CKI) δ/ε inhibitor known to lengthen the circadian period [21], caused a long-period phenotype in a dose-dependent manner at final concentrations of 3–20 μM. These results indicate the validity of the system for the screening of compounds affecting the circadian period.

Screening of LOPAC Chemical Library

The potential effect of 1280 compounds on the cellular circadian rhythm was investigated at final concentrations of 7 μM (Fig. 1). The screening was repeated

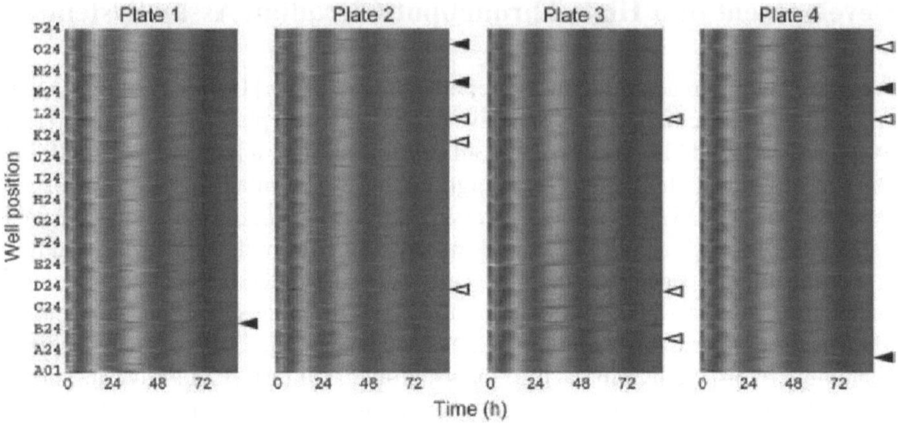

Fig. 1. Raster plot of the result of Library of Pharmacologically Active Compounds (LOPAC) chemical library screening. Luminescence rhythms of *Bmal1-dLuc* cells were monitored by using the ViewLux system in the presence of compounds (final concentration, 7 μM). One screening contained four 384-well plates, and profiles of one 384-well plate are represented in *each panel. Each horizontal raster line* represents a single well, with elapsed time plotted to the right. Luminescence intensity data from each well are normalized for amplitude, and then indicated by gray scale: *peak is white and trough is black.* The screening was repeated twice, and the result from the first experiment is shown. *Solid and open arrowheads* indicate the positions of long- and short-period compounds in Table 1, respectively. Note that there are many compounds that change the phase of the rhythm without affecting the period. (Modified from [18], with permission)

twice, and 13 primary "hits" were identified based on period change (lengthening or shortening) of 0.5 h or more in both screenings (Table 1). The effect of the 13 hit compounds was further investigated at various concentrations. Eleven of the compounds showed dose-dependent period lengthening or shortening, confirming the result of the screening. Interestingly, vincristine, an inhibitor of microtubule assembly, showed a bidirectional effect: period shortening at lower dose and lengthening at higher dose.

The hit compounds can generally be classified as inhibitors/activators of protein kinases, inhibitors of microtubule assembly, inhibitors of Ca^{2+} entry, and inducers of DNA damage (Table 1). Many of the compounds are related to the pathways already known to affect the circadian clock function in a variety of organisms and tissues, as follows. Roscovitine [a cyclin-dependent kinase (CDK) inhibitor], SP600125 [a c-Jun N-terminal kinase (JNK) inhibitor], and SB 203580 [an analog of the p38 mitogen-activated protein kinase (MAPK) inhibitor SB 202190] cause period lengthening in cultured *Aplysia* eye [22], mouse tissues [23], and chicken pineal gland [24], respectively. Decreased activity of casein kinase 2 (CK2) causes long-period behavioral rhythms in *Drosophila* [25, 26]. Furthermore, supporting the short-period phenotypes of phorbol myristate acetate (PMA) [an activator of protein kinase C (PKC)], etoposide, and mitoxantrone (DNA damage inducers) in our assay (Table 1), PMA treatment and DNA damage by γ-radiation have been reported to cause phase advances but not delays of the rhythms in cultured hamster

Table 1. Long- and short-period compounds identified from LOPAC chemical library screening[a]

Compound name	Function	Period change (h)		Confirmation
		Exp 1	Exp 2	
Roscovitine	Inhibitor of CDK	+1.1	+1.4	Long period
SP600125	Inhibitor of JNK	+1.0	+1.4	Long period
SB 202190	Inhibitor of p38 MAPK	+1.0	+1.2	Long period
DRB	Inhibitor of CK2	+0.7	+1.0	Long period
Vincristine	Inhibitor of microtubules	+0.6	+1.0	Long & short period
Etoposide	Inducer of DNA damage	−3.5	−1.5	Short period
Mitoxantrone	Inducer of DNA damage	−1.5	−2.0	Short period
PMA	Activator of PKC	−1.9	−1.6	Short period
SKF-96365	Inhibitor of Ca^{2+} entry	−1.3	−2.1	Short period
Indirubin-3′-oxime	Inhibitor of CDK and GSK-3	−1.0	−0.7	Short period
Kenpaullone	Inhibitor of CDK and GSK-3	−1.1	−0.5	Short period
Ethamivan	Respiratory stimulant	−2.5	−0.6	No period effect
2-Methoxyestradiol	Estrogen derivative	−1.2	−1.6	No period effect

[a] Modified from [18], with permission

Exp, experiment; GSK-3, glycogen synthase kinase 3; LOPAC, Library of Pharmacologically Active Compounds; CDK, cyclin-dependent kinase; JNK, c-Jun N-terminal kinase; MAPK, mitogen-activated protein kinase; CK2, casein kinase 2; PMA, phorbol myristate acetate; PKC, protein kinase C

Luminescence rhythms of *Bmal1-dLuc* cells were monitored by using the ViewLux system in the presence of compounds (final concentration, 7 μM). Each luminescence profile was fitted with a damped cosine curve to obtain the period parameter. The compounds that caused a period change of 0.5 h or more in both screenings were listed with a period change (h) relative to the median period of the dimethyl sulfoxide (DMSO) control wells. Positive and negative values represent period lengthening and shortening, respectively. Ethamivan and 2-methoxyestradiol did not show a period-changing effect in the confirmation assay

SCN and free-running mice, respectively [27, 28]. Taken together, the identification of these previously known compounds and pathways validates our screening paradigm and suggests a highly conserved mechanism of the circadian clock across species and tissue types.

Period Shortening by Inhibition of GSK-3β

The period-shortening compounds indirubin-3′-oxime and kenpaullone (Table 1) are known to inhibit both CDK and GSK-3 [29]. In contrast, roscovitine, which inhibits CDK but not GSK-3 [29], caused a long-period phenotype (Table 1). To clarify the difference of the effect of CDK inhibition and GSK-3 inhibition on the period length, we tested additional CDK and GSK-3 inhibitors. We found that GSK-3-directed inhibitors, including Chir99021 [30] and 1-azakenpaullone [31], shortened the period, but CDK-directed inhibitors such as purvalanol A [29] and NU6102 [32] did not. Interestingly, the period-shortening effect of GSK-3-directed inhibitors is opposite to the well-known period-lengthening effect of lithium [33], which has been proposed to act through GSK-3 inhibition [34]. We also observed prominent period lengthening by LiCl in our assay system.

To elucidate the role of GSK-3 in the period regulation of mammalian cells, we performed siRNA-mediated knock-down experiments. Transfection of GSK-3β siRNA strongly and specifically reduced endogenous GSK-3β mRNA levels and shortened the circadian period. We obtained similar period-shortening in mouse primary fibroblasts infected with a lentivirus vector encoding GSK-3β shRNA. Together, the chemical biology and genomic approaches clearly indicate a novel role of GSK-3β in the period regulation of the mammalian circadian clock.

Role of GSK-3β in the Circadian Clock Mechanism

In *Drosophila*, reduction of GSK-3 activity by genetic manipulation causes period lengthening [35], in contrast to our findings in mammalian cell culture. Of note, GSK-3 phosphorylates timeless (TIM) protein in *Drosophila* [35], but there is no tim ortholog in mammals [36]. Instead, GSK-3β is known to phosphorylate Period 2 (PER2), cryptochrome 2 (CRY2), and Rev-erbα in mammals [37–39]. GSK-3β-mediated phosphorylation leads to proteasomal degradation of CRY2 [38] and stabilization of Rev-erbα [39]. Considering that *Cry2* and *Rev-erbα* knockout mice show long- and short-period phenotypes, respectively [40–42], it is possible that the period shortening by GSK-3β inhibition is mediated at least in part by the regulation of CRY2 and Rev-erbα protein levels (i.e., stabilization of CRY2 and degradation of Rev-erbα). The role of GSK-3α is of interest and should be addressed in future studies.

Although lithium lengthens the period of circadian rhythms in a wide range of experimental systems such as unicellular organisms, insects, mice, and humans [33], the exact mode of action is still uncertain. Because lithium inhibits inositol monophosphatase and other phosphomonoesterases as well as GSK-3 [34], the long-period phenotype in mammals might be mediated by lithium-targeted protein(s) other than GSK-3.

Period-Changing Compounds from the LOPAC Chemical Library

The LOPAC chemical library contains many drugs currently on the market and in clinical trials. Therefore, our screening results obtained from human cells might have an important implication for the application of such drugs. Roscovitine, vincristine, etoposide, and mitoxantrone are in clinical trials against cancers, and they showed significant period-changing effects (Table 1) that may affect the circadian clock of normal, nondividing tissues.

The specificity of 20 kinase inhibitors including clinical drugs was reported by developing a small molecule-kinase interaction profiling method [43]. The kinase-interaction of roscovitine, SP600125, and SB 202190 was investigated, and interestingly, all of the compounds showed strong binding with CKIε [43], the inhibition of which caused a long-period phenotype. It is possible that the period-lengthening effect of roscovitine, SP600125, and SB 202190 (Table 1) is mediated via CKIε inhibition as well as by their primary target (CDK, JNK, or p38 MAPK). Further studies are necessary to determine the protein responsible for the circadian effect of these compounds.

Chemical Biology Approach for Circadian Clock Mechanism

We have successfully miniaturized the automated luminescence monitoring of cellular circadian rhythms to a 384-well format and set up data analysis tools to extract circadian parameters from thousands of datasets, enabling us to identify period-altering compounds from LOPAC chemical library screening. To discover novel clock components through the chemical biology approach, we need a more comprehensive, large-scale approach with many different types of compounds. Such diversity-oriented screening usually requires hundreds of thousands of chemical probes to identify a viable lead compound [44].

The 1536-well plate ultra-high throughput screening (uHTS) format allows screening of large compound libraries. We applied the GNF Automated Compound Profiling (ACP) System [20] to the circadian assay and enabled parallel recordings from 70 plates, achieving throughput of ~100 000 compounds per week in a 1536-well plate format. We are screening a nonproprietary compound collection containing ~650 000 diverse compounds with predicted drug-like properties. A wide variation of chemical structures has the advantage of probing many classes of potential targets, some of which may lead to the identification of new mechanisms regulating circadian clock functions in many tissues. The chemical probes identified will provide a basis for developing therapeutics based on alteration of circadian function.

References

1. Young MW, Kay SA (2001) Time zones: a comparative genetics of circadian clocks. Nat Rev Genet 2: 702–715
2. Takahashi JS, Hong HK, Ko CH et al (2008) The genetics of mammalian circadian order and disorder: implications for physiology and disease. Nat Rev Genet 9: 764–775
3. Toh KL, Jones CR, He Y et al (2001) An hPer2 phosphorylation site mutation in familial advanced sleep phase syndrome. Science 291: 1040–1043
4. Xu Y, Padiath QS, Shapiro RE et al (2005) Functional consequences of a CKIdelta mutation causing familial advanced sleep phase syndrome. Nature 434: 640–644
5. Reppert SM, Weaver DR (2002) Coordination of circadian timing in mammals. Nature 418: 935–941
6. Gallego M, Virshup DM (2007) Post-translational modifications regulate the ticking of the circadian clock. Nat Rev Mol Cell Biol 8: 139–148
7. Belden WJ, Dunlap JC (2008) SIRT1 is a circadian deacetylase for core clock components. Cell 134: 212–214

8. Welsh DK, Logothetis DE, Meister M et al (1995) Individual neurons dissociated from rat suprachiasmatic nucleus express independently phased circadian firing rhythms. Neuron 14: 697–706
9. Welsh DK, Yoo SH, Liu AC et al (2004) Bioluminescence imaging of individual fibroblasts reveals persistent, independently phased circadian rhythms of clock gene expression. Curr Biol 14: 2289–2295
10. Nagoshi E, Saini C, Bauer C et al (2004) Circadian gene expression in individual fibroblasts: cell-autonomous and self-sustained oscillators pass time to daughter cells. Cell 119: 693–705
11. Schibler U, Sassone-Corsi P (2002) A web of circadian pacemakers. Cell 111: 919–922
12. Aton SJ, Herzog ED (2005) Come together, right . . . now: synchronization of rhythms in a mammalian circadian clock. Neuron 48: 531–534
13. Liu AC, Welsh DK, Ko CH et al (2007) Intercellular coupling confers robustness against mutations in the SCN circadian clock network. Cell 129: 605–616
14. Liu AC, Lewis WG, Kay SA (2007) Mammalian circadian signaling networks and therapeutic targets. Nat Chem Biol 3: 630–639
15. Vollmers C, Panda S, DiTacchio L (2008) A high-throughput assay for siRNA-based circadian screens in human U2OS cells. PLoS ONE 3: e3457
16. Takahashi JS (2004) Finding new clock components: past and future. J Biol Rhythms 19: 339–347
17. Ding S, Schultz PG (2004) A role for chemistry in stem cell biology. Nat Biotechnol 22: 833–840
18. Hirota T, Lewis WG, Liu AC et al (2008) A chemical biology approach reveals period shortening of the mammalian circadian clock by specific inhibition of GSK-3beta. Proc Natl Acad Sci USA 105: 20 746–20 751
19. Liu AC, Tran HG, Zhang EE et al (2008) Redundant function of REV-ERBalpha and beta and non-essential role for Bmal1 cycling in transcriptional regulation of intracellular circadian rhythms. PLoS Genet 4: e1000023
20. Melnick JS, Janes J, Kim S et al (2006) An efficient rapid system for profiling the cellular activities of molecular libraries. Proc Natl Acad Sci USA 103: 3153–3158
21. Reischl S, Vanselow K, Westermark PO et al (2007) Beta-TrCP1-mediated degradation of PERIOD2 is essential for circadian dynamics. J Biol Rhythms 22: 375–386
22. Sankrithi N, Eskin A (1999) Effects of cyclin-dependent kinase inhibitors on transcription and ocular circadian rhythm of Aplysia. J Neurochem 72: 605–613
23. Chansard M, Molyneux P, Nomura K et al (2007) c-Jun N-terminal kinase inhibitor SP600125 modulates the period of mammalian circadian rhythms. Neuroscience 145: 812–823
24. Hayashi Y, Sanada K, Hirota T et al (2003) p38 mitogen-activated protein kinase regulates oscillation of chick pineal circadian clock. J Biol Chem 278: 25 166–25 171
25. Lin JM, Kilman VL, Keegan K et al (2002) A role for casein kinase 2alpha in the Drosophila circadian clock. Nature 420: 816–820
26. Akten B, Jauch E, Genova GK et al (2003) A role for CK2 in the Drosophila circadian oscillator. Nat Neurosci 6: 251–257
27. Schak KM, Harrington ME (1999) Protein kinase C inhibition and activation phase advances the hamster circadian clock. Brain Res 840: 158–161
28. Oklejewicz M, Destici E, Tamanini F et al (2008) Phase resetting of the mammalian circadian clock by DNA damage. Curr Biol 18: 286–291
29. Bain J, McLauchlan H, Elliott M et al (2003) The specificities of protein kinase inhibitors: an update. Biochem J 371: 199–204
30. Ring DB, Johnson KW, Henriksen EJ et al (2003) Selective glycogen synthase kinase 3 inhibitors potentiate insulin activation of glucose transport and utilization in vitro and in vivo. Diabetes 52: 588–595
31. Kunick C, Lauenroth K, Leost M et al (2004) 1-Azakenpaullone is a selective inhibitor of glycogen synthase kinase-3 beta. Bioorg Med Chem Lett 14: 413–416
32. Davies TG, Bentley J, Arris CE et al (2002) Structure-based design of a potent purine-based cyclin-dependent kinase inhibitor. Nat Struct Biol 9: 745–749

33. Engelmann W (1987) Effects of lithium salts on circadian rhythms. In: Halaris A (ed) Chronobiology and psychiatric disorders. Elsevier Science, Amsterdam
34. Quiroz JA, Gould TD, Manji HK (2004) Molecular effects of lithium. Mol Interv 4: 259–272
35. Martinek S, Inonog S, Manoukian AS et al (2001) A role for the segment polarity gene shaggy/GSK-3 in the Drosophila circadian clock. Cell 105: 769–779
36. Gotter AL, Manganaro T, Weaver DR et al (2000) A time-less function for mouse timeless. Nat Neurosci 3: 755–756
37. Iitaka C, Miyazaki K, Akaike T et al (2005) A role for glycogen synthase kinase-3beta in the mammalian circadian clock. J Biol Chem 280: 29 397–29 402
38. Harada Y, Sakai M, Kurabayashi N et al (2005) Ser-557-phosphorylated mCRY2 is degraded upon synergistic phosphorylation by glycogen synthase kinase-3 beta. J Biol Chem 280: 31 714–31 721
39. Yin L, Wang J, Klein PS et al (2006) Nuclear receptor Rev-erbalpha is a critical lithium-sensitive component of the circadian clock. Science 311: 1002–1005
40. Thresher RJ, Vitaterna MH, Miyamoto Y et al (1998) Role of mouse cryptochrome blue-light photoreceptor in circadian photoresponses. Science 282: 1490–1494
41. van der Horst GT, Muijtjens M, Kobayashi K et al (1999) Mammalian Cry1 and Cry2 are essential for maintenance of circadian rhythms. Nature 398: 627–630
42. Preitner N, Damiola F, Lopez-Molina L et al (2002) The orphan nuclear receptor REV-ERBalpha controls circadian transcription within the positive limb of the mammalian circadian oscillator. Cell 110: 251–260
43. Fabian MA, Biggs WH, 3rd, Treiber DK et al (2005) A small molecule-kinase interaction map for clinical kinase inhibitors. Nat Biotechnol 23: 329–336
44. Mullin R (2004) Drug discovery: as high-throughput sreening draws fire, researchers leverage science to put automation into perspective. Chem Eng News 82: 23–32

Cell Death and Autophagy

Yasuo Uchiyama[1], Masato Koike[1], and Masahiro Shibata[2]

Summary

Autophagy is a mechanism by which parts of a cell that are old and unneeded are segregated inside structures called "autophagosomes". The materials ingested by this autophagy are brought to cellular compartments called "lysosomes," which are specific intracellular compartments for degradation, and the degraded products are re-used for cell metabolism. We have shown that, in mice, deficiency in lysosomal proteinases such as cathepsin D or cathepsins B and L induces the accumulation of lysosomes containing ceroid-lipofuscin; the phenotypes of these mice resemble those of neuronal ceroid lipofuscinosis (NCL). In these mutant mice, the accumulation of abnormal lysosomal structures appears in accordance with an increase in the amount of membrane-bound microtubule associated protein 1 light chain 3 (LC3), a marker of "autophagosomes" in neurons. Such autophagosomes often contain granular osmiophilic deposits, a hallmark of NCL, together with part of the cytoplasm, which contains undigested materials. These data strongly argue for a major involvement of autophagy in the pathogenesis of NCL, although it remains largely unknown what signaling is essential for autophagosome formation.

Neonatal hypoxic/ischemic (H/I) brain injury causes neurological impairment, including cognitive and motor dysfunction, as well as seizures. However, the molecular mechanisms regulating neuron death after H/I injury are poorly defined and remain controversial. Here we show that *Atg7*, a gene essential for autophagy induction, is a critical mediator of H/I-induced neuron death. Neonatal mice subjected to H/I injury show dramatically increased autophagosome formation and extensive hippocampal neuron death that is regulated by both caspase-3-dependent and -independent execution. Mice deficient in Atg7 show nearly complete protection from both H/I-induced caspase-3 activation and neuron death, indicating that Atg7 is critically positioned upstream of multiple

[1]Department of Cell Biology and Neuroscience, Juntendo University Graduate School of Medicine, 2-1-1 Hongo, Bunkyo-ku, Tokyo 113-8421, Japan
[2]Division of Gross Anatomy and Morphogenesis, Department of Regenerative and Transplant Medicine, Niigata University Graduate School of Medical and Dental Sciences, Niigata 951-8510, Japan

N. Tamaki, Y. Kuge (eds.)
Molecular Imaging for Integrated Medical Therapy and Drug Development.
DOI: 10.1007/978-4-431-98074-2_19, ©Springer 2010

neuronal death executioner pathways. Adult H/I brain injury also produces a significant increase in autophagy, but, unlike neonatal H/I, neuron death is almost exclusively caspase-3-independent. These data suggest that autophagy plays an essential role in triggering neuronal death execution after H/I injury.

Although it has been considered that autophagy is essential for the maintenance of cellular metabolism, our data suggest that excess autophagy under pathological conditions may lead to cell death.

Key words Autophagy · Lysosomes · Cathepsins · LC3 · Brain ischemia

Introduction

Lysosomes are multifunctional membrane-bound organelles in mammalian cells and their internal environment is acidic, with pH ranging from 5.0 to 5.5. They contain various acid hydrolases and degrade excess, old, and unneeded intracellular substances and organelles, as well as extracellular materials, into biologically active monomers that may be recycled intracellularly. There are three intracellular trafficking routes for such macromolecules to lysosomes: (1) endocytosis of cell-surface-receptor proteins with bound ligands, which produces early endosomes; (2) heterophagocytosis of large extracellular materials, such as dead cells and bacteria, which produces heterophagosomes; and (3) autophagy of old and unneeded intracellular materials, which produces autophagosomes. Early endosomes, heterophagosomes, and autophagosomes then receive lysosomal enzymes by fusing with lysosomes or transporting vesicles from the *trans*-Golgi network, while degradation begins in these compartments that subsequently become late endosomes/lysosomes, heterolysosomes, and autolysosomes, respectively [1].

Stroke is one of the most common adult diseases that lead to death, while hypoxia/ischemia (H/I) in infants and children may be linked to subsequent neurological complications. This type of brain damage is related to cerebral palsy, mental retardation, and epilepsy [2]. Neurons are exposed to various stresses such as energy failure, oxidative stress, and unbalanced ion fluxes after H/I brain injury [2–4], and such stresses may induce autophagy in the neurons [5–7]. It has been shown that autophagy is highly induced in CA1 pyramidal neurons of the gerbil hippocampus after brief forebrain ischemia [8]. Autophagy is a highly regulated process involving the bulk degradation of cytoplasmic macromolecules and organelles in mammalian cells via the lysosomal system, and it is induced in response to various stresses [9, 10]. Autophagy is essential for the maintenance of cellular metabolism, and thus the induced autophagy in CA1 pyramidal neurons of the gerbil hippocampus may be neuroprotective. However, autophagy is also involved in neurodegenerative disorders and ,further, can trigger a form of cell death distinct from apoptosis in neurons [5, 11, 12]. The induction of autophagy has also been shown in neonatal and adult mouse cortex and striatum after H/I injury [13], when examined by western blotting for LC3, which is a marker protein for autophagy that is required for autophagosome formation via its conversion from cytosolic LC3-I to membrane-bound LC3-II [14]. Thus, it is important to decide whether autophagy participates as a pro- or anti-death factor in the execution of neuron death after ischemic brain insult. We therefore examined this issue using wild-type, Atg7-deficient, caspase-3-deficient, and caspase activated DNase (CAD)-deficient mice [5, 6].

Autophagy in Lysosome Storage Disorders Due to Deficiency in Lysosomal Cathepsins and Lamp2

Lysosomal cathepsins B and L (CB and CL), and D (CD) are representative cysteine and aspartic proteinases in cells of mammalian tissue, including the brain. The most common inherited neurodegenerative disease in childhood is neuronal ceroid-lipofuscinosis (NCL). NCL is known as a lysosomal storage disorder and is pathologically characterized by the accumulation of proteolipids, such as subunit c of mitochondrial ATP synthase and sphingolipid activator proteins in the lysosomes of neurons [15]. Because most lysosomal proteinases are cysteine proteinases, it has been thought that loss of cysteine proteinases such as CB and CL causes the pathogenesis of NCL. In fact, it has been shown that the administration of cysteine proteinase inhibitors such as leupeptin and E64 induces the accumulation of lysosomes with ceroid-lipofuscin in the brain cells of young rats [16]. However, the causative gene of NCLs that was first found in human cases was *CLN2* (late-infantile NCL), which encodes a lysosomal serine proteinase, tripeptidyl peptidase I (TPP-I) [17]. TPP-I is involved in the initial degradation of subunit c in lysosomes, and it leads to the sequential cleavage of tripeptides from the N-terminus of the peptide corresponding to the amino terminal sequence of subunit c. Therefore, loss of CLN2 leads to the accumulation of subunit c in the lysosomes of neurons [18, 19].

We have demonstrated that the central nervous system (CNS) neurons in CD-deficient mice show a new form of lysosomal accumulation disease with a phenotype resembling that of NCL; subunit c of mitochondrial ATP synthase accumulates in the lysosomes of the affected neurons. CD-deficient mice manifest neurological phenotypes such as seizures with trembling and stiff tails, and blindness during the terminal stages, because of massive neuron death in various brain regions, particularly in the thalamus, cerebral cortex, hippocampus, and retina [11, 20–22]. Such striking neuron death is executed by nitric oxide (NO) that is produced via inducible NO synthase activity in microglial cells [21, 22]. Similar to the CD activity in mouse brains, CD is also extremely important in human brains; it has been reported that mutations in the human *CD* gene are responsible for a novel autosomal recessive disorder, CD-deficient NCL, which can manifest in early childhood. Two missense mutations in the *CD* gene, F229I and W383C, have been identified and have been found to cause markedly reduced proteolytic activity and a diminished amount of CD in patient fibroblasts. The *CD* gene has also been shown to be responsible for a congenital form of NCL, which is currently denoted as CLN10 [23, 24].

Moreover, we have shown that the ultrastructural and morphometric features of CB/CL-double deficient neurons are very similar to those of CD-deficient neurons, indicating that mice deficient in CB and CL can be models of NCL [11]. The morphological hallmarks of NCL neurons are granular osmiophilic deposits (GRODs) and fingerprint profiles that can be seen in these mutant mouse neurons. Because CB and subunit c are detected in the GRODs of CD-deficient neurons [20], the GRODs and fingerprint profiles are lysosomes. Immunoreactivity for subunit c is granular in the neuronal perikarya of CD-deficient and CB/CL-double deficient mouse brains, while the localization pattern of LC3 is similar to that of subunit c in the neuronal perikarya and fibrous in the dendrites of cerebral and cerebellar cortical neurons [11]. Electron microscopic observations show that nascent autophagosomes containing part of the cytoplasm are frequently detected in CNS neurons of CD-deficient and CB/CL-double deficient mice that are near the terminal stages of the disease. Autophagosomes undergo stepwise maturation by fusing with endosomes and/or lysosomes [25], and two types of autophagosomes in this maturation process can be detected by electron microscopy [26, 27]: nascent or immature autophagosomes that are enwrapped by endoplasmic reticulum (ER)-like membrane saccules contain part of the morphologically intact cytoplasm, while autolysosomes enwrapped by a single membrane possess degraded but morphologically identifiable cytoplasmic materials and structures. Interestingly, double membranes, called the isolation membrane, in CD-deficient and CB/CL-deficient neurons often enwrap lysosome-like structures such as dense bodies, GROD-like inclusions, and autophagosome-like structures, along with part of the cytoplasm, that are morphological hallmarks of NCL. This indicates that autophagosome formation occurs frequently in these mutant neurons. In CD-deficient neurons, the volume densities of GRODs and autophagosomes and autolysosomes increase with days after birth when examined by morphometry. More interestingly, half of the nascent autophagosomes counted possess GRODs, and 20% of GRODs are detected in nascent autophagosomes. These data suggest that the presence of GROD-like inclusions and autophagosomes/autolysosomes with undigested materials such as GROD-like inclusions may be a potent inducer of autophagy in neuronal cells [11].

Loss of the lysosomal membrane protein, lamp2, has also been shown to induce lysosomal storage disorder, where increased autophagosome formation is found in various tissue cells of lamp2-deficient mice [28]. In such mice, autophagic degradation of long-lived proteins is severely impaired in hepatocytes, while cardiac myocytes are ultrastructurally abnormal and heart contractility is severely reduced, indicating that lamp2 is critical for autophagy. Danon's disease, which is associated with the accumulation of autophagic material in striated myocytes, has been shown to be caused by a deficiency of lamp2 in humans [28]. Thus, impairment of lysosomal functions due to the loss of lamp2 facilitates autophagosome formation, resulting in lysosomal storage disorder.

Until recently, however, it has remained largely unknown what signaling is essential for autophagosome formation. In conditional *Atg7*-knockout mice, specifically in liver or CNS tissue, numerous ubiquitinated aggregates are detected in the cytosol of hepatocytes or CNS neurons with the presence of functional proteasomes [29–31], indicating that protein ubiquitination may serve as a signal to the autophagic process, in addition to the proteasomal pathway. The presence of ubiquitin aggregates is one of the common pathological characteristics of neurodegenerative diseases, including lysosomal storage disorders [32–34]. Very interestingly, it has been shown that a multifunctional protein, p62/A170/SQSTM1 (p62) [35], which mediates diverse signaling pathways including cell stress, survival, and inflammation [36, 37], can bind both LC3 and ubiquitin [38]. In fact, p62 is degraded in lysosomes through autophagy, and accumulates in autophagy-deficient cells [38–40]. Moreover, the genetic ablation of p62 markedly suppresses the appearance of ubiquitin-positive protein aggregates in hepatocytes and suppresses the appearance of these aggregates to some extent in neurons, indicating that p62 plays an important role in inclusion body formation, although the pathological process associated with autophagic deficiency is cell type-specific.

H/I Injury-Induced Neonatal Neuron Death Is Executed by Both Caspase-Dependent and -Independent Pathways

Caspase-Dependent Neuron Death After H/I Brain Injury

Necrosis is known to occur accidentally and is a pathological process resulting from tissue damage and loss of energy [41]. During organogenesis, embryonic/fetal cells often undergo cell death to form tissues and organs, including CNS tissue. This type of cell death is called programmed cell death (PCD), and is a genetically controlled cell death [42, 43]. Apoptosis is a morphologically defined form of cell death, while PCD is a more physiologically defined event that occurs during development, and its dying form is apoptotic. Such cell death consists of cell and nuclear shrinkage with nuclear chromatin condensation. Dying cells are finally phagocytosed by phagocytic cells, such as macrophages and microglial cells [44] after fragmentation of the cells into apoptotic bodies. There are two major pathways of apoptosis in mammalian cells: death domain-mediated and mitochondrial pathways [45]. Apoptosis is known to occur in physiological and pathological situations following various death stimuli, but the final pathway of apoptosis is mainly mediated by the caspase family of proteinases, apoptotic proteinase activating factor-1 (Apaf-1), and DNA fragmentation factors (DFF40 and 45)/CAD and its inhibitor, ICAD [46–50]. In the spinal anterior horn, half of the neurons die by naturally occurring cell death/PCD during normal brain development [51]. It has been shown that the activation of apoptotic mechanisms contributes to neuronal death and brain injury after ischemia, which is more pronounced in the immature brain [13, 52–56]. The developing brain has a higher density of *N*-methyl-D-aspartate (NMDA) receptors [57]; thus, it has been demonstrated that the vulnerability to H/I injury and subsequent apoptotic neuronal degeneration peaks at postnatal day (P) 6, and diminishes thereafter [4, 58].

We have recently shown that the hippocampal pyramidal neurons of mice at P7 undergo pyknosis from the early stage after H/I injury, and the number of the pyknotic neurons increases thereafter [5]. This indicates that the death mode of pyramidal neurons in the hippocampus is distinct from necrosis. Approximately 35% of these dying pyramidal neurons are positive for activated caspase-3 at 8 h after H/I injury. In this situation, DEVDase activity is increased and an activated form of caspase-3, but not caspase-7, is detected by western blotting. The ratio of neurons that we found to have died in a caspase-dependent manner is consistent with previous data shown by the use of a caspase-3 inhibitor [59] and Bcl-xL-overexpressing mice [60]. From these data it seems likely that H/I injury-induced pyramidal neuron death is, in part, executed in a caspase-dependent manner. However, it is also puzzling that pyramidal neurons in the hippocampus 8 h after neonatal H/I brain injury are rarely costained for TUNEL and cleaved caspase-3. This may indicate that activated caspases, other than the activation of CAD, are involved in the degradation of cytosolic substrates.

Caspase-Independent Neuron Death After H/I Brain Injury

It has been shown that a great number of pyramidal neurons in the neonatal hippocampus after H/I injury die in a caspase-independent fashion [13]. To further verify this issue, we examined H/I injury-induced pyramidal neuron death in the neonatal hippocampus, using mice that were deficient in caspase-3 or CAD [61–64]. It is well known that genomic DNA fragmentation into oligonucleosomes is one of the hallmarks of apoptotic cell death and is mediated by CAD. In the early postnatal days around P7, the genomic DNA from the untreated hippocampus of neonatal brains is weakly but distinctly fragmented into oligonucleosomes, because PCD still occurs in this region [65]. Using CAD-deficient neonatal mice, H/I injury-induced alterations were analyzed in hippocampal pyramidal neurons; the degree of damage was found to be similar to that in the wild-type mice 24 h later, and DNA laddering was detectable in the ipsilateral hippocampus. These data indicate that nuclear DNA in the damaged neurons is fragmented by a DNase other than CAD. Moreover, no DNA laddering was observed in the contralateral hippocampus, suggesting that the DNA fragmentation of PCD is mediated in the wild-type neonatal hippocampus by CAD. Similar to the findings in wild-type and CAD-deficient mice, H/I injury-induced changes are abundant in each pyramidal layer of the hippocampus of caspase-3-deficient mice within 24 h. It has been suggested that the long-term inhibition of caspase-3 during development by its genetic ablation upregulates caspase-3-independent cell death pathways and increases the vulnerability of the developing brain to neonatal H/I injury. However, the DNA fragmentation in the ipsilateral and contralateral hippocampi of caspase-3-deficient mouse brains after H/I injury occurs in a manner similar to that in the wild-type mouse brains. This is because caspase-7, which is structurally and functionally similar to caspase-3, is activated in hippocampal pyramidal neurons and compensates for the lack of caspase-3, as previously reported [63, 66].

Ultrastructural Alterations in Nuclei of Pyramidal Neurons Damaged After H/I Insult

Interestingly, the morphological alterations in the nuclei of hippocampal pyramidal neurons after H/I injury are very similar in wild-type, caspase-3-deficient, and CAD-deficient mouse brains, being detected as shrunken nuclei with small patches of chromatin clumping. Even though apoptotic neuron death is also evident in the pyramidal layers of the wild-type hippocampus, a typical nuclear figure of apoptosis with chromatin condensation is not observed in the pyramidal layers of the hippocampus after H/I injury. Moreover, such chromatin clumping in nuclei is also found in the hippocampal pyramidal neurons of adult mice after H/I injury and in those of the gerbil after brief forebrain ischemia [8].

H/I Injury-Induced Pyramidal Neuron Death in Neonatal Hippocampus Is Prevented by Atg7 Deficiency

Since the time that Clarke [67] categorized physiological neuron death into three types that can be detected in CNS tissue during development: apoptotic, autophagic, and nonlysosomal vesiculate, numerous studies have shown the presence of autophagy-related cell death in various CNS tissues, peripheral tissues, and cultured cells [68–75]. In these criteria by Clarke [67], it is generally accepted that autophagic cell death is defined as follows: (1) autophagy is highly induced in cells that undergo death, (2) cell death is not prevented by the inhibition of caspases by pancaspase inhibitors, and (3) cell death is suppressed by autophagy inhibitors such as 3-methyladenin (3-MA) and wortmannin [74, 76]. According to these criteria, we have further investigated whether H/I injury-induced neuron death is autophagic, specifically in CNS tissue, using Atg7-deficient mice [30].

When the Atg7-deficient mice were subjected to H/I injury at P7, the neuronal damage in the hippocampal pyramidal layers of the mice was compared with that in control littermate mice, with the following results: (1) Approximately 80% of the Atg7-deficient mice showed intact morphology in the hippocampal pyramidal neurons, without pyknotic nuclei and with positive TUNEL staining 3 days after the H/I injury. (2) Although 22% of the littermate control mice escaped damage to the pyramidal neurons, the rest had severely damaged hippocampal neurons. Moreover, (3) when the damaged areas in all the pyramidal layers of the hippocampus were measured 3 days after the H/I injury, the mean value in Atg7-deficient mice was 0.32%, while it was 27.38% in the littermate controls. (4) The degree of damage was further analyzed 7 days after H/I injury and the area loss in the ipsilateral hippocampus, expressed as a percentage of the contralateral hippocampal area, was 14.0% in the Atg7-deficient mice, whereas it was 48.8% in the control littermate mice. Thus, the inhibition of hippocampal pyramidal neuron death after H/I injury is sustained at least until 7 days after H/I injury. These results strongly suggest that the Atg7 deficiency largely prevented both caspase-dependent and caspase-independent neuron death in hippocampi from the neonatal mouse brains after H/I injury. Our genetic study using Atg7-deficient mice is providing direct evidence for autophagy-induced neuron death following neonatal mouse H/I brain injury [5, 6].

H/I Injury-Induced Neuron Death in Adult Mice Is Executed by Caspase-Independent Pathways

Because neurodegeneration such as the loss of pyramidal neurons in the hippocampus and Purkinje cells in the cerebellum is detected in the mutant mice from 3 weeks of age [30], Atg7-deficient animals, specifically in regard to CNS tissue, cannot be used for adult brain H/I injury study. As far as we have examined, no activation of caspases-3 and -7 occurs in the hippocampal tissue of wild-type mouse brains at 8 weeks of age after H/I injury [5]. As in the neonatal mice, it is also evident that such pyramidal neuron death is not prevented in the hippocampus of the caspase-3- or CAD-deficient mouse brain, while DNA laddering in the ipsilateral hippocampi is detected in the brains each of these mouse types. These data indicate that the DNA fragmentation in H/I injury-induced adult pyramidal neuron death is mediated by an unknown DNase, other than CAD. Autophagy is highly induced in the pyramidal neurons of the adult hippocampus after H/I injury; the ratio of the amounts of LC3-II to LC3-I is significantly higher in the ipsilateral side at each time point than in the contralateral side, while intense granular LC3-imunoreactivity is observed in hippocampal pyramidal neurons with pyknotic nuclei after H/I injury. Electron microscopy shows that pyramidal neurons having irregularly-shaped nuclei with small patches of chromatin clumping often become shrunken and contain numerous vacuolar structures, including autophagosomal and lysosomal structures, indicating that the morphological features of these degenerating neurons resemble the features of type-2 neuron death [67]. Taken together, these data suggest that autophagy also plays an important role in the H/I-induced pyramidal neuron death in the adult hippocampus.

Autophagic Cell Death Detected in Peripheral Tissue

Liver transplantation has become an established therapy for child and adult patients with end-stage liver disease. However, it is also true that primary graft dysfunction still occurs in 5%-25% of these recipient patients [77, 78]. Among the many factors that induce primary graft dysfunction, cold ischemia-warm reperfusion (CI/WR) is a major cause of poor initial graft function [79, 80]. Moreover, the severity of the dysfunction depends largely on the time of cold storage, this being associated with the survival of individual patients. Better ways to prevent the CI/WR injury associated with liver transplantation are needed, and many investigations have focused on the molecular mechanisms of this injury. However, the mechanisms reported to date are controversial and no improvement in therapy has resulted.

We have noted that autophagy is involved not only in CI but also in WR [81, 82]. By 15 min after the start of WR, small masses of hepatocytes that possess abundant autophagosomes and autolysosomes frequently dissociate from the hepatic cords and obstruct the sinusoid, causing massive necrosis of hepatocytes within 2 h. The cell masses include TUNEL-positive nuclei without caspase-3 and -7 activation. Autophagy suppression with the phosphatidylinositol 3-kinase (PI3K) inhibitors, wortmannin (WM) or LY294002, markedly reduced both liver damage and the mortality rate of recipient rats: the survival rate of rats receiving a liver graft that was not treated with PI3K inhibitors was 10.0% on the tenth day after transplantation, whereas the mortality rates of rats that received LY294002-treated or WM-treated liver grafts were 75.0% and 90.0%, respectively. The prolonged survival of rats with the WM-treated grafts corresponded well with the markedly decreased liver damage. When examined by light and electron microscopy, liver sections showing hepatocytes in the perivenous and midzonal regions of the untreated liver grafts were frequently swollen, and the plasma membranes and cytoplasmic structures of the damaged hepatocytes were disrupted. In contrast, hepatocytes in the WM-treated liver grafts were largely intact.

To elucidate the downstream mechanisms of this autophagic pathway, liver grafts were treated with the aspartic proteinase inhibitor pepstatin A, alone or with leupeptin, a cysteine proteinase inhibitor. These treatments also improved the survival rate of recipient rats. These data suggest that autophagy-associated hepatocyte death triggers liver graft dysfunction. The protective effects of suppressing autophagy may suggest new ways to prevent CI/WR injury of the liver.

Development of Autophagy Inhibitors for Therapeutic Use

It is known that specific inhibitors of caspases attenuate H/I-induced pyramidal neuron death in the neonatal hippocampus [59, 83]. Thus far, chemical and genetically manipulated suppressors of autophagy that are suitable for therapeutic use have not been generated. Because our data strongly suggest that autophagy regulates H/I-induced pyramidal neuron death in the neonatal hippocampus, and hepatocyte death in liver grafts during the early phases of WR followed by prolonged CI [5, 81], the development of chemical tools that prevent autophagic cell death by specifically blocking each step of autophagy could be important.

Acknowledgments This work was supported by a Grant-in-Aid for Creative Scientific Research from the Japan Society for the Promotion of Science (16GS0315).

References

1. Uchiyama Y, Shibata M, Koike M et al (2008) Autophagy—physiology and pathophysiology. Histochem Cell Biol 129: 407–420
2. Blomgren K, Leist M, Groc L (2007) Pathological apoptosis in the developing brain. Apoptosis 12: 993–1010
3. Blomgren K, Hagberg H (2006) Free radicals, mitochondria, and hypoxia-ischemia in the developing brain. Free Radic Biol Med 40: 388–397
4. Blomgren K, Zhu C, Hallin U et al (2003) Mitochondria and ischemic reperfusion damage in the adult and in the developing brain. Biochem Biophys Res Commun 304: 551–559
5. Koike M, Shibata M, Tadakoshi M et al (2008) Inhibition of autophagy prevents hippocampal pyramidal neuron death after hypoxic-ischemic injury. Am J Pathol 172: 454–469
6. Uchiyama Y, Koike M, Shibata M (2008) Autophagic neuron death in neonatal brain ischemia/hypoxia. Autophagy 4: 404–408
7. Uchiyama Y, Koike M, Shibata M et al (2009) Autophagic neuron death. Methods Enzymol 453: 33–51
8. Nitatori T, Sato N, Waguri S et al (1995) Delayed neuronal death in the CA1 pyramidal cell layer of the gerbil hippocampus following transient ischemia is apoptosis. J Neurosci 15: 1001–1011
9. Mizushima N (2007) Autophagy: process and function. Genes Dev 21: 2861–2873
10. Shintani T, Klionsky DJ (2004) Autophagy in health and disease: a double-edged sword. Science 306: 990–995
11. Koike M, Shibata M, Waguri S et al (2005) Participation of autophagy in storage of lysosomes in neurons from mouse models of neuronal ceroid-lipofuscinoses (Batten disease). Am J Pathol 167: 1713–1728
12. Nixon RA (2006) Autophagy in neurodegenerative disease: friend, foe or turncoat? Trends Neurosci 29: 528–535
13. Zhu C, Wang X, Xu F et al (2005) The influence of age on apoptotic and other mechanisms of cell death after cerebral hypoxia-ischemia. Cell Death Differ 12: 162–176
14. Kabeya Y, Mizushima N, Ueno T et al (2000) LC3, a mammalian homologue of yeast Apg8p, is localized in autophagosome membranes after processing. EMBO J 19: 5720–5728
15. Haltia M (2006) The neuronal ceroid-lipofuscinoses: from past to present. Biochim Biophys Acta 1762: 850–856
16. Ivy GO, Schottler F, Wenzel J et al (1984) Inhibitors of lysosomal enzymes: accumulation of lipofuscin-like dense bodies in the brain. Science 226: 985–987
17. Sleat DE, Donnelly RJ, Lackland H et al (1997) Association of mutations in a lysosomal protein with classical late-infantile neuronal ceroid lipofuscinosis. Science 277: 1802–1805
18. Ezaki J, Takeda-Ezaki M, Kominami E (2000) Tripeptidyl peptidase I, the late infantile neuronal ceroid lipofuscinosis gene product, initiates the lysosomal degradation of subunit c of ATP synthase. J Biochem 128: 509–516
19. Ezaki J, Tanida I, Kanehagi N et al (1999) A lysosomal proteinase, the late infantile neuronal ceroid lipofuscinosis gene (CLN2) product, is essential for degradation of a hydrophobic protein, the subunit c of ATP synthase. J Neurochem 72: 2573–2582
20. Koike M, Nakanishi H, Saftig P et al (2000) Cathepsin D deficiency induces lysosomal storage with ceroid lipofuscin in mouse CNS neurons. J Neurosci 20: 6898–6906
21. Koike M, Shibata M, Ohsawa Y et al (2003) Involvement of two different cell death pathways in retinal atrophy of cathepsin D-deficient mice. Mol Cell Neurosci 22: 146–161

22. Nakanishi H, Zhang J, Koike M et al (2001) Involvement of nitric oxide released from microglia-macrophages in pathological changes of cathepsin D-deficient mice. J Neurosci 21: 7526–7533
23. Siintola E, Partanen S, Stromme P et al (2006) Cathepsin D deficiency underlies congenital human neuronal ceroid-lipofuscinosis. Brain 129: 1438–1445
24. Steinfeld R, Reinhardt K, Schreiber K et al (2006) Cathepsin D deficiency is associated with a human neurodegenerative disorder. Am J Hum Genet 78: 988–998
25. Dunn WA, Jr. (1994) Autophagy and related mechanisms of lysosome-mediated protein degradation. Trends Cell Biol 4: 139–143
26. Dunn WA Jr (1990) Studies on the mechanisms of autophagy: maturation of the autophagic vacuole. J Cell Biol 110: 1935–1945
27. Liou W, Geuze HJ, Geelen MJ et al (1997) The autophagic and endocytic pathways converge at the nascent autophagic vacuoles. J Cell Biol 136: 61–70
28. Tanaka Y, Guhde G, Suter A et al (2000) Accumulation of autophagic vacuoles and cardiomyopathy in LAMP-2-deficient mice. Nature 406: 902–906
29. Hara T, Nakamura K, Matsui M et al (2006) Suppression of basal autophagy in neural cells causes neurodegenerative disease in mice. Nature 441: 885–889
30. Komatsu M, Waguri S, Chiba T et al (2006) Loss of autophagy in the central nervous system causes neurodegeneration in mice. Nature 441: 880–884
31. Komatsu M, Waguri S, Ueno T et al (2005) Impairment of starvation-induced and constitutive autophagy in Atg7-deficient mice. J Cell Biol 169: 425–434
32. Ardley HC, Hung CC, Robinson PA (2005) The aggravating role of the ubiquitin-proteasome system in neurodegeneration. FEBS Lett 579: 571–576
33. Settembre C, Fraldi A, Jahreiss L et al (2008) A block of autophagy in lysosomal storage disorders. Hum Mol Genet 17: 119–129
34. Zhan SS, Beyreuther K, Schmitt HP (1992) Neuronal ubiquitin and neurofilament expression in different lysosomal storage disorders. Clin Neuropathol 11: 251–255
35. Bjørkøy G, Lamark T, Brech A et al (2005) p62/SQSTM1 forms protein aggregates degraded by autophagy and has a protective effect on huntingtin-induced cell death. J Cell Biol 171: 603–614
36. Moscat J, Diaz-Meco MT, Albert A et al (2006) Cell signaling and function organized by PB1 domain interactions. Mol Cell 23: 631–640
37. Wooten MW, Hu X, Babu JR et al (2006) Signaling, polyubiquitination, trafficking, and inclusions: sequestosome 1/p62's role in neurodegenerative disease. J Biomed Biotechnol 2006: 62–79
38. Komatsu M, Waguri S, Koike M et al (2007) Homeostatic levels of p62 control cytoplasmic inclusion body formation in autophagy-deficient mice. Cell 131: 1149–1163
39. Nakai A, Yamaguchi O, Takeda T et al (2007) The role of autophagy in cardiomyocytes in the basal state and in response to hemodynamic stress. Nat Med 13: 619–624
40. Wang QJ, Ding Y, Kohtz DS et al (2006) Induction of autophagy in axonal dystrophy and degeneration. J Neurosci 26: 8057–8068
41. Walker NI, Harmon BV, Gobe GC et al (1988) Patterns of cell death. Methods Achiev Exp Pathol 13: 18–54
42. Lockshin RA, Williams CM (1964) Programmed cell death. II. Endocrine potentiation of the breakdown of the intersegmental muscles of silkmoths. J Insect Physiol 10: 643–649
43. Lockshin RA, Zaleri Z (1991) Programmed cell death and apoptosis. Cold Spring Harbor Laboratory, New York
44. Kerr JF, Wyllie AH, Currie AR (1972) Apoptosis: a basic biological phenomenon with wide-ranging implications in tissue kinetics. Br J Cancer 26: 239–257
45. Salvesen GS, Dixit VM (1999) Caspase activation: the induced-proximity model. Proc Natl Acad Sci U S A 96: 10 964–10 967
46. Enari M, Sakahira H, Yokoyama H et al (1998) A caspase-activated DNase that degrades DNA during apoptosis, and its inhibitor ICAD. Nature 391: 43–50

47. Li P, Nijhawan D, Budihardjo I et al (1997) Cytochrome c and dATP-dependent formation of Apaf-1/caspase-9 complex initiates an apoptotic protease cascade. Cell 91: 479–489
48. Nicholson DW, Thornberry NA (1997) Caspases: killer proteases. Trends Biochem Sci 22: 299–306
49. Sakahira H, Enari M, Nagata S (1998) Cleavage of CAD inhibitor in CAD activation and DNA degradation during apoptosis. Nature 391: 96–99
50. Sakahira H, Enari M, Ohsawa Y et al (1999) Apoptotic nuclear morphological change without DNA fragmentation. Curr Biol 9: 543–546
51. Raff MC, Barres BA, Burne JF et al (1993) Programmed cell death and the control of cell survival: lessons from the nervous system. Science 262: 695–700
52. Blomgren K, Zhu C, Wang X et al (2001) Synergistic activation of caspase-3 by m-calpain after neonatal hypoxia-ischemia: a mechanism of "pathological apoptosis"? J Biol Chem 276: 10 191–10 198
53. Gill R, Soriano M, Blomgren K et al (2002) Role of caspase-3 activation in cerebral ischemia-induced neurodegeneration in adult and neonatal brain. J Cereb Blood Flow Metab 22: 420–430
54. Hu BR, Liu CL, Ouyang Y et al (2000) Involvement of caspase-3 in cell death after hypoxia-ischemia declines during brain maturation. J Cereb Blood Flow Metab 20: 1294–1300
55. Liu CL, Siesjo BK, Hu BR (2004) Pathogenesis of hippocampal neuronal death after hypoxia-ischemia changes during brain development. Neuroscience 127: 113–123
56. Zhu C, Qiu L, Wang X et al (2003) Involvement of apoptosis-inducing factor in neuronal death after hypoxia-ischemia in the neonatal rat brain. J Neurochem 86: 306–317
57. McDonald JW, Silverstein FS, Johnston MV (1988) Neurotoxicity of N-methyl-D-aspartate is markedly enhanced in developing rat central nervous system. Brain Res 459: 200–203
58. Ikonomidou C, Mosinger JL, Salles KS et al (1989) Sensitivity of the developing rat brain to hypobaric/ischemic damage parallels sensitivity to N-methyl-aspartate neurotoxicity. J Neurosci 9: 2809–2818
59. West T, Atzeva M, Holtzman DM (2006) Caspase-3 deficiency during development increases vulnerability to hypoxic-ischemic injury through caspase-3-independent pathways. Neurobiol Dis 22: 523–537
60. Parsadanian AS, Cheng Y, Keller-Peck CR et al (1998) Bcl-xL is an antiapoptotic regulator for postnatal CNS neurons. J Neurosci 18: 1009–1019
61. Kawane K, Fukuyama H, Yoshida H et al (2003) Impaired thymic development in mouse embryos deficient in apoptotic DNA degradation. Nat Immunol 4: 138–144
62. Kuida K, Zheng TS, Na S et al (1996) Decreased apoptosis in the brain and premature lethality in CPP32-deficient mice. Nature 384: 368–372
63. Lakhani SA, Masud A, Kuida K et al (2006) Caspases 3 and 7: key mediators of mitochondrial events of apoptosis. Science 311: 847–851
64. Leonard JR, Klocke BJ, D'Sa C et al (2002) Strain-dependent neurodevelopmental abnormalities in caspase-3-deficient mice. J Neuropathol Exp Neurol 61: 673–677
65. Reznikov KY (1991) Cell proliferation and cytogenesis in the mouse hippocampus. Adv Anat Embryol Cell Biol 122: 1–74
66. Houde C, Banks KG, Coulombe N et al (2004) Caspase-7 expanded function and intrinsic expression level underlies strain-specific brain phenotype of caspase-3-null mice. J Neurosci 24: 9977–9984
67. Clarke PG (1990) Developmental cell death: morphological diversity and multiple mechanisms. Anat Embryol (Berl) 181: 195–213
68. Bursch W (2001) The autophagosomal-lysosomal compartment in programmed cell death. Cell Death Differ 8: 569–581
69. Canu N, Tufi R, Serafino AL et al (2005) Role of the autophagic-lysosomal system on low potassium-induced apoptosis in cultured cerebellar granule cells. J Neurochem 92: 1228–1242
70. Isahara K, Ohsawa Y, Kanamori S et al (1999) Regulation of a novel pathway for cell death by lysosomal aspartic and cysteine proteinases. Neuroscience 91: 233–249

71. Ohsawa Y, Isahara K, Kanamori S et al (1998) An ultrastructural and immunohistochemical study of PC12 cells during apoptosis induced by serum deprivation with special reference to autophagy and lysosomal cathepsins. Arch Histol Cytol 61: 395–403
72. Shibata M, Kanamori S, Isahara K et al (1998) Participation of cathepsins B and D in apoptosis of PC12 cells following serum deprivation. Biochem Biophys Res Commun 251: 199–203
73. Telbisz A, Kovacs AL, Somosy Z (2002) Influence of X-ray on the autophagic-lysosomal system in rat pancreatic acini. Micron 33: 143–151
74. Uchiyama Y (2001) Autophagic cell death and its execution by lysosomal cathepsins. Arch Histol Cytol 64: 233–246
75. Yu L, Alva A, Su H et al (2004) Regulation of an ATG7-beclin 1 program of autophagic cell death by caspase-8. Science 304: 1500–1502
76. Bursch W, Ellinger A, Kienzl H et al (1996) Active cell death induced by the anti-estrogens tamoxifen and ICI 164 384 in human mammary carcinoma cells (MCF-7) in culture: the role of autophagy. Carcinogenesis 17: 1595–1607
77. Ploeg RJ, D'Alessandro AM, Knechtle SJ et al (1993) Risk factors for primary dysfunction after liver transplantation—a multivariate analysis. Transplantation 55: 807–813
78. Strasberg SM, Howard TK, Molmenti EP et al (1994) Selecting the donor liver: risk factors for poor function after orthotopic liver transplantation. Hepatology 20: 829–838
79. Calmus Y, Cynober L, Dousset B et al (1995) Evidence for the detrimental role of proteolysis during liver preservation in humans. Gastroenterology 108: 1510–1516
80. Furukawa H, Todo S, Imventarza O et al (1991) Effect of cold ischemia time on the early outcome of human hepatic allografts preserved with UW solution. Transplantation 51: 1000–1004
81. Gotoh K, Lu Z, Morita M et al (2009) Participation of autophagy in the initiation of graft dysfunction after rat liver transplantation. Autophagy 5: 351–360
82. Lu Z, Dono K, Gotoh K et al (2005) Participation of autophagy in the degeneration process of rat hepatocytes after transplantation following prolonged cold preservation. Arch Histol Cytol 68: 71–80
83. Cheng Y, Deshmukh M, D'Costa A et al (1998) Caspase inhibitor affords neuroprotection with delayed administration in a rat model of neonatal hypoxic-ischemic brain injury. J Clin Invest 101: 1992–1999

Bioluminescent Imaging for Assessing Heterogeneous Cell Functions in the Mammalian Central Circadian Clock

Sato Honma[1,2], Tomoko Yoshikawa[2], Shin-ya Nishide[1], Daisuke Ono[1], and Ken-ichi Honma[1]

Summary

The mammalian circadian system is composed of a master clock in the suprachiasmatic nucleus (SCN) of the hypothalamus, and peripheral clocks in organs and tissues. Light is the most dominant time cue which adjusts the intrinsic circadian period to a 24-h day-night cycle. The master clock in the SCN is the only clock which acts as an interface between our body and diurnal and seasonal cycles in the environment, while the peripheral clocks drive tissue-specific rhythms. The SCN is composed of a number of cell-autonomous oscillator neurons. Over the past decade, our understanding of the molecular clock mechanisms has advanced tremendously. Of special note is a bioluminescent cell-imaging technique which enables us to monitor each clock cell's ticking for days and weeks, using luciferase reporters of clock genes. By using bioluminescent imaging, we found the multiple oscillators in the SCN that encode seasonal changes in the photoperiod.

Key words Luciferase reporter · Clock gene · Circadian rhythm · Photoperiod · Bioluminescent imaging

Introduction

The physiological functions of the body exhibit various rhythms, the periods of which vary from a few milliseconds to years. Pacemakers have been identified for many of these rhythms, but the mechanisms for rhythm generation differ. Some

[1]Department of Physiology, Hokkaido University Graduate School of Medicine, North 15 West 7, Kita-ku, Sapporo 060-8638, Japan
[2]Department of Chronomedicine, Hokkaido University Graduate School of Medicine, North 15 West 7, Kita-ku, Sapporo 060-8638, Japan

N. Tamaki, Y. Kuge (eds.)
Molecular Imaging for Integrated Medical Therapy and Drug Development.
DOI: 10.1007/978-4-431-98074-2_20,

rhythms are generated by neural networks, while others are generated by the combination of ion channels. Among them, rhythms with a period of about 24 h (circadian rhythm) are generated by an intrinsic circadian clock, in which a set of genes called "clock genes" and their protein products form interlocking transcriptional and translational autoregulatory feedback loops, and one turn of the loops takes about 24 h [1]. Bioluminescent reporter techniques have enabled us to monitor the clock gene expression rhythms in real time [2, 3]. In mammals, the master circadian clock resides in the suprachiasmatic nucleus (SCN) of the hypothalamus. Total lesion of the SCN eradicates overt circadian rhythms, while transplantation of the SCN from newborn or fetal animals recovers circadian behavioral rhythms with the period of the donor animals [4]. Surprisingly, not only peripheral tissues but also cell lines that have been frozen and thawed for generations have been shown to have a circadian clock [3, 5]. These findings suggest that the SCN is not the only circadian clock that drives circadian rhythms in bodily functions. The SCN receives photic signals to entrain the intrinsic periodicity to an environmental light-dark cycle, and it orchestrates peripheral clocks throughout the body to exhibit temporally coordinated rhythms in physiology and behavior. The SCN is composed of a number of cell-autonomous oscillator neurons [6, 7]. Therefore, synchronization among these cellular oscillators is critical for rhythm expression. However, the mechanisms of how these clock cells synchronize to form a master clock adjustable to changing environments are still poorly understood. By using bioluminescent reporters, we examined the functions of heterogeneous SCN cells whose circadian phase can be dynamically changed depending on the day length.

Bioluminescence Reporters to Assess Clock Functions

In mammals, the intracellular clock machinery comprises a core feedback loop for the rhythmic expression of the clock gene *Pers* and associated loops which interlock with the core loop, such as loops for rhythmic *Bmal1*, differentiated embryo chondrocytes (*Dec*), and *Rev-Erbα* expressions [8] (Fig. 1). In the core loop, basic Helix-loop-helix PER-ARNT-SIM (PAS) transcription factors, CLOCK and BMAL1, form a heterodimer which binds to the enhancer called E-box (CACGTG) or E′ box (CACGTT), which resides upstream of the *Per1* or *Per2* genes, respectively, to activate transcription. The protein products PER1 and PER2 translocate into the nucleus and bind to CLOCK/BMAL1 heterodimers to repress their own transcription. *Per1* and *Per2* expression in the SCN exhibits robust circadian rhythms with a peak at the subjective noon and afternoon, respectively [9,10]. *Bmal1* is also expressed rhythmically with a peak, antiphasically to the rhythms in *Per1* and *Per2* expressions, at the subjective night [11]. For rhythmic *Bmal1* expression, the transcription factors ROR and REVERBα bind to the RORE site upstream of the *Bmal1* gene to activate and to repress *Bmal1* transcription, respectively [12,13]. Because *RevErbα* expression is regulated by its upstream E-box

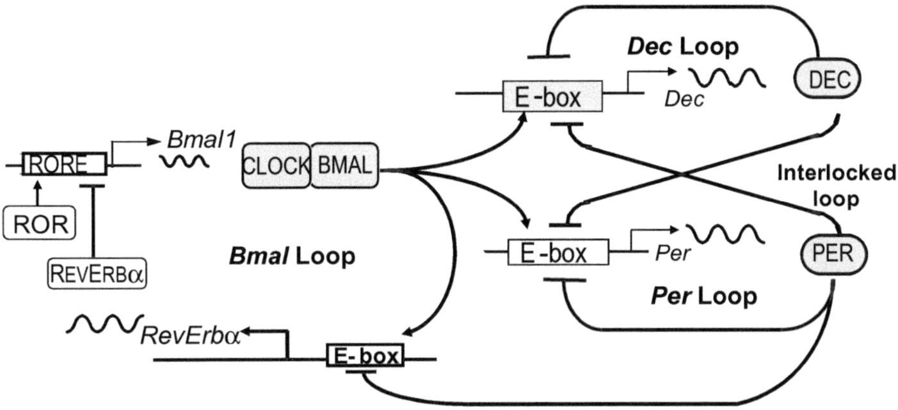

Fig. 1. Translational and transcriptional feedback model of intracellular circadian clock. A core feedback loop for cyclic *Per* expression and interlocking loops for cyclic *Bmal1* and *Dec* expression are shown. *Rectangles*, responsive elements; *oval enclosures*, proteins; *wavy lines*, rhythmic mRNA transcription; *arrows*, transcription activation; ⊣, transcription inhibition; ↱, transcription start site

enhancer, BMAL1 indirectly suppresses its own transcription during the subjective day. Rhythmic gene transcription is observed in cells and tissues throughout the body. Therefore, a reporter technique has become a powerful tool for monitoring a cellular clock's "tick." Especially, luciferase reporters have advantages over fluorescent ones for long-term and continuous monitoring due to their low toxicity and wide dynamic range for quantification. We have developed several transgenic mouse lines to monitor *Per1* and *Bmal1* expressions (Fig. 2a, b). We also aimed to monitor the expression of multiple genes simultaneously from the same cells and tissues (Fig. 2b, c). The two strategies were applied in collaboration with Drs. Ohmiya and Nakajima at the Institute of Advanced Industrial Science and Technology; we used a combination of secretory and nonsecretory luciferases [14] and luciferases emitting red and green lights [15].

Using a transgenic mouse carrying a reporter for *Per1* promoter activity with secretory luciferase, *Vargula hilgendorfii* luciferase (VL), and a reporter for *Bmal1* promoter activity with firefly luciferase (FL), we simultaneously monitored VL and FL activities from a cultured SCN (Fig. 2b) [14]. We observed a time lag of about 6–8 h for VL, while there was a time lag of 4–6 h for FL when compared with the peak phase of gene expression rhythms measured by in-situ hybridization at 4-h intervals. These time lags may be due to the processes of luciferase synthesis and secretion.

Red- and green-light-emitting luciferases which catalyze the same substrate are also excellent tools for monitoring the multiple genes that constitute a cellular clock [15]. By splitting the emission from red and green luciferases, using a longpass filter, we monitored the antiphasic expression of *Per2* and *Bmal1* (Fig. 2c).

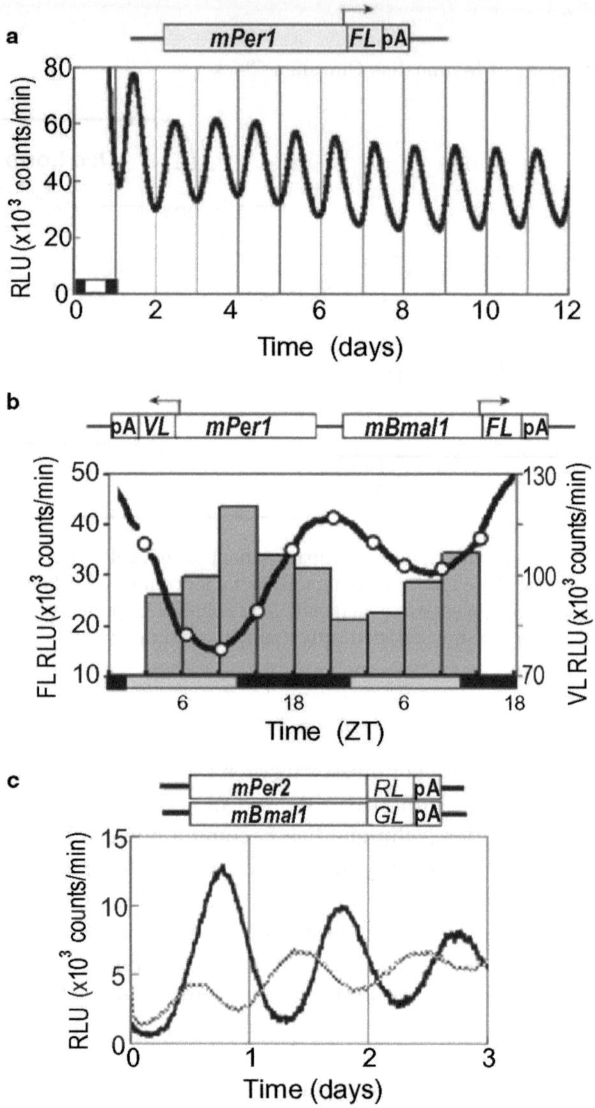

Fig. 2a–c. Bioluminescent reporter constructs and clock gene expression rhythms monitored by luciferase reporters. **a** A construct of mouse *Per1* expression reporter (*upper scheme*) and a representative bioluminescent rhythm of a cultured suprachiasmatic nucleus (SCN) of the hypothalamus from a *Per1-luc* transgenic mouse (*lower panel*); 6.1 kb upstream of *Per1*, including five conserved E-box enhancers, were inserted into pGL3.1 basic vector. *Vertical lines in the graph* indicate 24:00. *Black and white horizontal bars on the abscissa* indicate the light and dark phases on the day of brain sampling for the culture. **b** Dual reporter construct for simultaneous monitoring of *Per1* and *Bmal1* expressions by Vargula luciferase (*VL*) and firefly luciferase (*FL*), respectively (*upper scheme*), and representative simultaneous monitoring of VL and FL activity rhythms of the SCN from a transgenic mouse on the second day of the culture. *Histograms* indicate *Per1-VL* activity measured every 4 h and *the curve* indicates *Bmal1-FL* activity measured every minute. *Circles on the curve* are the times of medium sampling and refilling of the new medium. *Black and gray horizontal bars on the abscissa* indicate the light and dark phase of the day of slice preparation. **c** Constructs of *Per2* and *Bmal1* expression reporters with red-luciferase (*RL*) and green-luciferase (*GL*), respectively (*upper scheme*) and representative bioluminescent rhythms of *Per2* (*dotted line*) and *Bmal1* (*solid line*) expression (*lower graph*) monitored simultaneously. Two reporter vectors were transfected simultaneously to Rat-1 fibroblasts. On the following day, cells were treated with 100 nM dexamethasone for 2 h, and the bioluminescence was monitored with and without a longpass filter for 1 min each time at 18-min intervals. *Abscissa* indicates time after the washout of dexamesasone. *pA*, SV40 poly A; *RLU*, relative light unit; Zeitgeber Time (ZT) where ZTO is lights-on time in a light dark cycle.

Multiple Clocks Composed of Heterogeneous Clock Cells

The mammalian circadian clock in the SCN is composed of multiple oscillatory cells. By monitoring the neuronal activity rhythms of single SCN neurons, we found that the neuronal rhythms in an organotypic slice culture were mostly synchronized and the periods varied between 22 and 26 h. In a dispersed cell culture, on the other hand, circadian firing rhythms were desynchronized and the periods varied between 20 and 32 h [16]. Important to note is that the circadian periods followed a Gaussian distribution in both cultures and the averaged circadian periods were not significantly different in the two culture systems. These findings suggest that the SCN cell assemblage, with its close cell-to-cell apposition, plays an important role in synchronizing the constitutional oscillators in the SCN.

Clocks Coding Photoperiods

For animals living at middle and high latitudes, adaptation to seasonal changes is critical for survival. The behavioral rhythms of animals in the field change dramatically depending on the seasons. In nocturnal rodents, the behaviorally active time is compressed during the short summer nights, while it is decompressed during the long winter nights. This photoperiodic response of behavioral rhythm has been explained by a model of two separate, but mutually coupled, circadian oscillators, an evening (E) oscillator, which drives the activity onset and entrains to dusk, and a morning (M) oscillator, which drives the end of activity and entrains to dawn [17]. However, neither the localization nor the oscillation mechanism has been elucidated. Using *Per1-luc* transgenic mice, we compared their behavioral rhythms and *Per1-luc* rhythms in various SCN regions. We exposed *Per1-luc* mice that had been kept in a regular 12-h light and 12-h dark cycle (LD12 : 12) to either a long day of LD18 : 6 or to a short day of LD 6 : 18 for 3 weeks. Two serial anterior and posterior coronal SCN slices were made to monitor bioluminescence from whole SCN explants with a photomultiplier tube. We found that, irrespective of the photoperiod, the *Per1-luc* peak of the anterior SCN phase-locked to the end-of-activity rhythm, whereas that of the posterior SCN phase-locked to the activity-onset, suggesting the localization of M and E oscillators, respectively [18]. In addition, in LD18 : 6, the anterior SCN exhibited two *Per1-luc* peaks a day, one phase-locked to the activity-end, while the other was phase-locked neither to the activity-onset nor to the activity-end (Fig. 3a). Single-cell bioluminescent imaging with a charge-coupled device (CCD) camera revealed that all SCN cells with detectable bioluminescence showed circadian rhythms in both the anterior and posterior SCN, and there were two separate oscillatory cell groups with different circadian phases in the anterior SCN. In horizontal SCN slices, we could identify three regionally specific circadian oscillators (Fig. 3b). By collaborating with Drs. Ueda and Takagi of the Research Institute for Electronic Science, Hokkaido University, we developed spatial mappings of the cellular clocks in the SCN [19]. Fast

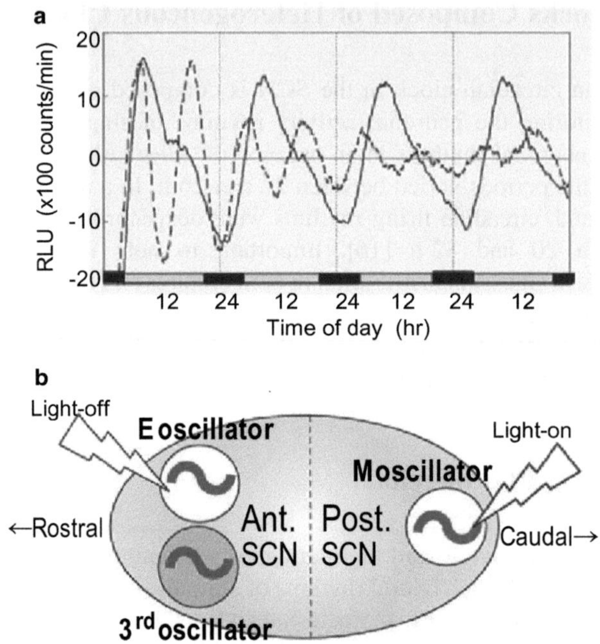

Fig. 3a,b. Three oscillatory cell networks in the SCN. **a** Detrended waveforms of *Per1-luc* rhythms of anterior (*broken line*) and posterior (*solid line*) SCN from a mouse exposed to a long day of 18-h light and 6-h dark (LD18 : 6). Data were obtained by subtracting 12 h of moving average values from the original luminescence curves. The anterior SCN exhibits a bimodal *Per1-luc* rhythm. Bioluminescent cell imaging revealed two oscillatory cell groups with different circadian phases in the anterior SCN. **b** Scheme for three oscillatory cell networks. One in the posterior (*Post.*) SCN is entrained by the light-on signal and regulates the activity end (*M oscillator*). Of the two oscillatory cell networks in the anterior (*Ant.*) SCN, one with a peak at the late afternoon is entrained by the light-off signal and regulates the activity onset (*E oscillator*), while the function of the other has not yet been identified (*3rd oscillator*) *E*, evening; *M*, morning

Fourier Transform (FFT) was applied to identify the phase, period, and relative amplitude of the circadian rhythms at each pixel of time-series images. The resultant frequency-, amplitude-, and phase-maps showed the spatial distribution of rhythmic phenomena in the SCN (Fig. 4b–d). The rhythm amplitude of *Per1-luc* was high in the posterior SCN in both the short and long photoperiods. The phase-map in Fig. 4d shows differentially phasing cells in a single pseudo color map. By using this phase-map, we could further separate the regions of the three differentially phasing oscillatory cell groups with circadian peaks in the subjective morning, afternoon, and late afternoon of the long photoperiod (Fig. 4e–g). The *Per1-luc* rhythms of the cells in the posterior SCN were phase-advanced compared with those in the anterior SCN, in both LD18 : 6 and LD6 : 18. However, the phase-difference was more than 12 hours for the cells in the posterior SCN, and only a couple of hours for those in the anterior SCN. With bioluminescent imaging,

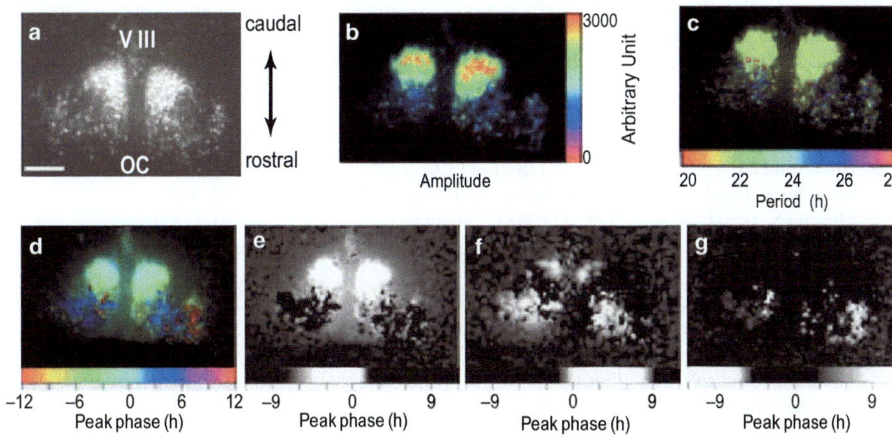

Fig. 4a–g. Bioluminescent cell imaging of cultured mouse SCN and spatial mapping of the circadian parameters. **a** A representative bioluminescent image of the cultured SCN from a *Per1-luc* mouse. Bioluminescence from a 100-μm-thick horizontal SCN slice was measured every hour with a 59-min exposure and at 1-min intervals with an electron multiplying (EM)-charge-coupled device (CCD) camera (ImageEM; Hamamatsu Photonics, Hamamatsu, Japan). Optic chiasm (OC) VIII: the third ventricle; *scale bar* indicates 200 μm. **b** An amplitude-map analyzed by Fast Fourier Transform (FFT). Amplitude is shown in arbitrary units. **c** A frequency-map, showing that most of the cells exhibit the same frequency. Slight differences were observed in the anterior SCN. A pseudo color scale is expressed in the circadian period. **d** A phase-map of the SCN showing phase-relations in pseudo color. The *positive values* indicate phase advance, and the *negative ones*, phase delay. **e–g** Phase-maps with enhanced bioluminescence in the *phase indicated below each image*

together with FFT-based spatial mapping, we are now able to visualize the heterogeneity of our master circadian clock in the SCN.

References

1. Reppert SM, Weaver DR (2002) Coordination of circadian timing in mammals. Nature 418: 935–941
2. Yamazaki S, Numano R, Abe M et al (2000) Resetting central and peripheral circadian oscillators in transgenic rats. Science 288: 682–685
3. Yoo SH, Yamazaki S, Lowrey PL et al (2004) PERIOD2::LUCIFERASE real-time reporting of circadian dynamics reveals persistent circadian oscillations in mouse peripheral tissues. Proc Natl Acad Sci USA 101: 5339–5346
4. Ralph MR, Foster RG, Davis FC et al (1990) Transplanted suprachiasmatic nucleus determines circadian period. Science 247: 975–978
5. Balsalobre A, Damiola F, Schibler U (1998) A serum shock induces circadian gene expression in mammalian tissue culture cells. Cell 93: 929–937
6. Welsh DK, Logothetis DE, Meister M et al (1995) Individual neurons dissociated from rat suprachiasmatic nucleus express independently phased circadian firing rhythms. Neuron 14: 697–706

7. Honma S, Shirakawa T, Katsuno Y et al (1998) Circadian periods of single suprachiasmatic neurons in rats. Neurosci Lett 250: 157–160
8. Ueda HR, Hayashi S, Chen W et al (2005) System-level identification of transcriptional circuits underlying mammalian circadian clocks. Nat Genet 37: 187–192
9. Tei H, Okamura H, Shigeyoshi Y et al (1997) Circadian oscillation of a mammalian homologue of the Drosophila period gene. Nature 389: 512–516
10. Shearman LP, Zylka MJ, Weaver DR et al (1997) Two period homologs: circadian expression and photic regulation in the suprachiasmatic nuclei. Neuron 20: 1103–1110
11. Honma S, Ikeda M, Abe H et al (1998) Circadian oscillation of BMAL1, a partner of a mammalian clock gene Clock, in rat suprachiasmatic nucleus. Biochem Biophys Res Commun 250: 83–87
12. Sato TK, Panda S, Miraglia LJ et al (2004) A functional genomics strategy reveals Rora as a component of the mammalian circadian clock. Neuron 43: 527–537
13. Preitner N, Damiola F, Lopez-Molina L, et al (2002) The orphan nuclear receptor REV-ERBalpha controls circadian transcription within the positive limb of the mammalian circadian oscillator. Cell 110: 251–260
14. Nishide S, Honma S, Nakajima Y, et al (2006) New reporter system for Per1 and Bmal1 expressions revealed self-sustained circadian rhythms in peripheral tissues. Genes Cells 11: 1173–1182
15. Nakajima Y, Ikeda M, Kimura T et al (2004) Bidirectional role of orphan nuclear receptor RORalpha in clock gene transcriptions demonstrated by a novel reporter assay system. FEBS Lett 565: 122–126
16. Honma S, Nakamura W, Shirakawa T et al (2004) Diversity in the circadian periods of single neurons of the rat suprachiasmatic nucleus depends on nuclear structure and intrinsic period. Neurosci Lett 358: 173–176
17. Pittendrigh C, Daan S (1976) A functional analysis of circadian pacemakers in nocturnal rodents. V. Clock for all seasons. J Comp Physiol A 14: 345–350
18. Inagaki N, Honma S, Ono D, et al (2007) Separate oscillating cell groups in mouse suprachiasmatic nucleus couple photoperiodically to the onset and end of daily activity. *P*roc Natl Acad Sci USA 104: 7664–7669
19. Fricker MD, Tlalka M, Bebber D et al (2007) Fourier-based spatial mapping of oscillatory phenomena in fungi. Fungal Genet Biol 44: 1077–1084

Early Postnatal Stress and the Serotonergic System in the Brain

Kohtarou Konno and Mitsuhiro Yoshioka

Summary

Recent studies have focused on the serotonergic mechanism, mediated via serotonin (5-hydroxytryptamine; 5-HT) receptors, underlying the regulation of emotional stress during the developmental period. The present study was undertaken to elucidate whether early postnatal stress affects rat brain development and influences serotonergic function in the midbrain raphe nuclei in the adult, focusing on the response to unconditioned fear stress assessed with the elevated plus maze test. Rats that received aversive foot shock (FS) stimuli in the third week of the postnatal period (3wFS), but not those that received the aversive FS stimuli in the second week (2wFS), spent an increased percentage of time on the open arms of the elevated plus maze in the postadolescent period (age, 10–12 weeks). The number of 5-HT neurons in the raphe nuclei was significantly decreased by FS administered when animals were 3 weeks old. Fluvoxamine treatment (10 mg/kg, p.o.) given for 14 days immediately after the aversive stress normalized the behavioral patterns in the elevated plus maze test. These findings suggest that exposure to aversive stress during the early postnatal period may affect serotonergic development, and thereby may affect emotional responses to fear stimuli in the post-adolescent period.

Key words raphe nuclei · unconditioned fear · elevated plus maze · fluvoxamine

Introduction

Several lines of evidence have shown that traumatic events during the early postnatal stage precipitate long-lasting alterations in the functional properties of the brain underlying emotional expression [1, 2]. It is well known that the serotonin

Department of Neuropharmacology, Hokkaido University Graduate School of Medicine, North 15 West 7, Kita-ku, Sapporo 060-8638, Japan

N. Tamaki, Y. Kuge (eds.)
Molecular Imaging for Integrated Medical Therapy and Drug Development.
DOI: 10.1007/978-4-431-98074-2_21, ©Springer 2010

(5-hydroxytryptamine; 5- HT) neuronal system is involved in the regulation of emotional stress [3, 4]. The midbrain raphe nuclei (the main origin of ascending serotonergic projections to the forebrain), in particular, are considered to play a critical role in the regulation of stress responsiveness [5, 6]. The serotonergic system develops early [7, 8] and eventually becomes one of the most widely distributed systems in the brain. 5-HT is also known to influence the maturation of many other cells [9, 10], i.e., 5-HT modulates developmental processes such as dendritic elaboration [11–13], neurogenesis [14, 15], synaptogenesis [16], and organization of the cortex [17, 18]. Furthermore, 5-HT can regulate the maturation of its own serotonergic neurons, termed autoregulation of development [9]. Given the association with the development of the serotonergic system and its prominent role in maturation, early life stress may lead to persistent changes in the serotonergic system in the raphe nuclei and alter the response to emotional stress later in life.

Recent studies have focused on the relationship between certain stresses and the characteristics of 5-HT receptors during brain development. For instance, in studies in rats, maternal separation, i.e., deprivation of the pups from the dam, increased the number of hippocampal 5-HT1A receptors and cortical 5-HT2A receptors [19, 20], whereas prenatal stress produced decreases in the hippocampal 5-HT1A receptors in male offspring [21]. Gartside et al. [22] reported that maternally deprived pups had desensitization of 5-HT1A autoreceptors located on the raphe nuclei. Few studies, however, have addressed whether early postnatal stress affects serotonergic function in the midbrain raphe nuclei. In the present study, we examined whether early postnatal stress influenced the serotonergic mechanism(s) underlying the regulation of emotional stress in adult rats, focusing on 5-HT neuronal development in the midbrain raphe nuclei.

For this purpose, pups were exposed to aversive footshock (FS) stimuli in the early postnatal period, during the third week (3w). In the postadolescent period (age, 10–12 weeks), a behavioral study was undertaken to analyze fear-related behavior, assessed by the elevated plus maze (EPM) test [23]. An immunohistochemical study was carried out to determine the 5-HT-immunoreactive cells in the raphe nuclei.

Methods

Animals

The rats used for experiments with juvenile stress models were bred in our own laboratory, except for the first breeder adult rats (two females and two males) supplied by Nippon SLC. Weaning was carried out on postnatal day (PND) 21. Rats were housed in a room with lights on from 19:00 to 07:00 h under a constant temperature (21 ± 2 °C). All handling of animals was performed in accordance with the guidelines for the Care and Use of Laboratory Animals of the Animal Research Committee of the Hokkaido University Graduate School of Medicine.

Aversive Stress in Early Postnatal Period

Pups were divided into two groups as previously described [24]. They were acclimated in an FS box for 5 min and subjected to five shocks (shock intensity, 0.5 mA; intershock interval, 30 s; shock duration, 2 s). Each rat remained in the box for 5 min after the last FS stimulation and was then returned to the home cage. This FS stimulation was carried out for 5 days. Rats exposed to FS stimulation during PNDs 21 to 25 (3 weeks) were denoted the 3wFS groups. Non-FS controls remained in the FS box for 12.5 min without FS stimuli. Rats subjected to FS and the non-FS controls in the colony were separately housed after weaning. In the postadolescent period (age, 10–12 weeks), the behavioral experiment was performed.

Behavioral Study Using the Elevated Plus Maze (EPM) Test

The EPM test, which has been established as an animal model of anxiety [23], was performed in the postadolescent period (age, 10–12 weeks) in the rats subjected to FS and the non-FS controls. The behavioral experiments were performed between 11:00 h and 14:00 h to minimize circadian influence. The EPM had two open arms (50 × 10 cm), perpendicular to two closed arms of equal dimensions, surrounded by 40-cm walls. The apparatus was elevated 50 cm from the floor and illumination was set at 400 lx. The behavioral parameters measured during 10 min were the number of entries into, and time spent on, the open and closed arms. A camera was set up to photograph rat behavior. It was considered that a rat had entered an arm only when all four paws were in the respective arm. The numbers of entries into all four arms (total arm entry counts) were estimated as measures of locomotor activity. The time spent on the open arms was quantified as a percentage of the total time spent on any arm. The camera signal was relayed to a monitor and behavior was recorded on a digital versatile disc for analysis. Behavioral parameters were measured and scored with computer software developed by LimeLight (Actimetrics, Wisconsin, Elmwood, USA).

Immunohistochemical Study: Staining of 5-HT-Immunoreactive Cells

An immunohistochemical study was performed to elucidate the developmental processes of the raphe nuclei in normal rats and rats subjected to early postnatal stress. In the postadolescent period, rats (non-FS and 3wFS) were anesthetized with sodium pentobarbital (60 mg/kg, i.p.) and perfused transcardially via the left ventricle with 0.9% saline, followed by 4% paraformaldehyde in 0.1 M phosphate buffer (PB). After perfusion, the brain was removed and immersed overnight in a postfixative consisting of 4% paraformaldehyde in PB at 4 °C. The brain was then placed in PB containing 30% sucrose at 4 °C until it sank. Each brain was frozen in liquid nitrogen and sectioned at 30 μm in a cryostat. Sections were collected in 0.1 M phosphate-buffered saline (PBS) containing 0.1% Triton X-100 (PBSTx) and stored for at least 1 h before immunohistochemical staining. The specimens were always treated as free-floating sections dipped in solution until the 3, 3′-diaminobenzidine (DAB) coloration step. Endogenous peroxidase activity was blocked by preincubation in 3% hydrogen peroxide in methanol. Sections were incubated with normal goat serum (Histofine; Nichirei, Tokyo, Japan) for 1 h, followed by rabbit anti-5-HT (Zymed Laboratories, South San Francisco, CA) diluted 1: 500 for 24 h. Then, biotinylated goat anti-rabbit IgG secondary antiserum (Histofine; Nichirei.) was applied for 1 h. Incubation was carried out with peroxidase-conjugated streptavidin (Histofine; Nichirei) for 1 h. During incubations, sections were rinsed with PBSTx three times for 5 min each time. Finally, these sections were reacted with DAB as a chromogen. Nickel sulfate was used to intensify the reaction product. The sections were then mounted on gelatin-chromalum-coated slides and dehydrated and coverslipped. Immunoreactive 5-HT cells were counted, according to the atlas of Paxinos and Watson [25], by observers blind to previous treatments.

Statistics

Experimental values are given as means ± SEM. Results obtained for the EPM were expressed as the percentage of time spent on the open arm determined for a 10-min period. For comparisons of the experimental groups, post-hoc multiple comparison was performed with Fisher's test after one-way analysis of variance (ANOVA). Probability values of less than 5% were considered significant.

Results

Changes in Fear-Related Behavior in Rats Subjected to Early Postnatal Stress

As shown in Fig. 1B, the percentage of time spent on the open arms during the EPM test was significantly increased in the 3wFS group when compared to that in

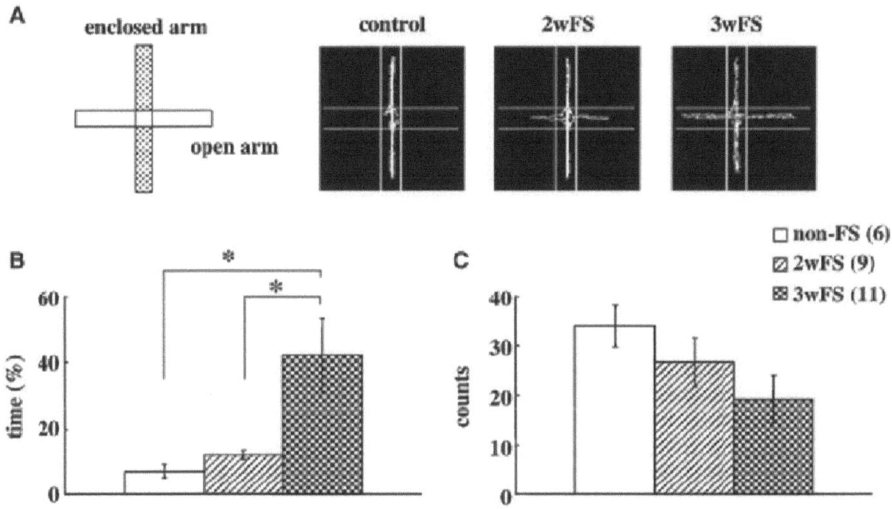

Fig. 1A–C. Effects of early postnatal stress on fear-related behavior. Behavioral changes are shown by the typical locus (**A**) and the percentage of time spent on the open arms (**B**) of the elevated plus maze. Total arm entries (counts/10 min), as an assessment of locomotor activity, are shown in **C**. Behavioral experiments were performed in the postadolescent period (age, 10–12 weeks). *Figures in parentheses* show the numbers of rats tested. Each value in the graphs represents the mean ± SEM. *Non-FS*, non-footshock (*FS*) control; *2wFS* and *3wFS*, pups exposed to FS stimuli in the second and third postnatal weeks, respectively. $*P < 0.05$

the non-FS controls. There was no significant difference in the time spent on the open arms between the 2wFS group and the controls. The total arm entries (counts/10 min), as an assessment of locomotor activity, were not significantly different among the groups (Fig. 1C).

Changes in 5-HT-Immunoreactive Cells in the Raphe Nuclei in Rats Subjected to Early Postnatal Stress

As shown in Fig. 2B, the number of 5-HT-immunoreactive cells in the median raphe nucleus (MRN) in the 3wFS group was obviously reduced compared to the numbers in the non-FS control and 2wFS groups. Significant decreases in 5-HT-immunoreactive cells in the MRN were observed in the 3wFS group when compared to those in the non-FS controls. There were no statistically significant differences in 5-HT-immunoreactive cells in the dorsolateral dorsal raphe nucleus (DRN) between rats subjected to early postnatal stress and controls (Fig. 2B).

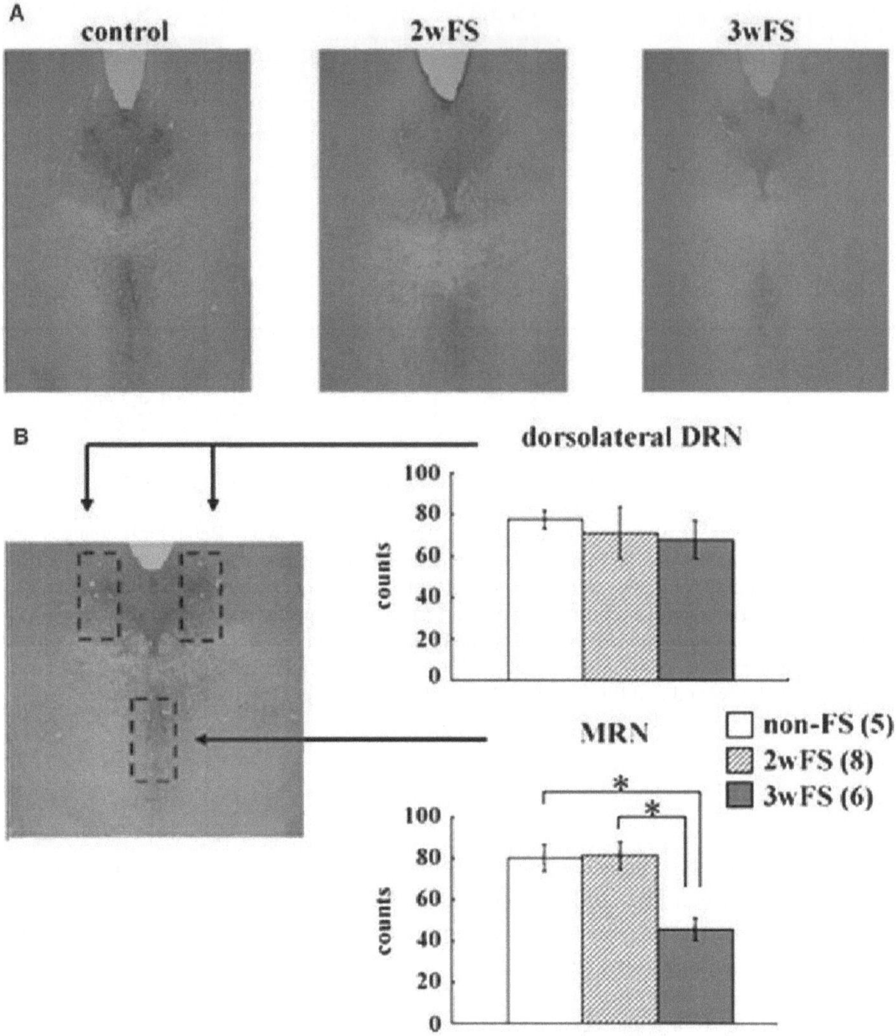

Fig. 2A,B. Effects of early postnatal stress on serotonin (5-hydroxytryptamine; 5-HT)-immunoreactive cells in the median raphe nucleus (*MRN*) and dorsal raphe nucleus (*DRN*) in the postadolescent period. Following the criteria, morphological changes in the raphe nuclei (**A**) and changes in 5-HT-immunoreactive cells in the dorsolateral area of the DRN and MRN (**B**) in the postadolescent period (10–12 weeks old) in rats subjected to early postnatal stress. The numbers of 5-HT-immunoreactive cells in the dorsolateral area of the DRN and MRN were determined as described in "Methods". *Figures in parentheses* show the numbers of rats tested. Each value represents the mean ± SEM. * $P < 0.05$

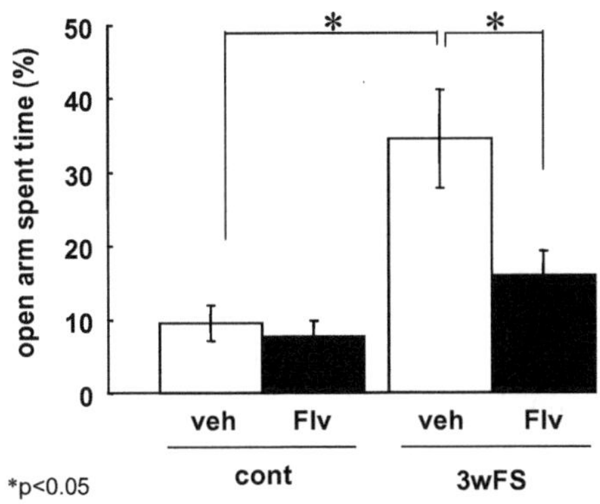

Fig. 3. Effects of fluvoxamine (*Flv*) on fear-related behavior in rats subjected to early postnatal stress. *veh*, vehicle; *cont*, control

Effects of Fluvoxamine on Fear-Related Behavior in Rats Subjected to Early Postnatal Stress

As shown in Fig. 3, the results on the EPM test were normalized by fluvoxamine treatment. Namely, the increase in the time spent on the open arm was decreased to the control level.

Discussion

In this study, exposure to aversive footshock (FS) stimuli during the third week (3w) of the postnatal period (3wFS) increased the time spent on the open arms of the elevated plus maze (EPM) test. In general, an increase in the time spent on the open arms is considered to show a decrease in anxiety, because clinically effective anxiolytics increase, but anxiogenic agents reduce, the percentage of time spent on the open arms [23]. The present data, therefore, indicate that the aversive stress exposure during the third postnatal week decreased anxiety in the postadolescent period. This hypothesis was supported by the present findings that 5-HT-immunoreactive cells in the raphe nucleus were markedly reduced in the post-adolescent period in the 3wFS group. In other words, aversive stimuli during the third postnatal week disrupt 5-HT neurons in the raphe nucleus, and consequently may cause abnormal responses to emotional stress in later life.

Numerous studies have demonstrated the correlation between anxiety and hypofunction of the somatodendritic 5-HT1A receptors located on the raphe nuclei [26, 27]. Furthermore, developmental factors affecting anxiety-related behavior can alter the expression of 5-HT1A receptors not only in the raphe nuclei [22] but also in the projection areas such as the hippocampus [19, 20]. Further studies, however, are required to clarify the precise mechanism of the serotonergic system in the raphe nuclei.

The interesting finding in the present study is that decreases in the number of 5-HT-immunoreactive cells in the median raphe nucleus (MRN), but not in the dorsolateral dorsal raphe nucleus (DRN), were observed in the 3wFS group. Although we cannot sufficiently explain why such marked behavioral and immunohistochemical changes were observed in the 3wFS group, one possibility may depend on the developmental process of the serotonergic system in the rodent brain, i.e., 5-HT levels continue to increase and 5-HT axons reach target areas, and the final innervation occurs robustly up until the third postnatal week [28]. During development, the 5-HT1A receptors in the raphe nuclei play a role in the inhibitory autoregulation of 5-HT neuronal development [29]. The present finding that the number of 5-HT-immunoreactive cells in the MRN was reduced in the 3wFS group suggests that the number of somatodendritic 5-HT1A receptors was also decreased. In turn, the negative feedback role of the serotonergic neurons in the MRN mediated via 5-HT1A receptors may be disrupted by early postnatal stress. In other words, the attenuation of fear-related behavior could be considered to be the result of a loss of autoinhibition via somatodendritic 5-HT1A autoreceptors in the MRN.

Interestingly, Béïque et al [30] have demonstrated that 5-HT1A receptor-induced depolarization in the neonatal prefrontal cortex occurs in the early postnatal period. Thus, 5-HT elicits depolarization during the second postnatal week and exhibits a gradual shift from depolarization to hyperpolarization beginning in the third postnatal week. It is plausible, therefore, that not only disruption but also persistent changes in the naive characteristics of the 5-HT1A receptors occur as a result of early postnatal stress and, consequently, may precipitate long-lasting alterations in the functional properties of the serotonergic projection areas underlying emotional expression. Another possibility is that a hyper-responsive limbic-hypothalamic-pitutary-adrenal axis, along with prolonged glucocorticoid responses, might affect the serotonergic system in the MRN during this period. Indeed, numerous studies have shown that elevated corticosterone levels, resulting from exposure to aversive stress, directly impair 5-HT receptor-mediated responses [31, 32]. Long-term elevation of endogenous corticosteroid also causes downregulation of 5-HT1A receptors in the hippocampus [33–35]. Therefore, it is possible that sustained corticosterone levels induced by early postnatal stress impair the serotonergic system in the MRN and/or the MRN-hippocampal circuit. More speculatively, insufficient levels of 5-HT in the MRN might lead to long-lasting changes in the neural circuits underlying emotional regulation in the postadolescent period. These previous reports strongly support the immunohistochemical evidence observed in the present study; in other words, dysfunction of the serotonergic system in the MRN seems to be associated with the behavioral changes seen in the 3wFS group.

Our immunohistochemical study showed that 5-HT-immunoreactive cells in the MRN were decreased in the 3wFS group. These results suggest that the developmental changes of the serotonergic system in the MRN likely contribute to altered behavioral responses to emotional stress in adults. Thus, early postnatal stress appears to affect the serotonergic system in the MRN and DRN in different ways. In other words, the MRN is a critical region underlying the serotonergic modulation of anxiety during brain development, and variations in 5-HT neurons and/or 5-HT receptors in the MRN during early life appear to be important in preventing emotional disturbances in adults. These findings further underline the pivotal role played by the serotonergic mechanism in the homeostasis of anxiety-like behavior during early postnatal development for normal anxiety-like behavior throughout life.

References

1. Modell S, Lauer CJ, Schreiber W et al (1998) Hormonal response pattern in the combined DEX-CRH test is stable over time in subjects at high familial risk for affective disorders. Neuropsychopharmacology 18: 253–262
2. Heim C, Nemeroff CB (2001) The role of childhood trauma in the neurobiology of mood and anxiety disorders: preclinical and clinical studies. Biol Psychiatry 49: 1023–1039
3. Yoshioka M, Matsumoto M, Togashi H et al (1995) Effects of conditioned fear stress on 5-HT release in the rat prefrontal cortex. Pharmacol Biochem Behav 51: 515–519
4. Roche M, Commons KG, Peoples A et al (2003) Circuitry underlying regulation of the serotonergic system by swim stress. J Neurosci 23: 970–977
5. Maes M, Meltzer HY (1995) The serotonin hypothesis of major depression. In: Bloom FE, Kupfer DJ (eds) Psychopharmacology: the fourth generation of progress. Lippincott, Philadelphia, pp 933–944
6. Sandford JJ, Argyropoulos SV, Nutt DJ (2000) The psychobiology of anxiolytic drugs: part 1. Basic neurobiology. Pharmacol Ther 88: 197–212
7. Aitken AR, Tork I (1988) Early development of serotonin-containing neurons and pathways as seen in wholemount preparations of the fetal rat brain. J Comp Neurol 274: 32–47
8. Lauder JM (1990) Ontogeny of the serotonergic system in the rat: serotonin as a developmental signal. Ann N Y Acad Sci 600: 297–313
9. Whitaker-Azmitia PM, Lauder JM, Shemmer A et al (1987) Postnatal changes in serotonin receptors following prenatal alterations in serotonin levels: further evidence for functional fetal serotonin receptors. Brain Res 430: 285–289
10. Zhou FC, Sari Y, Zhang JK (2000) Expression of serotonin transporter protein in developing rat brain. Dev Brain Res 119: 33–45
11. Mazer C, Muneyyirci J, Taheny K et al (1997) Serotonin depletion during synaptogenesis leads to decreased synaptic density and learning deficits in the adult rat. A model of neurodevelopmental disorders with cognitive deficits. Brain Res 760: 68–73
12. Faber KM, Haring JH (1999) Synaptogenesis in the postnatal rat fascia dentata is influenced by 5-HT1a receptor activation. Dev Brain Res 114: 245–252
13. Kondoh M, Shiga T, Okado N (2004) Regulation of dendrite formation of Purkinje cells by serotonin through serotonin1A and serotonin2A receptors in culture. Neurosci Res 48: 101–109
14. Lauder JM, Wallace JA, Krebs H (1981) Roles for serotonin in neuroembryogenesis. Adv Exp Med Biol 133: 477–506
15. Lauder JM, Wallace JA, Wilkie MB et al (1983) Roles for serotonin in neurogenesis. Monogr Neural Sci 9: 3–10
16. Okado N, Cheng L, Tanatsugu Y et al (1993) Synaptic loss following removal of serotoninergic fibers in newly hatched and adult chickens. J Neurobiol 24: 687–698

17. Cases O, Vitalis T, Seif I et al (1996) Lack of barrels in the somatosensory cortex of monoamine oxidase A-deficient mice: role of a serotonin excess during the critical period. Neuron 16: 297–307
18. Janusonis S, Gluncic V, Rakic P (2004) Early serotonergic projections to Cajal–Retzius cells: relevance for cortical development. J Neurosci 24: 1652–1659
19. Vazquez DM, Lopez JF, Van Hoers H et al (2000) Maternal deprivation regulates serotonin 1A and 2A receptors in the infant rat. Brain Res 855: 76–82
20. Vazquez DM, Eskandari R, Zimmer CA et al (2002) Brain 5-HT receptor system in the stressed infant rat: implications for vulnerability to substance abuse. Psychoneuroendocrinology 27: 245–272
21. Van den Hove DL, Lauder JM, Scheepens A et al (2006) Prenatal stress in the rat alters 5-HT1A receptor binding in the ventral hippocampus. Brain Res 1090: 29–34
22. Gartside SE, Johnson DA, Leitch MM et al (2003) Early life adversity programs changes in central 5-HT neuronal function in adulthood. Eur J Neurosci 17: 2401–2408
23. Pellow S, Chopin P, File SE et al (1985) Validation of open: closed arm entries in an elevated plus-maze as a measure of anxiety in the rat. J Neurosci Methods 14: 149–167
24. Matsumoto M, Higuchi K, Togashi H et al (2005) Early postnatal stress alters the 5-HTergic modulation to emotional stress at postadolescent periods of rats. Hippocampus 15: 775–781
25. Paxinos G, Watson C (1986) The rat brain in stereotaxic coordinates, 2nd edn. Academic, New York
26. Andrews N, Hogg S, Gonzalez LE et al (1994) 5-HT1A receptors in the median raphe nucleus and dorsal hippocampus may mediate anxiolytic and anxiogenic behaviours respectively. Eur J Pharmacol 264: 259–264
27. Cervo L, Mocaër E, Bertaglia A et al (2000) Roles of 5-HT1A receptors in the dorsal raphe and dorsal hippocampus in anxiety assessed by the behavioral effects of 8-OH-DPAT and S 15535 in a modified Geller–Seifter conflict model. Neuropharmacology 39: 1037–1043
28. Chugani DC, Muzik O, Rothermel R et al (1997) Altered serotonin synthesis in the dentatothalamocortical pathway in autistic boys. Ann Neurol 42: 666–669
29. Lauder JM, Liu J, Grayson DR (2000) In utero exposure to serotonergic drugs alters neonatal expression of 5-HT1A receptor transcripts: a quantitative RT-PCR study. Int J Dev Neurosci 18: 171–176
30. Béïque JC, Campbell B, Perring P et al (2004) Serotonergic regulation of membrane potential in developing rat prefrontal cortex: coordinated expression of 5-hydroxytryptamine (5-HT)1A, 5-HT2A, and 5-HT7 receptors. J Neurosci 24: 4807–4817
31. Young AH, Dow RC, Goodwin GM et al (1993) The effects of adrenalectomy and ovariectomy on the behavioral and hypothermic responses of rats to 8-hydroxy-2 (di-n-propylamino) tetralin (8-OH-DPAT). Neuropharmacology 32: 653–657
32. Meijer OC, de Kloet ER (1998) Corticosterone and serotonergic neurotransmission in the hippocampus: functional implications of central corticosteroid receptor diversity. Crit Rev Neurobiol 12: 1–20
33. Watanabe Y, Sakai RR, McEwen BS (1993) Stress and antidepressant effects on hippocampal and cortical 5-HT1A and 5-HT2 receptors and transport sites for serotonin. Brain Res 615: 87–94
34. McKittrick CR, Blanchard DC, Blanchard RJ et al (1995) Serotonin receptor binding in a colony model of chronic social stress. Biol Psychiatry 37: 383–393
35. Lopez JF, Chalmers DT, Little KY et al (1998) A.E. Bennett Research Award. Regulation of serotónin1A, glucocorticoid, and mineralocorticoid receptor in rat and human hippocampus: implications for the neurobiology of depression. Biol Psychiatry 43: 547–573

Neural Circuit Development and Plasticity Shaped by Glutamate Transporters

Taisuke Miyazaki, Chihiro Takasaki, and Masahiko Watanabe

Summary

Glutamate receptors and ion channels that mediate Ca^{2+} influx and release play a critical role in the activity-dependent pruning of immature synaptic circuitry, as shown for *N*-methyl-D-aspartate (NMDA) receptors in the whisker-related patterning of thalamocortical (TC) synapses and mGluR1 and P/Q-type Ca^{2+} channels in the elimination of surplus climbing fibers (CFs) onto cerebellar Purkinje cells (PCs). Recently we have identified unique roles played in synapse development by the glutamate transporters GLT1 and GLAST, two astrocytic transporters predominant in the cerebrum and cerebellum, respectively. In the somatosensory cortex of GLT1-knockout mice, whisker-related patterning of TC synapses and their critical period termination occurred normally in the first postnatal week. However, when a lesion was given to row-C whiskers during the critical period (postnatal days 0–3) in GLT1-knockout mice, the shrinkage of lesioned row-C barrels and the reciprocal expansion of intact row-B/D barrels were significantly diminished. Thus, GLT1 magnifies the lesion-induced plasticity of TC synapses during the critical period. In the cerebellum of GLAST-knockout mice, the territorial innervation of PC dendrites, i.e., the innervation of proximal dendrites by CFs and distal dendrites by parallel fibers, was normally structured. However, dendritic innervation by single major CFs was significantly regressed and, instead, their aberrant wiring to neighboring PC dendrites was induced conspicuously and caused multiple innervation. This aberrant innervation was infrequent during the first 3 postnatal weeks (when surplus CFs are normally eliminated in wild-type rodents), and became progressively exacerbated thereafter. Moreover, in a similar time course, the synapse-enwrapping processes of Bergmann glia were progressively retracted, resulting in incomplete glial sealing of PC synapses. Presumably through glutamate uptake and synapse-sealing functions, GLAST thus promotes and maintains CF monoinnervation by consolidating single major CFs innervating the PCs and preventing their aberrant wiring to neighboring PCs. These phenotypes collectively suggest that

Department of Anatomy, Hokkaido University School of Medicine, North 15 West 7, Kita-ku, Sapporo 060-8638, Japan

N. Tamaki, Y. Kuge (eds.)
Molecular Imaging for Integrated Medical Therapy and Drug Development.
DOI: 10.1007/978-4-431-98074-2_22, ©Springer 2010

glutamate transporters operate as an *activity discriminator* in competitive synaptic wiring; with this operation, major afferents to postsynaptic targets can further expand their innervation, whereas innervation by minor afferents is diminished or suppressed. This molecular function provides neural circuits with a *"winner-takes-more"* strategy, by which activity-dependent remodeling is facilitated in the developing somatosensory cortex and monoinnervation by single CFs is maintained in the adult cerebellum.

Key words Glutamate transporter · Synapse development · Plasticity · Cerebellum · Somatosensory cortex

Introduction

Synaptic circuits attain final maturation through activity-dependent refinement and remodeling during the early postnatal period. This process has been best investigated using the thalamocortical (TC) synapse in the somatosensory and visual cortices and the climbing fiber-Purkinje cell (CF-PC) synapse in the cerebellum. In both types of synaptic circuits, connection with major afferents that cause strong activation to postsynaptic neurons is strengthened, whereas that with minor afferents is weakened and eliminated [1–4].

Whiskers are tactile hairs on the face in nonhuman mammals that are topographically represented as modular patterns at each somatosensory station [5]. Whisker-related patterns are referred to as "barrels" in the somatosensory cortex, "barreloids" in the thalamus, and "barrelettes" in the brainstem. These patterns are plastic during the early neonatal days known as the critical or sensitive period, and later, they are consolidated as permanent structures. If sensory nerves innervating whisker follicles are completely severed during the critical period, central patterns are blurred or fused [6–9]. When particular whiskers are lesioned during the critical period, the lesioned barrels are diminished and adjacent intact barrels expand reciprocally in the somatosensory cortex; this process is termed lesion-induced plasticity [10–12]. *N*-Methyl-D-aspartate (NMDA)-type and metabotropic mGluR5 glutamate receptors play an essential role in the formation of these whisker-related central patterns [13–18].

A remarkable feature of PC circuits is territorial innervation by two distinct excitatory afferents. A single CF innervates proximal PC dendrites, whereas hundreds of thousands of parallel fibers (PFs), the axons of granule cells, innervate distal PC dendrites. This mature circuitry is formed and maintained through homosynaptic competition among CFs and heterosynaptic competition between CFs and PFs [19–22]. These competitions are mediated by a number of signaling molecules, including the glutamate receptors GluRδ2 [20, 23–30] and mGluR1 [31–35], the P/Q-type Ca^{2+} channel α1A [21], and the secretory protein precerebellin (or Cbln1) [36–37]. The ablation of these genes in mice leads to aberrant circuitry, such as shifted territorial innervation, impaired PF synapse formation, and persistent multiple innervations by CFs. Thus, in both the cerebrum and cerebellum, the activation of glutamate receptors and ion channels that mediate Ca^{2+} influx and release plays a key role in activity-dependent synapse refinement.

Extracellular glutamate concentration has to be kept low enough to terminate glutamate receptor activation and to protect neurons from excitotoxicity, by means of high-affinity glutamate transporters on the cell membrane [38–39]. GLT1 and GLAST are astrocytic glutamate transporters that are predominant in the cerebrum and cerebellum, respectively, of adult animals [40]. Previous reports have demonstrated physiological roles of these transporters in neuronal differentiation and survival, synaptic transmission and plasticity, and neuron-glial metabolic crosstalk [41–47]. Little is known, however, about their role in activity-dependent synapse development. In this chapter, our recent data are presented to show how glutamate transporters regulate TC patterning in the somatosensory cortex [48] and CF innervation to PCs in the cerebellum (T. Miyazaki et al., article in preparation).

Materials and Methods

Animals

C57BL/6J wild-type mice, GLT1-KO mice [41], and GLAST-KO mice [42] were used and treated according to the Guidelines for the Care and Use of Laboratory Animals of the Hokkaido University School of Medicine. The day after overnight mating was counted as embryonic day 0 (E0), and the day of birth (~24 h) was designated as postnatal day 0 (P0). Genotyping of genomic tail DNA by polymerase chain reaction and phenotypic analyses of barrel development were performed in a blind fashion. Under deep pentobarbital anesthesia (100 mg/kg body weight; i.p.), mice were fixed by transcardial perfusion with 4% paraformaldehyde in 0.1 M sodium phosphate buffer (pH 7.2; PB) for light microscopy and 2% paraformaldehyde/2% glutaraldehyde in PB for electron microscopy. Sections (40- to 50-μm-thick) were prepared using a microslicer.

Barrel Analyses

Barrels were visualized by cytochrome oxidase (CO) histochemistry [49]. The right infraorbital nerve (ION) was transected to determine when the critical period was terminated. With the mice under hypothermia-induced anesthesia, a vertical slit was made just behind the mystacial pad to expose the right ION. The ION was cut with a pair of iridectomy scissors, the cut edge was subjected to electrical cautery to prevent nerve regeneration, and then pups were returned to their mothers. After 8 days, mice were anesthetized with pentobarbital (100 mg/kg of body weight) and killed for CO histochemistry. To compare genotypic differences, we estimated the age when barrels appeared in 50% of mice (DA_{50}) and the age when the critical period was terminated in 50% of mice (CPT_{50}). To obtain the optimal DA_{50} and CPT_{50} values and their 95% confidence intervals, probit regression analysis was used [50]; the fraction of mice that had formed barrels or completed the critical period was converted into probits at each postnatal age. Probit-fitted linear regression lines were used for calculation. Differences between control and mutant DA_{50} or CPT_{50} values were considered statistically significant if there was no overlap between their confidence intervals. Fisher's exact probability test was used for statistical evaluation.

Row-C whiskers were lesioned to compare critical period plasticity. Hair follicles of the right row-C whiskers were cauterized on P0, P2, or P4; CO histochemistry and immunofluorescence were performed on P15. Whisker pads were examined stereoscopically and histologically to verify the absence of regrowth. Area measurements were performed with IPLab software (Scananalytics, Rockville, MD, USA) for a pair of barrels in each row (A2 and A3, B2 and B3, C2 and C3, D2 and D3, or E2 and E3) including their intervening septa. From the scores, we calculated the ratio of row-C or row-B/D to the total measured area and the map plasticity index. Statistically significant differences were determined with the Mann-Whitney *U*-test. Data values are presented as means ± SEM. Differences were considered statistically significant for $P < 0.05$.

Climbing Fiber Labeling

With mice under anesthesia with chloral hydrate (350 mg/kg body weight; i.p.), a glass pipette (G-1.2; Narishige, Tokyo, Japan) filled with 2–3 μl of 10% solution of dextran Alexa 594 (DA594; Invitrogen, Carlsbad, CA, USA) in phosphate-buffered saline (pH 7.4) was inserted stereotaxically to the inferior olive by the dorsal approach. The anterograde tracer was injected by air pressure at 20 psi at 5-s intervals for 1 min (Pneumatic Picopump; World Precision Instruments, Sarasota, FL, USA). After 4 days of survival, mice were anesthetized and fixed by transcardial perfusion. For combined anterograde and immunofluorescence labelings, DA594-labeled microslicer sections were incubated overnight with a mixture of goat calbindin (1 μg/ml) [51] and rabbit VGluT2 (0.5 μg/ml) [52] antibodies, and then with Alexa Fluor 488- or indodicarbocyanine (Cy5)-labeled species-specific secondary antibodies for 2 h at a dilution of 1: 200 (Invitrogen; Jackson ImmunoResearch, West Grove, PA, USA). Images were taken with a confocal laser scanning microscope (FV1000; Olympus Optical, Tokyo, Japan) and analyzed with Metamorph software (Molecular Devices, Downingtown, PA, USA).

Electron Microscopy

For conventional electron microscopy, ultrathin sections cut in the parasagittal plane were prepared from the straight portion of lobules 4/5, and stained with 2% uranyl acetate for 5 min and mixed lead solution for 2 min. Twelve electron micrographs, at the original magnification of ×10 000 were taken randomly from the neuropil region of the molecular layer to evaluate synaptic enwrapping with Bergmann glia (BG). Films of the electron micrographs were scanned and analyzed with MetaMorph software, and all data values were expressed as means ± SEM.

Results

Role of GLT1 in Barrel Development

The role of GLT1 in TC patterning in the somatosensory cortex was examined by comparing the stage of barrel appearance, the stage of critical period termination, and the magnitude of lesion-induced plasticity, using GLT1-KO mice and control littermates. Normal body growth and brain development were indicated from the findings of no significant differences in the body weight and cortical histology between GLT1-KO and control neonates [48].

CO histochemistry in flattened tangential cortical sections revealed that no barrels were visible at P2 and none were segregated at P3 in any GLT1-KO or control mice (Fig. 1A). The percentage of mice having segregated barrels was increased to 64% in GLT1-KO mice and 75% in control mice at P4, and reached 100% in both types of mice at P5. From the scores, a probit regression analysis was used to calculate the age when barrels appeared in 50% of mice (DA_{50}). DA_{50} values were P3.8 (95% confidence interval [CI], P3.1-P4.4) in GLT1-KO mice and P3.8 (95% CI, P3.3-P4.1) in control mice, showing no significant difference (Fisher's exact probability test; $P > 0.05$).

To compare the stage of critical period termination, transection of the ION was conducted, and lesioned (contralateral) and intact (ipsilateral) cortices were examined by CO histochemistry 8 days later. In all GLT1-KO and control mice lesioned on P2, CO intensity was lowered in the lesioned barrels, and the barrels were blurred and fused into rows: thus, the critical period was complete in 0% of these mice (Fig. 1B). The percentage of mice that had completed the critical period was increased from those with the P3 lesion (30% in GLT1-KO and 25% in control) to those with the P4 lesion (93% in GLT1-KO and 92% in control mice), and attained 100% in both types of mice lesioned on P5. The age when the critical period was complete in 50% of mice (CPT_{50}) was P3.2 (95% CI, P2.8-P3.5) in GLT1-KO mice and P3.3 (95% CI, P2.9-P3.7) in control mice, showing no significant difference between the two groups ($P > 0.05$).

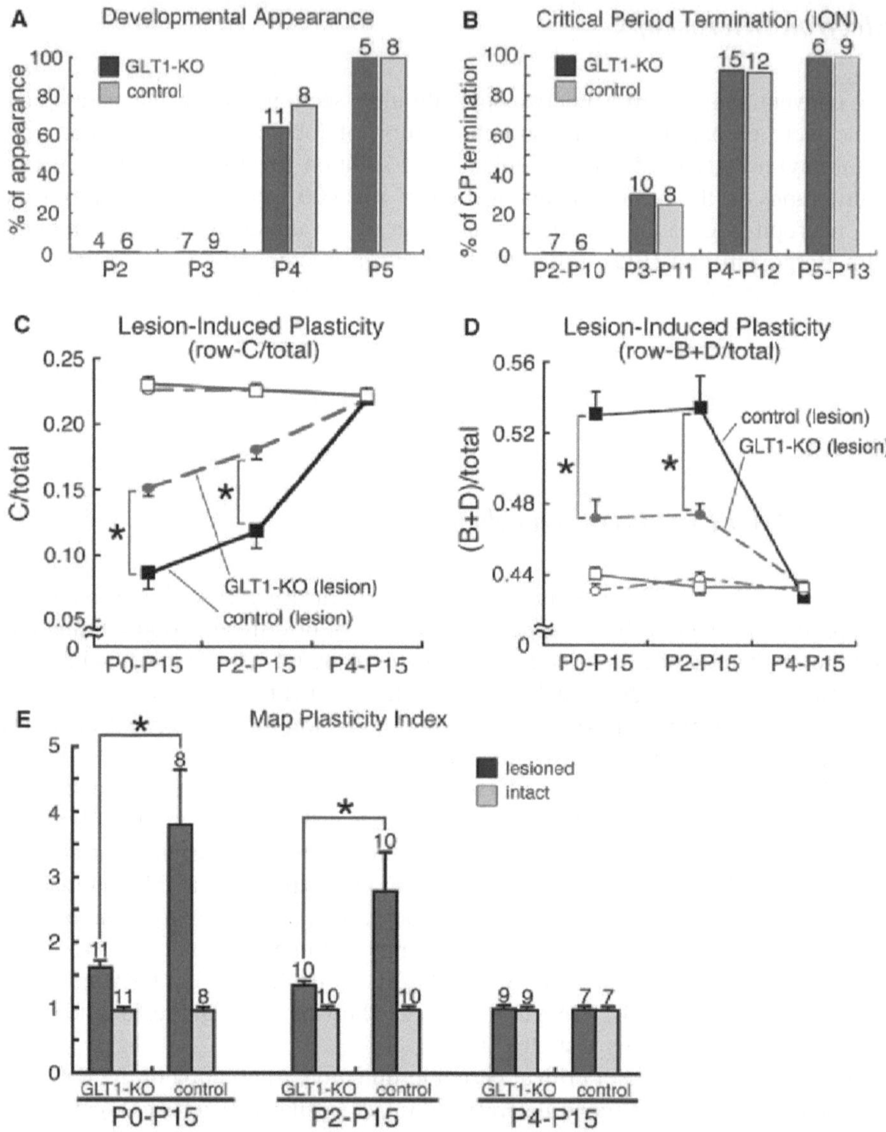

Fig. 1A–E. Quantitative comparison of the appearance of cortical barrels (**A**), critical period (*CP*) termination (**B;** *ION*, infraorbital nerve), and critical period plasticity (**C–E**). **A, B** The percentage of mice with cortical barrels (**A**) and the percentage of mice in which the critical period was complete (**B**) increased with age in both GLT1-knockout (*KO*) (*black bars*) and control (*gray bars*) mice. **C, D** The ratio of the row-C (C2+C3) area (**C**) or row-B/D (B2+B3+D2+D3) area (**D**) relative to the total measured area (A2+A3+B2+B3+C2+C3+D2+D3+E2+E3) in GLT1-KO (*circles*) and control (*squares*) mice. *Filled and open symbols* indicate the lesioned and intact sides, respectively. **E** Map plasticity index in GLT1-KO mice and their littermate controls. *Black bars and gray bars* indicate the lesioned and intact sides, respectively. *Error bars* represent SD. *Numbers above the bars* indicate the total number of mice examined at each stage; *Asterisks* indicate $P < 0.01$. *P on horizontal axes,* postnatal day. This figure is modified from [48], with permission

To examine the plasticity during the critical period, hair follicles of right row-C whiskers were electrocauterized on P0, P2, and P4, and changes in barrel map structure were examined by CO histochemistry on P15 (Fig. 2). When row-C hair follicles were lesioned on P0 and P2, row-C barrels in the lesioned cortex were diminished in size and fused in a continuous row, and adjacent row-B and row-D barrels expanded reciprocally in both GLT1-KO (Fig. 2C) and control mice (Fig. 2B, D). Reproducibly, the reduction of lesioned barrels and reciprocal expansion of intact barrels was milder in GLT1-KO mice than in control mice (compare Fig. 2A1 with Fig. 2B1, and Fig. 2C1 with Fig. 2D1). No changes in histochemical patterns were observed by producing a P4 lesion in either mouse type (Fig. 2E1, F1). This phenotypic change was quantitated by measuring the area of A2-A3, B2-B3, C2-C3, D2-D3, or E2-E3 barrels. Following lesion on P0 or P2, the ratio of the C2-C3 area relative to the total measured area was significantly greater in GLT1-KO mice than in control mice ($P < 0.01$ for each; Mann-Whitney *U*-test; Fig. 1C). A similar effect was not observed in mice lesioned on P4 ($P > 0.05$). Conversely, the ratio of the summed B2-B3 and D2-D3 areas relative to the total measured area was significantly lower in GLT1-KO mice than in control mice lesioned on P0 or P2 ($P < 0.01$ for each), but this was not the case in mice lesioned on P4 ($P > 0.05$; Fig. 1D). Consequently, the map plasticity index (MPI), as obtained from (B2+B3+D2+D3)/2(C2+C3), was significantly lower in GLT1-KO

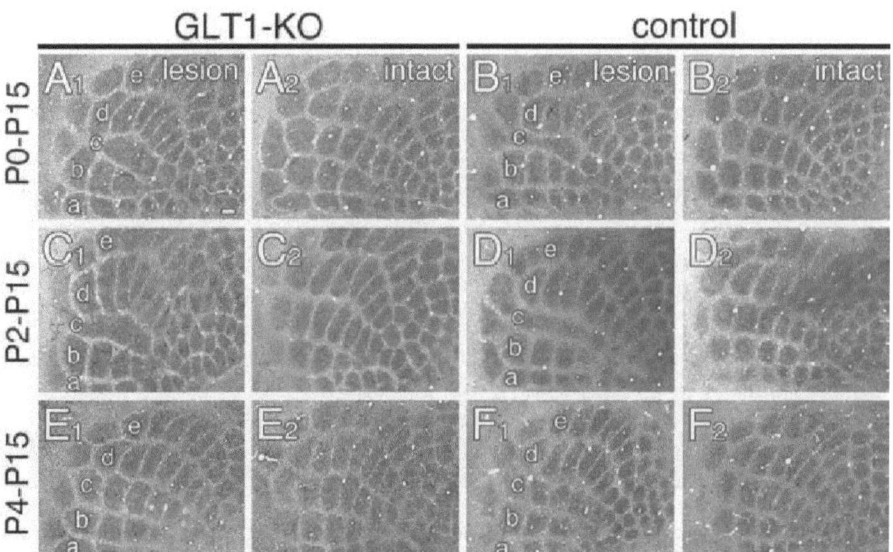

Fig. 2A–F. Lesion of row-C whiskers reveals diminished critical period plasticity in GLT1-KO mice (**A, C, E**) and control (**B, D, F**) mice. **A–F** Cytochrome oxidase (CO) histochemistry was performed on P15 after row-C whisker follicle cautery on P0 (**A, B**), P2 (**C, D**), or P4 (**E, F**). *Scale bar,* 100 μm. *a–e*; rows a through e of barrels. This figure is modified from [48], with permission

mice than in control mice lesioned on P0 or P2: 1.61 ± 0.34 (n = 11; mean ± SD) vs 3.80 ± 2.21 (n = 8) for P0 lesion and 1.34 ± 0.21 (n = 10) vs 2.79 ± 1.79 (n = 10) for P2 lesion, respectively ($P < 0.01$ for each; Fig. 1E). No significant differences in the MPI between GLT1-KO and control mice were observed in the cortices lesioned on P4 (0.99 ± 0.05; n = 9 vs 0.99 ± 0.02; n = 7; $P > 0.05$). Therefore, the magnitude of lesion-induced critical period plasticity was significantly reduced in GLT1-KO mice. Normal temporal closing of the critical period was also confirmed by this row-C lesion experiment (Fig. 1C-E).

Role of GLAST in CF Innervation to PCs

Patterns of dendritic wiring by CFs were compared between GLAST-KO and control mice by anterograde tracer labeling with dextran Alexa 594 (DA594, Fig. 3; red) and by immunofluorescence for vesicular glutamate transporter VGluT2 (green) and calbindin (blue). In the control mice, DA594- or VGluT2-labeled terminals on ascending CFs were associated with dendrites until the transition between proximal shaft dendrites and distal spiny branchlets (Fig. 3A). From the ascending CFs, DA594-labeled collateral branches were occasionally emitted in the transverse plane (white arrows in Fig. 3B), but they rarely formed VGluT2-positive terminals (white arrowhead in Fig. 3B). In such PCs wired by DA594-labeled CFs, DA594

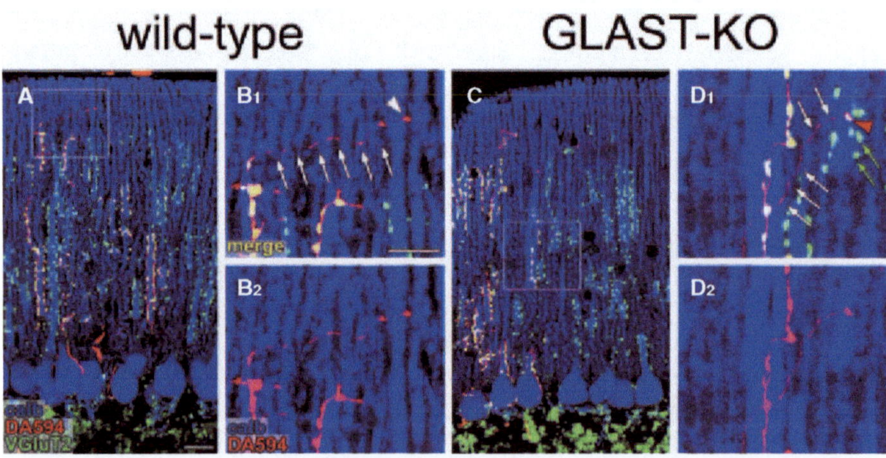

Fig. 3A–D. Multiple climbing fiber (CF) innervation in GLAST-KO mice. Triple labeling for calbindin (*blue*), vesicular glutamate transporter (VGluT2; *green*), and dextran Alexa 594 (DA594; *red*) in control (**A, B**) and GLAST-KO (**C, D**) mice. *Boxed regions* in **A** and **C** are shown in **B** and **D**, respectively. *White arrows* indicate transverse collaterals extending from ascending CFs. *White and red arrowheads* indicate VGluT2-negative and VGluT2-positive terminals, respectively. In GLAST-KO mice, a VGluT2-positive terminal on a transverse collateral contacts an adjacent Purkinje cell (PC) dendrite, which is mainly innervated by DA594-unlabeled CFs *(green arrows)*, thus causing multiple innervation. *Scale bars*, **A** 20 μm; **B** 10 μm

and VGluT2 were overlapped completely along shaft dendrites of given PCs, and DA594-negative/VGluT2-positive terminals were barely found. This represented the monoinnervation pattern of CF wiring in most PCs of the control mice.

In the GLAST-KO mice, the reach of ascending CFs apparently became lower in the molecular layer (Fig. 3C). CF innervation often stopped in the middle of shaft dendrites, leaving the rest of the shaft dendrites free of CF innervation. Despite regressive innervations by ascending CFs, transverse collaterals carried VGluT2-positive terminals and innervated neighboring PC dendrites (red arrowhead in Fig. 3D). Because the dendrites were also innervated by DA594-unlabeled ascending CFs (green arrows in Fig. 3D), this aberrant innervation thus caused multiple innervation to the PCs. Such aberrant innervation by CFs was infrequent during the first 3 postnatal weeks (when surplus CFs are eliminated and monoinnervation is achieved in wild-type rodents), and was progressively exacerbated after this period. Therefore, CFs in GLAST-KO mice display aberrant local wiring to neighboring PC dendrites after the stage of CF elimination, and this causes multiple innervation in the GLAST-KO mice.

Because GLAST is strongly expressed in the BG [53], the morphology of the BG was investigated by electron microscopy. In control mice, lamellate processes of the BG completely enwrapped PC synapses and sealed the synaptic cleft (Fig. 4A). In GLAST-KO mice, glial synaptic sealing was incomplete and the synaptic cleft was often open to the extracellular space (arrowheads; Fig. 4B). The extent of synapse enwrapping was evaluated by counting the number of synaptic edges that were devoid of BG coverage. Surprisingly, 35.8 ± 7.8 % of synaptic edges were devoid of glial coverage in the GLAST-KO mice, compared to 1.2 ± 0.3 % in the control mice, showing a significant difference (mean ± SEM; $n = 3$, $P = 0.047$ by t-test). Impaired synaptic enwrapping in the GLAST-KO mice was rare in the first week, but progressively exacerbated thereafter until adulthood.

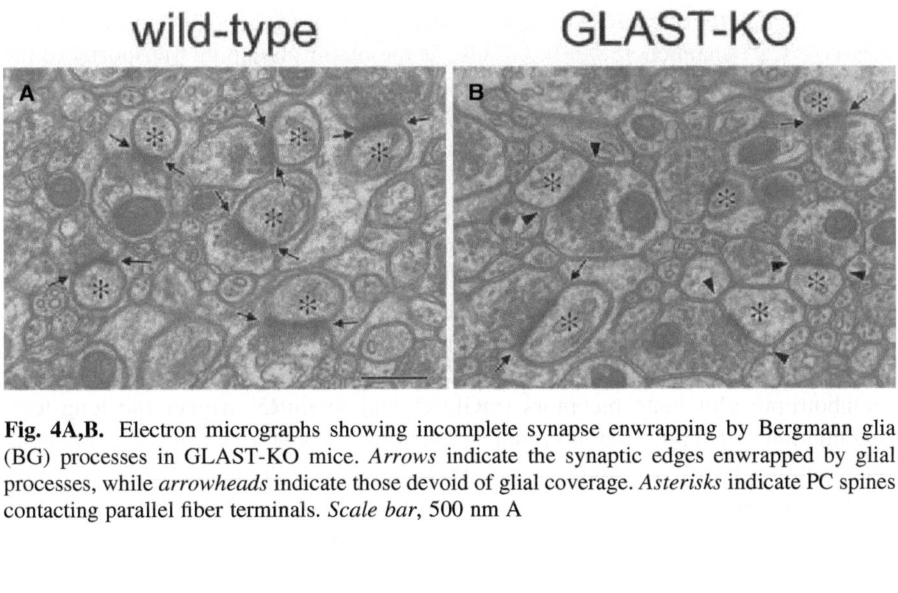

Fig. 4A,B. Electron micrographs showing incomplete synapse enwrapping by Bergmann glia (BG) processes in GLAST-KO mice. *Arrows* indicate the synaptic edges enwrapped by glial processes, while *arrowheads* indicate those devoid of glial coverage. *Asterisks* indicate PC spines contacting parallel fiber terminals. *Scale bar*, 500 nm A

Discussion

In the present study, we pursued the role of glutamate transporters in the process of activity-dependent synapse development. We found here diminished lesion-induced plasticity in TC projections in GLT1-KO mice, and aberrant CF innervation that caused multiple innervation to GLAST-KO PCs. Furthermore, glial enwrapping of PC synapses was progressively impaired in GLAST-KO mice. GLT1-KO mice were normal in body weight, cortical histoarchitecture, and the stages of barrel appearance and critical period termination. Furthermore, GLAST-KO mice are known to normally grow and mate [42]. Thus, the observed phenotypes in GLT1-KO and GLAST-KO mice likely resulted from specific defects in the development of TC and CF-PC projections, rather than from general defects in body growth or brain development.

Astrocytic glutamate transporters play a major role in the rapid and bulk uptake of extracellular glutamate [45]. This transporter function is further facilitated by the co-expression in astrocytes of glutamine synthetase, which converts glutamate to glutamine. Despite normal cortical expression of GLAST in the cortex of GLT1-KO neonates, extracellular glutamate levels are significantly elevated there [48]. Chronic elevation of ambient glutamate levels could affect presynaptic release function and postsynaptic activation. Indeed, the blocking of glutamate transporters in the immature neocortex and hippocampus generates recurrent NMDA receptor-mediated currents associated with synchronous Ca^{2+} oscillations in the entire neuronal population [54–57]. Elevated glutamate levels could also affect synaptic competition so that inactive afferents can even activate postsynaptic targets, as active afferents do. In this regard, the mechanisms that maximize such *activity discriminator* function could magnify lesion-induced plasticity, i.e., by promoting advantaged TC circuits to further expand and disadvantaged ones to shrink (Fig. 5A). Any factors that shift from the optimal point would blunt the plasticity. This notion may be also true for GLAST. CF activity causes strong depolarization that triggers Ca^{2+} influx through P/Q-type Ca^{2+} channels [58–60]. GLAST is the major glutamate transporter in the cerebellum, being expressed at levels sixfold higher than those of GLT-1 [61, 62]. The ablation of GLAST elevates extracellular glutamate levels [46], and impaired synaptic enwrapping further accelerates glutamate spillover to the extracellular space. Such a situation will accelerate cross-talk among CFs and blunt the *activity discriminator* function, resulting in failure to promote major ascending CFs to expand their innervation and to prevent their ectopic wiring to adjacent PCs (Fig. 5B).

Intriguingly, these phenotypes in mice that are defective in glutamate transporters differ from those that are defective in glutamate receptors. Through the activity-dependent rise of intracellular Ca^{2+} levels, NMDA receptors and group I metabotropic glutamate receptors (mGluR1 and mGluR5) trigger the long-term potentiation and depression of synaptic transmission [63–66]. During development, the ablation of NMDA receptors and mGluR5 impairs the formation of whisker-related patterning without affecting lesion-induced plasticity [13, 16–17, 18, 48].

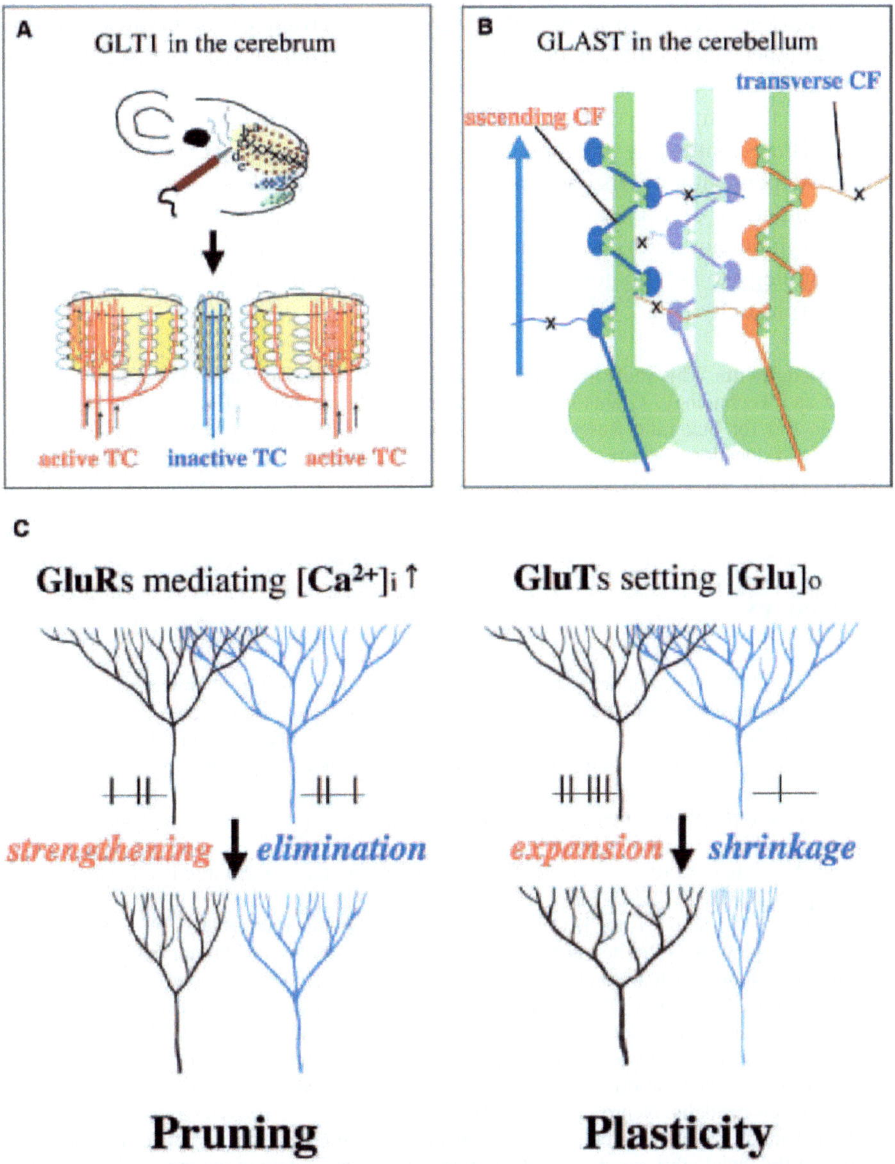

Fig. 5A–C. Schematic illustrations of the roles played by glutamate transporters. **A** GLT1 helps major thalamocortical (*TC*) projections expand and minor ones shrink during the critical period. **B** GLAST helps main ascending CFs extend their innervation along PC dendrites and suppresses their ectopic innervation to nearby PCs. **C** Distinct roles of glutamate receptors (*GluRs*) and transporters (*GluTs*) in activity-dependent circuit development. Activation of GluRs leading to Ca^{2+} influx and release strengthens the connection with major afferents and eliminates that with minor afferents. On the other hand, GluTs may set ambient glutamate levels so that they are optimal to expand the innervation of major afferents and to shrink or suppress the innervation of minor afferents. *[Ca^{2+}]i*, intracellular Ca^{2+} level; *[Glu]o*, extracellular glutamate level

In mutant mice defective in mGluR1 signaling, aberrant CF innervation causing multiple innervation emerges during the early postnatal period when PF synaptogenesis proceeds and surplus CFs are eliminated [31–34]. In GLAST-KO mice, however, aberrant innervation by CFs mainly appeared after the synaptogenic period. These contrasting phenotypes suggest that molecular machineries mediating activity-dependent Ca^{2+} rise are involved in the activity-dependent strengthening and elimination of immature synapses (i.e., the pruning of synaptic circuits), whereas glutamate transporters may provide neural circuits with a *"winner-takes-more"* strategy to help major inputs expand their territory in an activity-dependent manner (plasticity; Fig. 5C).

Acknowledgments This study was supported through Special Coordination Funds for Promoting Science and Technology, Grants-in-Aid for Scientific Research (S) (19100005 to M.W.) and Grants-in-Aid for Scientific Research on Priority Area (17023001 to M.W.) provided by the Ministry of Education, Culture, Sports, Science and Technology of Japan. M.W. was also supported in part by the Uehara Foundation, the Naito Foundation, and the Akiyama Foundation.

References

1. Crepel F (1982) Regression of functional synapses in the immature mammalian cerebellum. Trends Neurosci 5: 266–269
2. O'Leary DD, Ruff NL, Dyck RH (1994) Development, critical period plasticity, and adult reorganizations of mammalian somatosensory systems. Curr Opin Neurobiol 4: 535–544
3. Singer W (1995) Development and plasticity of cortical processing architectures. Science 270: 758–764
4. Crair MC, Malenka RC (1995) A critical period for long-term potentiation at thalamocortical synapses. Nature 375: 325–328
5. Woolsey TA, Van der Loos H (1970) The structural organization of layer IV in the somatosensory region (SI) of mouse cerebral cortex. The description of a cortical field composed of discrete cytoarchitectonic units. Brain Res 17: 205–242
6. Weller WL, Johnson JI (1975) Barrels in cerebral cortex altered by receptor disruption in newborn, but not in 5-day-old mice (Cricetidoe and Muridae). Brain Res 83: 504–508
7. Chiaia NL, Bennett-Clarke CA, Rhoades RW (1992) Differential effects of peripheral damage on vibrissa-related patterns in trigeminal nucleus principalis, subnucleus interpolaris, and subnucleus caudalis. Neuroscience 49: 141–156
8. Yamakado M (1995) Remodelling in the array of cell aggregates in somatotopic representation of the facial vibrissae through the trigeminal sensory system of the mouse. Neurosci Res 23: 399–413
9. Toki S, Watanabe M, Ichikawa R et al (1999) Early establishment of lesion-insensitive mature barrelettes corresponding to upper lip vibrissae in developing mice. Neurosci Res 33: 9–15
10. Van der Loos H, Woolsey TA (1973) Somatosensory cortex: structural alterations following early injury to sense organs. Science 179: 395–398
11. Woolsey TA, Wann JR (1976) Areal changes in mouse cortical barrels following vibrissal damage at different postnatal ages. J Comp Neurol 170: 53–66
12. Jeanmonod D, Rice FL, Van der Loos H (1977) Mouse somatosensory cortex: Development of the alterations in the barrel field which are caused by injury to the vibrissal follicles. Neurosci Lett 6: 151–156

13. Li Y, Erzurumlu RS, Chen C, Jhaveri S et al (1994) Whisker-related neuronal patterns fail to develop in the trigeminal brainstem nuclei of NMDAR1 knockout mice. Cell 76: 427–437
14. Fox K, Schlaggar BL, Glazewski S et al (1996) Glutamate receptor blockade at cortical synapses disrupts development of thalamocortical and columnar organization in somatosensory cortex. Proc Natl Acad Sci USA 93: 5584–5589
15. Kutsuwada T, Sakimura K, Manabe T et al (1996) Impairment of suckling response, trigeminal neuronal pattern formation, and hippocampal LTD in NMDA receptor ε2 subunit mutant mice. Neuron 16: 333–344
16. Iwasato T, Erzurumlu RS, Huerta PT et al (1997) NMDA receptor-dependent refinement of somatotopic maps. Neuron 19: 1201–1210
17. Iwasato T, Datwani A, Wolf AM et al (2000) Cortex-restricted disruption of NMDA R1 impairs neuronal patterns in the barrel cortex. Nature 406: 726–731
18. Hannan AJ, Blakemore C, Katsnelson A et al (2001) PLC-β1, activated via mGluRs, mediates activity-dependent differentiation in cerebral cortex. Nat Neurosci 4: 282–288
19. Bravin M, Morando L, Vercelli A et al (1999) Control of spine formation by electrical activity in the adult rat cerebellum. Proc Natl Acad Sci USA 96: 1704–1709
20. Ichikawa R, Miyazaki T, Kano M et al (2002) Distal extension of climbing fiber territory and multiple innervation caused by aberrant wiring to adjacent spiny branchlets in cerebellar Purkinje cells lacking glutamate receptor GluR™2. J Neurosci 22: 8487–8503
21. Miyazaki M, Hashimoto K, Shin HS et al (2004) P/Q-type Ca^{2+} channel α1A regulates synaptic competition on developing cerebellar Purkinje cells. J Neurosci 24: 1734–1743
22. Watanabe M (2008) Molecular mechanisms governing competitive synaptic wiring in cerebellar Purkinje cells. Tohoku J Exp Med 214: 175–190
23. Guastavino JM, Sotelo C, Damez-Kinselle I (1990) Hot-foot murine mutation: behavioral effects and neuroanatomical alterations. Brain Res, 523: 199–210
24. Kashiwabuchi N, Ikeda K, Araki K et al (1995) Impairment of motor coordination Purkinje cell synapse formation and cerebellar long-term depression in GluR™2 mutant mice. Cell 81: 245–252
25. Kurihara H, Hashimoto K, Kano M et al (1997) Impaired parallel fiber-Purkinje cell synapse stabilization during cerebellar development of mutant mice lacking the glutamate receptor δ2 subunit (GluR™2). J Neurosci 17: 9613–9623
26. Hashimoto K, Ichikawa R, Takechi H et al (2001) Roles of glutamate receptor δ2 subunit (GluR™2) and metabotropic glutamate receptor subtype1 (mGluR1) in climbing fiber synapse elimination during postnatal cerebellar development. J Neurosci 21: 9701–9712
27. Lalouette A, Lohof A, Sotelo C et al (2001) Neurobiological effects of a null mutation depend on genetic context: comparison between two hotfoot alleles of the delta-2 ionotropic glutamate receptor. Neuroscience 105: 443–455
28. Takeuchi T, Miyazaki T, Watanabe M et al (2005) Control of synaptic connection by glutamate receptor δ2 in the mature cerebellum. J Neurosci 25: 2146–2156
29. Uemura T, Kakizawa T, Yamasaki M et al (2007) Regulation of long-term depression and climbing fiber territory by GluRδ2 at parallel fiber synapses through its carboxyl terminal domain in cerebellar Purkinje cells. J Neurosci 27: 12 096–12 108
30. Kakegawa W, Miyazaki T, Emi K et al (2008) Differential regulation of synaptic plasticity and cerebellar motor learning by the C-terminal PDZ-binding motif of GluRδ2. J Neurosci 28: 1460–0468
31. Kano M, Hashimoto K, Chen C et al (1995) Impaired synapse elimination during cerebellar development in PKCγ mutant mice. Cell 83: 1223–1231
32. Kano M, Hashimoto K, Kurihara H et al (1997) Persistent multiple climbing fiber innervation of cerebellar Purkinje cells in mice lacking mGluR1. Neuron 18: 71–79
33. Kano M, Hashimoto K, Watanabe M et al(1998) Phospholipase C®4 is specifically involved in climbing fiber synapse elimination in the developing cerebellum. Proc Natl Acad Sci USA 95: 15724–15729

34. Offermanns S, Hashimoto K, Watanabe M et al (1997) Impaired motor coordination and persistent multiple climbing fiber innervation of cerebellar Purkinje cells in mice lacking Gαq. Proc Natl Acad Sci USA 94: 14089–14094
35. Ichise T, Kano M, Hashimoto K et al (2000) mGluR1 in cerebellar Purkinje cells essential for long-term depression, synapse elimination, and motor coordination. Science 288: 1832–1835
36. Hirai H, Pang Z, Bao D et al (2005) Clbn1 is essential for synaptic integrity and information processing in the cerebellum. Nature Neurosci 8: 1534–1541
37. Ito-Ishida A, Miura E, Emi K et al (2008) Cbln1 regulates rapid formation and maintenance of excitatory synapses in mature cerebellar Purkinje cells in vitro and in vivo. J Neurosci 28: 5920–5930
38. Hertz L (1979) Functional interactions between neurons and astrocytes I. Turnover and metabolism of putative amino acid transmitters. Prog Neurobiol 13: 277–323
39. Choi DW (1992) Excitotoxic cell death. J Neurobiol 23: 1261–1276
40. Danbolt NC (2001) Glutamate uptake. Prog Neurobiol 65: 1–105
41. Tanaka K, Watase K, Manabe T et al (1997) Epilepsy and exacerbation of brain injury in mice lacking the glutamate transporter GLT-1. Science 276: 1699–1702
42. Watase K, Hashimoto K, Kano M et al (1998) Motor discoordination and increased susceptibility to cerebellar injury in GLAST mutant mice. Eur J Neurosci 10: 976–988
43. Katagiri H, Tanaka K, Manabe T (2001) Requirement of appropriate glutamate concentrations in the synaptic cleft for hippocampal LTP induction. Eur J Neurosci 14: 547–553
44. Voutsinos-Porche B, Bonvento G, Tanaka K et al (2003) Glial glutamate transporters mediate a functional metabolic crosstalk between neurons and astrocytes in the mouse developing cortex. Neuron 37: 275–286
45. Takayasu Y, Iino M, Kakegawa W et al (2005) Differential roles of glial and neuronal glutamate transporters in Purkinje cell synapses. J Neurosci 25: 8788–8793
46. Takayasu Y, Iino M, Shimamoto K et al (2006) Glial glutamate transporters maintain one-to-one relationship at the climbing fiber-Purkinje cell synapse by preventing glutamate spillover. J Neurosci 26: 6563–6572
47. Matsugami TR, Tanemura K, Mieda M et al (2006) Indispensability of the glutamate transporters GLAST and GLT1 to brain development. Proc Natl Acad Sci USA 103: 12 161–12 166
48. Takasaki C, Okada R, Mitani A et al (2008) Glutamate transporters regulate lesion-induced period plasticity in the developing somatosensory cortex. J Neurosci 28: 4995–5006
49. Wong-Riley M (1979) Changes in the visual system of monocularly sutured or enucleated cats demonstrable with cytochrome oxidase histochemistry. Brain Res 171: 11–28
50. Finney DJ (1971) Probit analysis, 3rd edn. Cambridge University Press, Cambridge, UK, p 333
51. Miura E, Fukaya M, Sato T et al (2006) Expression and distribution of JNK/SAPK-associated scaffold protein JSAP1 in developing and adult mouse brain. J Neurochem 97: 1431–1446
52. Miyazaki T, Fukaya M, Shimizu H et al (2003) Subtype switching of vesicular glutamate transporters at parallel fibre-Purkinje cell synapses in developing mouse cerebellum. Eur J Neurosci 17: 2563–2572
53. Yamada K, Fukaya M, Shibata T et al (2000) Dynamic transformation of Bergmann glial fibers proceeds in correlation with dendritic outgrowth and synapse formation of cerebellar Purkinje cells. J Comp Neurol 418: 106–120
54. Kidd FL, Isaac JT (2000) Glutamate transport blockade has a differential effect on AMPA and NMDA receptor-mediated synaptic transmission in the developing barrel cortex. Neuropharmacology 39: 725–732
55. Demarque M, Villeneuve N, Manent JB et al (2004) Glutamate transporters prevent the generation of seizures in the developing rat neocortex. J Neurosci 24: 3289–3294
56. Cattani AA, Bonfardin VD, Represa A et al (2007) Generation of slow network oscillations in the developing rat hippocampus after blockade of glutamate uptake. J Neurophysiol 98: 2324–2323

57. Milh M, Becq H, Villeneuve N et al (2007) Inhibition of glutamate transporters results in a "suppression-burst" pattern and partial seizures in the newborn rat. Epilepsia 48: 169–174
58. Kano M, Rexhausen U, Dreessen J et al (1992) Synaptic excitation produces a long-lasting rebound potentiation of inhibitory synaptic signals in cerebellar Purkinje cells. Nature 356: 601–604
59. Konnerth A, Dreessen J Augustine G.J (1992) Brief dendritic calcium signals initiate long-lasting synaptic depression in cerebellar Purkinje cells. Proc Natl Acad Sci USA 89: 7051–7055
60. Regehr WG Mintz IM (1994) Participation of multiple calcium channel types in transmission at single climbing fiber to Purkinje cell synapses. Neuron 12: 605–613
61. Lehre KP, Danbolt NC (1998) The number of glutamate transporter subtype molecules at glutamatergic synapses: chemical and stereological quantification in young adult rat brain. J Neurosci 18: 8751–8757
62. Takatsuru Y, Takayasu Y, Iino M et al (2006) Roles of glial glutamate transporters in shaping EPSCs at the climbing fiber-Purkinje cell synapses. Neurosci Res 54: 140–148
63. Bliss TV, Collingridge GL (1993) A synaptic model of memory: long-term potentiation in the hippocampus. Nature 361: 31–39
64. Aiba A, Kano M, Chen C et al (1994) Deficient cerebellar long-term depression and impaired motor learning in mGluR1 mutant mice. Cell 79: 377–388
65. Conquet F, Bashir ZI, Davies CH et al (1994) Motor deficit and impairment of synaptic plasticity in mice lacking mGluR1. Nature 372: 237–243
66. Bear MF (1996) A synaptic basis for memory storage in the cerebral cortex. Proc Natl Acad Sci USA 93: 13453–13459

Molecular and Functional Imaging for Drug Development and Elucidation of Disease Mechanisms Using Positron Emission Tomography (PET)

Manabu Tashiro[1], Toshihiko Fujimoto[2], Nobuyuki Okamura[4], Ren Iwata[3], Hiroshi Fukuda[5], and Kazuhiko Yanai[1,4]

Summary

Tohoku University has a more than 30-year-long history of molecular and functional imaging research using radiopharmaceuticals. This article provides a brief overview of various achievements in molecular and functional imaging at Tohoku University. It is noteworthy that many of these early studies were associated with positron emission tomography (PET) studies in oncology and neuroscience. Later, new application to sports sciences was initiated, and PET and [^{18}F]fluorodeoxyglucose ([^{18}F]FDG) has been used for exercise physiology and psychology studies. This technique, similar to the performance of autoradiography, allows subjects to be scanned just after carrying out exercise tasks. We have observed the metabolic effects of exercise on brain and skeletal muscles.

One of the important contributions we have made in neuroscience has been associated with the histaminergic neuronal system in the brain. The histaminergic system is associated with various autonomic functions such as the sleep-wake cycle and appetite control. Using PET and [^{11}C]doxepin, a ligand for histamine H1 receptors, various studies have been conducted regarding physiological changes such as those occurring in aging, and pathological changes such as those occurring in Alzheimer's disease (AD), depression, schizophrenia, and anorexia nervosa. In addition, PET and [^{11}C]doxepin has also been used for the evaluation of side effects due to histamine H1 receptor antagonists (antihistamines). Antihistamines are frequently used for the treatment of allergic disorders such as seasonal rhinitis, but these drugs can induce sedative side effects that can sometimes result in serious

[1]Division of Cyclotron Nuclear Medicine, Cyclotron and Radioisotope Center, Tohoku University, 6-3 Aoba, Aramaki, Aoba-ku, Sendai 980-8578, Japan
[2]Division of Radiopharmaceutical Chemistry, Cyclotron and Radioisotope Center, Tohoku University, 6-3 Aoba, Aramaki, Aoba-ku, Sendai 980-8578, Japan
[3]Center for the Advancement of Higher Education, Tohoku University, 41 Kawauchi, Aoba-ku, Sendai 980-8576, Japan
[4]Department of Pharmacology, Tohoku University Graduate School of Medicine, 2-1 Seiryo-machi, Aoba-ku, Sendai 980-8575, Japan
[5]Institute of Development, Aging and Cancer, Tohoku University, 4-1 Seiryo-machi, Aoba-ku, Sendai 980-8575, Japan

N. Tamaki, Y. Kuge (eds.)
Molecular Imaging for Integrated Medical Therapy and Drug Development.
DOI: 10.1007/978-4-431-98074-2_23, ©Springer 2010

traffic accidents. Objective measurement of the sedative property of antihistamines was established using histamine H1 receptor occupancy as a reliable index. Our additional functional imaging studies using [^{18}F]FDG and [^{15}O]H_2O have revealed the brain mechanisms of these sedative side effects in the brains of allergic patients.

Finally, we have been involved in the development of novel tracers for amyloid deposits in the brains of patients with AD and mild cognitive disorder. We have been conducting clinical studies of an ^{11}C-labelled tracer, [^{11}C]2-(2-[2-dimethylaminothiazol-5-yl]ethenyl)-6-(2-[fluoro]ethoxy)benzoxazole (BF-227). Our early evaluation has demonstrated that [^{11}C]BF-227 is a promising tracer for the differentiation of aged normal volunteers and AD patients. In future, PET will undoubtedly be used more frequently in drug development and for the early diagnosis of various diseases.

Key word [^{18}F]fluorodeoxyglucose (FDG) · sports medicine · [^{11}C]doxepin · histamine H1 receptor · [^{11}C]BF-227 · amyloid imaging

Introduction to Molecular and Functional Imaging

Molecular and functional imaging with nuclear medicine techniques has been useful to investigate endophenotypic alterations in a living human or animal body externally without perturbing biological phenomena. This nuclear medicine technique was founded in the early twentieth century, when a tracer technique was developed by Dr. George von Hevesy, a Nobel laureate in chemistry, in 1943. This technique was later applied to scintigraphy for human subjects, and was fused with computed tomography technology, positron emission tomography (PET), and single photon emission computed tomography (SPECT) established in the late twentieth century. Now this field, established as nuclear medicine, has been actively used for molecular and functional imaging.

Using PET as a tool for molecular and functional imaging, we can measure cerebral energy (glucose) consumption by injecting [^{18}F]fluorodeoxyglucose ([^{18}F]FDG). Increased regional brain activity may result in increased demands for glucose and oxygen, inducing dilation of brain capillaries. This can be accompanied by an increase in regional cerebral perfusion, which can be measured using PET and radiolabeled water ([^{15}O]H_2O), and other methods such as functional magnetic resonance imaging (fMRI) and near-infrared spectroscopy (NIRS). Nowadays, PET is more often used for measuring regional brain glucose consumption and for evaluating neurotransmission function. In the human brain, neurotransmitters can exert their actions even in very small amounts. It is not easy to visualize the actions of neurotransmitters in the living human brain externally without using a highly sensitive technique such as PET. In addition, PET enables the quantification of interactions between neurotransmitters and neuroreceptors. PET has been used for more than 30 years as a tool for molecular and functional imaging at Tohoku University.

Research and Education in Molecular and Functional Imaging at Tohoku University

In 1977, the Cyclotron and Radioisotope Center (CYRIC) was founded at Tohoku University for the multipurpose use of a cyclotron, including nuclear physics, nuclear chemistry, solid-state physics, and radioisotope production for biology and medicine. Clinical studies were started at CYRIC in 1982 following the first installation of a PET scanner in 1981. In 1976, the production of [^{18}F]FDG had already been successfully done, initially for brain research at CYRIC, and the use of [^{18}F] FDG was later applied to oncology research. The challenge of using [^{18}F]FDG for cancer imaging was based on Matsuzawa's concept of "fishing for cancer with baits." He claimed that "cancer cells were always hungry and almost blind, therefore they try to eat anything including false sugars, amino acids, and nucleic acids." Soon, Fukuda and colleagues reported the applicability of [^{18}F]FDG PET for diagnosing metastatic liver cancer, in 1982 [1]. Later, in 1983, [^{11}C]methionine was proven to be useful for lung cancer detection [2]. In addition, [^{18}F]fluorodeoxygalactose was also demonstrated to be useful for cancer imaging, in 1986 [3] (Fig. 1). The first international symposium on PET oncology was held in 1985, and soon after PET oncology became known worldwide. Recently, a new tracer for oncology imaging, [^{18}F]O-fluoro-methyl-tyrosine (FMT), has been examined for its

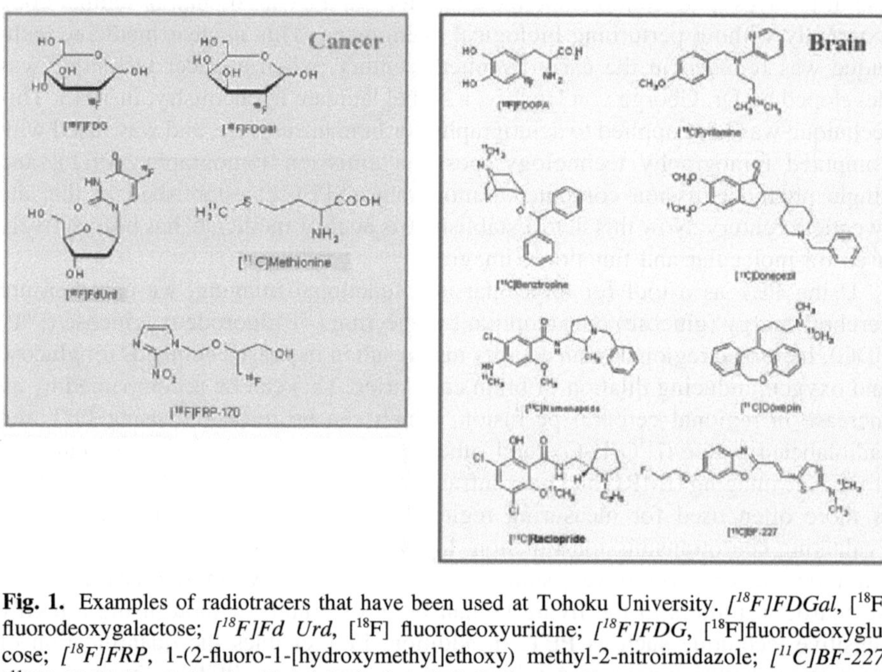

Fig. 1. Examples of radiotracers that have been used at Tohoku University. *[^{18}F]FDGal*, [^{18}F] fluorodeoxygalactose; *[^{18}F]Fd Urd*, [^{18}F] fluorodeoxyuridine; *[^{18}F]FDG*, [^{18}F]fluorodeoxyglucose; *[^{18}F]FRP*, 1-(2-fluoro-1-[hydroxymethyl]ethoxy) methyl-2-nitroimidazole; *[^{11}C]BF-227*, [^{11}C]2-(2-[2-dimethylaminothiazol-5-yl]ethenyl)-6-(2-[fluoro]ethoxy)benzoxazole. Tohoku University has carried out human positron emission tomography (PET) studies for more than 25 years, since 1982

usefulness for oncology diagnosis [4]. [^{18}F]1-(2-fluoro-1-[hydroxymethyl]ethoxy) methyl-2-nitroimidazole ([^{18}F]FRP-170) has been synthesized for imaging hypoxic cells resistant to radiation therapy; this tracer has been applied to cancer patients to determine its usefulness for diagnosing hypoxic malignant cells that tend to be resistant to radiation therapy [5].

For brain research, we have mainly used [^{18}F]fluoro-L-DOPA (dopamine metabolism) [6, 7], [^{11}C]YM9151–2 and [^{11}C]nemonapride (dopamine D2 receptors) [6, 8], [^{11}C]benztropine (muscarinic acetylcholine receptors) [9], and [^{11}C]doxepin (histamine H1 receptors; H1R) [10] for human brain studies. Brain activation studies during running [11] and car-driving [12] have also been initiated. Various new tracers have been introduced, such as [^{11}C]BF-227, for imaging beta-amyloid deposition [13], and [^{11}C]donepezil, for evaluating the function of acetylcholinergic nerves [14]. [^{11}C]BF-227 has recently attracted the attention of many Japanese investigators as the first domestic probe for the clinical examination of AD patients. [^{11}C]donepezil has also been used previously in clinical examinations at Tohoku University [14] (Fig. 1).

Since 2006, Tohoku University has managed the "Molecular Imaging Program," including an educational course for postgraduate students. This program has been organized in collaboration with the National Institute of Radiological Science (Chiba, Japan). In this program, CYRIC has been playing important roles in performing experiments and in further education. Below, we outline our recent achievements in molecular and functional imaging research (Fig. 2).

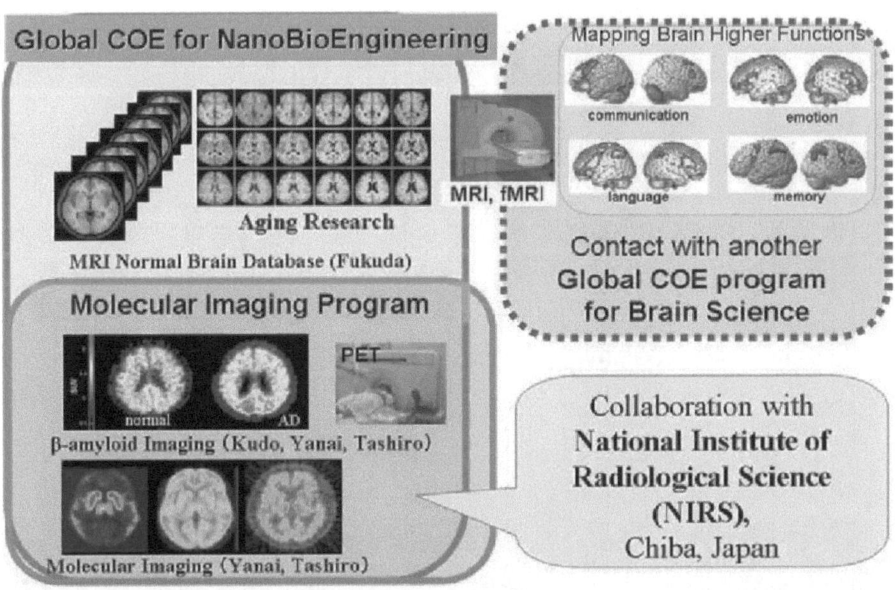

Fig. 2. Organizations and projects of Tohoku University's current work in molecular and functional imaging and brain research. *COE*,; *MRI*, magnetic resonance imaging; *fMRI*, functional magnetic resonance imaging; *AD*, Alzheimer's disease; *SUV*, standardized uptake value

Exercise and Brain Imaging Related to Physical and Mental Health

In the scope of applying PET to health promotion science, we have performed imaging studies in subjects during various forms of exercise such as running and bicycle riding. We have scanned not only the brain [11, 15] but also the skeletal [15, 16] and cardiac muscles [17]. By conducting whole-body scanning, we can obtain a whole-body map of energy metabolism in the living human body [16] (Fig. 3). Regional cerebral metabolic changes induced by exercise have been examined in animals with an autoradiography technique using [^{14}C]deoxyglucose ([^{14}C]2-DG) as a tracer, because it did not require the simultaneous scanning of subjects during exercise [18, 19]. These studies provided the first functional indices of brain activity with respect to exercise. Using this technique, Sharp [20] demonstrated a regional increase in glucose uptake in the cerebellar vermis of swimming rats. Human studies were later conducted by Herholz and coworkers [21] in the late 1980s, first using a ^{133}Xe clearance method for studying regional changes in brain activity. Later, Fink and coworkers [22] demonstrated regional activation during and immediately after an ergometer task by PET using [^{15}O]H_2O. In addition, Mishina and coworkers [23] applied [^{18}F]FDG PET to the neuropathological evalu-

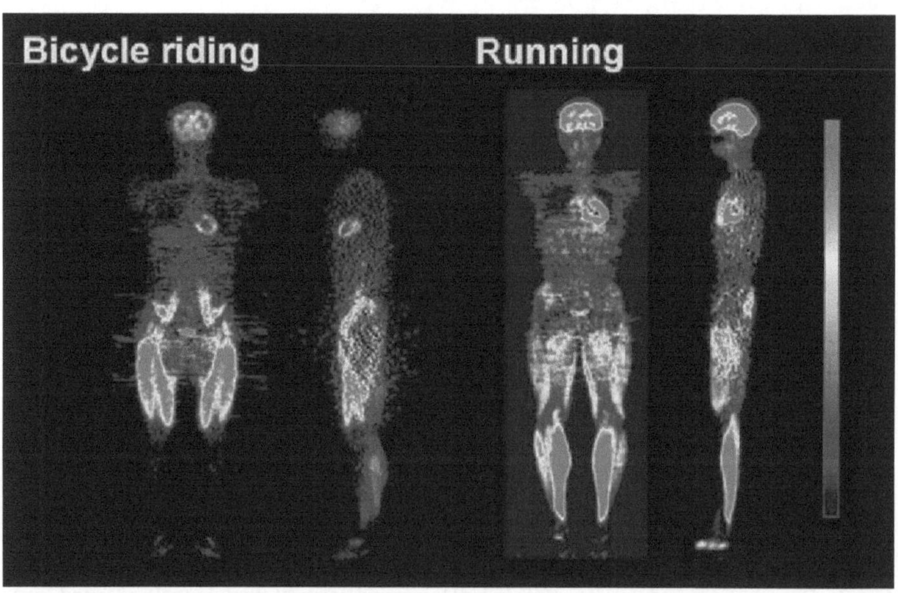

Fig. 3. An example of [^{18}F]FDG PET use in exercise physiology. A clear difference can be seen between bicycle riding (*left*) and running (*right*).

ation of patients with olivo-pontine-cerebellar atrophy manifesting gait disturbances; these patients exhibited a decreased response to a walking task in the cerebellar vermis compared with normal subjects.

We first applied [^{18}F]FDG PET to healthy human subjects undertaking a running task in the upright posture (Fig. 3). We demonstrated a relative increase in glucose uptake in the temporo-parietal association cortex, occipital cortex, premotor cortex, primary sensorimotor cortex, and the cerebellar vermis [24]. This was probably due to the higher energy consumption necessary for integrating multimodal sensory inputs. A relative reduction in glucose uptake was detected in the prefrontal cortex, temporal cortex, cerebellar hemisphere, brain stem, and striatum. The mean value of global brain glucose uptake was relatively lower in the runners than in resting controls [24]. Kemppainen and coworkers [25] later demonstrated a significant reduction of the regional glucose metabolic rate in all cortical regions in correlation with exercise intensity. They also pointed out that exercise could be associated with adaptive metabolic changes in the frontal cortex [25]. Thus, global and regional brain metabolic decline, especially in the limbic and frontal regions, was observed using [^{18}F]FDG PET [24, 25]. It is easy to explain the metabolic increase in the regions directly associated with the execution of exercise tasks, while it is not so easy to explain the mechanism of the relative decrease in the regions not involved in exercise. Previous imaging studies in patients with anxiety disorders demonstrated increased glucose metabolism in the limbic and frontal regions [26, 27]. We speculated that the metabolic reduction in the frontal and limbic regions in runners was associated with emotional changes, including the phenomenon called "runner's high" [28].

Dietrich and Sparling [29] reported that endurance exercise impaired prefrontal-dependent cognitive ability in healthy volunteers. Dietrich [30] later proposed a new theory (transient hypofrontality theory; THT) to explain the metabolic reduction in the prefrontal region, where the prefrontal activity is suppressed indirectly due to the limitation in energy supply to the brain. Interestingly, this theory also explains the neural mechanism regarding the mental health benefits of exercise [29, 30].

Imaging Study for Improvement of Quality of Life (QOL) in Patients with Cancer

Functional imaging such as PET has also been used for studies of psychiatric disorders in patients with physical diseases such as malignant tumors. Because cancer and cancer treatments have various effects on the central nervous system, the diagnosis of psychiatric symptoms in cancer patients is, in part, problematic. Molecular and functional imaging could be used as a supplementary diagnostic tool. Previously, we have proposed the use of [^{18}F]FDG PET in the neuropsychiatric evaluation of cancer patients, and we have performed a series of studies to examine whether or not the images of a cancer patient's brain are normal. It is now widely accepted that psychological factors are as important as external factors in disease progression. Thus, psychological evaluation and patient care are very important, not only for improving quality of life (QOL) but also for prolonging survival. If we assume that a certain psychological disturbance truly exists in a patient, to the extent of affecting systemic function, it is also possible for their brain activity to have significant alterations. Thus, it seems reasonable to think that cancer patients may manifest a corresponding abnormality in their functional brain images. Our preliminary work has demonstrated regional hypometabolic findings, mainly in the prefrontal cortex, limbic structures, and striatum, in Japanese cancer patients [31, 32]. Common findings in these studies were hypometabolism in the prefrontal cortex, anterior and posterior cingulate, insular cortex, and striatum. These regions were similar to those demonstrated in previous neuroimaging studies of patients with major depression showing commonly repeated findings of hypometabolism in the prefrontal cortex, anterior cingulate gyrus, and basal ganglia. Our additional cross-sectional study in Japanese cancer patients showed that the depth of hypometabolic findings tended to fluctuate in some regions and constantly decreased in other regions [33]. Finally, we demonstrated that regional hypometabolism in the prefrontal cortex was negatively correlated with the subjective measure of depression in cancer patients. Nowadays, mild cognitive impairment after chemotherapy, so-called chemo-fog, has been studied intensively in association with QOL [34].

Molecular and Functional Imaging of the Histaminergic Nervous System

Psychosocial stress has been associated with increased rates of various psychiatric disorders, such as major depression, schizophrenia, anxiety, and eating disorders in our daily lives. The incidence of cognitive disorders is also increasing. We have recently conducted a variety of studies to elucidate the pathophysiological mechanisms of psychiatric disorders mentioned above [35–39], putting emphasis on alterations in neural transmission in the histaminergic neuronal system [35]. For this purpose, [^{11}C]doxepin is the tracer of choice for imaging histamine H1 receptors (H1Rs). Using PET and [^{11}C]doxepin, a large amount of evidence has been accumulated regarding the role of the histaminergic neuronal system in the pathophysiology of these disorders. In our recent studies, histamine H1 receptor binding was measured, by PET and [^{11}C]doxepin, in 10 normal male subjects and 10 patients with schizophrenia [36], 10 patients with major depression [37], and 10 patients with AD [39], respectively, as well as in 10 normal female subjects and 12 female patients with anorexia nervosa [38]. In these studies, significant reductions in H1 receptor binding were observed in the patients with schizophrenia, major depression, and AD, while a significant increase was observed in the patients with anorexia nervosa [35–39].

In addition, PET and [^{11}C]doxepin is also useful for the evaluation of drug-induced side effects and the elucidation of their mechanism [35]. Some of the most frequently used therapeutic drugs for allergies such as seasonal pollinosis, or hay fever, are H1R antagonists (antihistamines). There are many available antihistamines, but some of them have sedative side effects. Therefore, it is important to develop an objective and reliable method for measuring the strength of such sedative side effects [35, 40, 41]. To date, we have studied the mechanism of functional suppression in signal transmission through H1Rs in the brain. Usually, antihistamines are used to suppress the actions of mast cells in the peripheral blood and to control allergic reactions. However, some of these drugs may enter the brain and suppress the signal transmission of intracerebral H1Rs. As a result, it becomes difficult to maintain arousal (sedative effects), and sometimes these drugs may cause us to make mistakes during work or while driving, resulting in decreased work efficiency or traffic accidents. Considering such a background, the objective measurement of the sedative effects of these drugs becomes very important.

We have succeeded in quantifying the strength of the sedative effects of antihistamines in terms of H1R occupancy (rate) in the brain using PET, and we have measured this clinically. Previously, to evaluate drug sedative effects, investigators have used macroscopic behavioral techniques such as the measurement of psychomotor performance, including psychomotor speed and accuracy, as well as the measurement of subjective sleepiness, using many volunteer subjects [35, 40]. Recently, we conducted a clinical test to evaluate the sedative profile of bepotastine besilate, a new antihistamine developed in Japan [40]. The basic pharmaceutical classification of this antihistamine has been "a mildly sedative antihistamine". We succeeded in obtaining supporting data regarding this classification using PET [35] (Fig. 4).

From the viewpoint of cognitive neuroscience, the category of cognitive function (psychomotor performance) measured in antihistamine studies is mainly "vigilance and attention." The sedative side effects of antihistamines have been recognized to be potentially dangerous in our daily tasks such as car driving, but the mechanism underlying these effects has not yet been elucidated well. We, therefore, attempted to elucidate the brain mechanism of impaired performance by using a car-driving simulator and [^{15}O]H_2O PET [42]. We examined regional cerebral blood flow (rCBF) responses during a simulated car-driving task following the oral administration of D-chlorpheniramine, using [^{15}O]H_2O PET. The results of the performance evaluation revealed that part of the driving performance (lane deviation) significantly increased in the D-chlorpheniramine condition compared with the placebo condition, while subjective sleepiness was not significantly different between the two drug conditions. In addition, brain imaging analysis suggested that D-chlorpheniramine tended to suppress the regional brain activities associated with visuo-spatial cognition and visuo-motor coordination [42].

Thus, nuclear medicine techniques are very useful for the objective evaluation of the intensities and mechanisms of the effects of various drugs on the brain. These data can be used for the development of new drugs with reduced side effects.

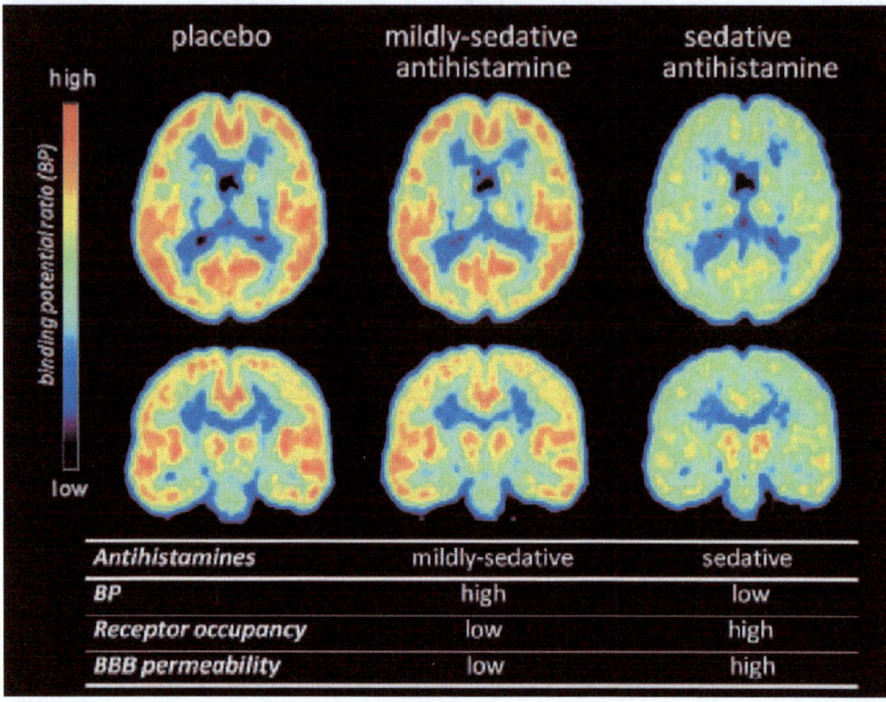

Fig. 4. Binding potential ratio (BPR) images of [^{11}C]doxepin anatomically-normalized for the standard brain space. BPR of [^{11}C]doxepin reflects permeability of the blood-brain barrier (BBB) and histamine H1 receptor occupancy. A mildly-sedative antihistamine tends to manifest high BPR, low receptor occupancy and low BBB permeability. A sedative antihistamine tends to manifest low BPR, high receptor occupancy and high BBB permeability.

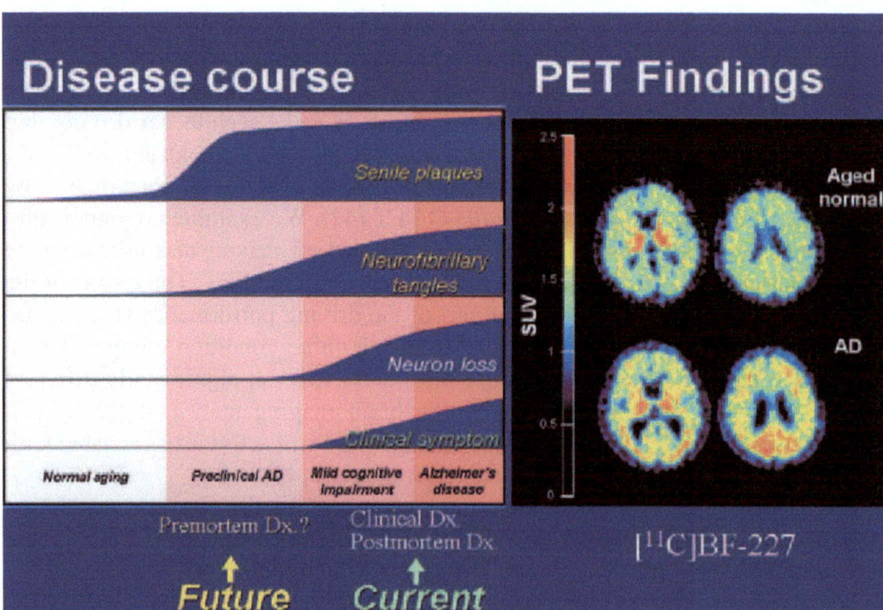

Fig. 5. Diagram showing the disease course of Alzheimer's disease (*AD*) and PET findings of [^{11}C]BF-227 in an aged normal subject and an AD patient. The aim of the amyloid imaging project is to establish a system to predict a possible converter from mild cognitive impairment to AD. *Dx.*, diagnosis

Imaging for Pathophysiological Evaluation and Early Diagnosis of Alzheimer's Disease (AD)

In recent years, the number of patients with dementia has been increasing. The deposition of senile plaques (SPs) in the brain has been demonstrated to represent the neuropathological characteristics of AD [13]. The progressive accumulation of SPs has been considered to be fundamental to the initial development of AD. The deposition of SPs in the brain is present even in very mild AD and precedes the initiation of cognitive impairment at the clinical level [43, 44]. That is why noninvasive imaging of SP deposition would be useful for the early detection of AD. In addition, several anti-amyloid drugs are already under development for the treatment and prevention of AD [45]. This could be another good motivation for the promotion of the noninvasive imaging of SP deposition. Though several imaging techniques have been developed for this purpose, PET has been the most advanced and practical method for the in vivo evaluation of SP deposition. At present, 2-(4′methylaminophenyl)benzothiazole (6OH-BTA-1) or Pittsburgh Compound B (PIB) is the most successful PET agent for in vivo amyloid imaging and it has been proven to be useful for the early diagnosis of AD [46–48].

We have demonstrated a novel series of benzoxazole derivatives as promising candidates for in vivo imaging probes of SPs [49, 50]. These derivatives have shown comparatively high blood-brain barrier (BBB) permeability, high binding affinity for Aβ aggregates, and high specificity for fibrillar Aβ deposits. These characteristics have been suggested to be potential advantages for the early detection of AD-related pathological changes. We have introduced a new derivative, 2-(2-[2-dimethylaminothiazol-5-yl]ethenyl)-6-(2-[fluoro]ethoxy)benzoxazole (BF-227), as a new candidate for a PET probe for the in vivo detection of dense amyloid deposits in AD patients. The binding property of BF-227 to SP was evaluated by neuropathological staining of AD brain sections, and by in vivo measurement in 11 normal healthy subjects and in 10 AD patients, using PET and [^{11}C]BF-227. The clinical PET study demonstrated the retention of [^{11}C]BF-227 in the cerebral cortices of AD patients but not in those of the normal subjects. All AD patients were clearly distinguishable from healthy subjects, using the temporal standardized uptake value (SUV) ratio normalized by the uptake in the cerebellum (Fig. 5). Voxel-by-voxel analysis of the PET images revealed that cortical [^{11}C]BF-227 retention in the AD patients was distributed primarily to the posterior association area of the brain and corresponded well with the preferred site for SP deposition. These findings suggest that [^{11}C]BF-227 is a promising PET probe for the in vivo detection of dense amyloid deposits in AD patients.

Thus, PET is a promising tool for the evaluation and early diagnosis of various neurological disorders.

Acknowledgments The authors thank all the staff of the Cyclotron and Radioisotope Center, Tohoku University, for their support during the study. The molecular imaging course and some of the studies outlined here were partly supported by a Japan Science and Technology Agency (JST) Grant for Research and Education in Molecular Imaging.

References

1. Fukuda H, Matsuzawa T, Abe Y et al (1982) Experimental study for cancer diagnosis with positron-labeled fluorinated glucose analogs: [18F]-2-fluoro-2-deoxy-D-mannose: a new tracer for cancer detection. Eur J Nucl Med 7: 294–297
2. Kubota K, Ito M, Fukuda H et al (1983) Cancer diagnosis with positron computed tomography and carbon-11-labelled L-methionine. Lancet 2: 1192
3. Fukuda H, Matsuzawa T, Tada M et al (1986) 2-Deoxy-2-[18F]fluoro-D-galactose: a new tracer for the measurement of galactose metabolism in the liver by positron emission tomography. Eur J Nucl Med 11: 444–448
4. Suzuki M, Yamaguchi K, Honda G et al (2005) An experimental study on O-[18F]fluoromethyl-L-tyrosine for differentiation between tumor and inflammatory tissues. Ann Nucl Med 19: 589–595
5. Kaneta T, Takai Y, Iwata R et al (2007) Initial evaluation of dynamic human imaging using 18F-FRP170 as a new PET tracer for imaging hypoxia. Ann Nucl Med 21: 101–107
6. Nagasawa H, Kogure K, Ido T (1993) Simultaneous demonstration of neurotransmitter and receptor systems of the rat brain using in vivo double autoradiography. Tohoku J Exp Med 169: 87–89
7. Meguro K, Yamaguchi S, Itoh M et al (1997) Striatal dopamine metabolism correlated with frontotemporal glucose utilization in Alzheimer's disease: a double-tracer PET study. Neurology 49: 941–945
8. Nakajima T, Nimura T, Yamaguchi K et al (2003) The impact of stereotactic pallidal surgery on the dopamine D2 receptor in Parkinson disease: a positron emission tomography study. J Neurosurg 98: 57–63
9. Ono S, Kawashima R, Ito H et al (1996) Regional distribution of the muscarinic cholinergic receptor in the human brain studied with 11C-benztropine and PET using an anatomical standardization technique (in Japanese). Kaku Igaku (The Japanese Journal of Nuclear Medicine) 33: 721–727
10. Yanai K, Watanabe T, Yokoyama H et al (1992) Histamine H1 receptors in human brain visualized in vivo by [11C]doxepin and positron emission tomography. Neurosci Lett 137: 145–148
11. Tashiro M, Ota H, Fujimoto T et al (1998) Functional analysis of runner's brain by 18F-FDG and PET during field running. In: Nose H, Nadel E, Morimoto T (eds) The 1997 Nagano Symposium on Sports Sciences., Cooper Publishing, Carmel, USA, pp 178–184
12. Jeong M, Tashiro M, Singh LN et al (2006) Functional brain mapping of actual car-driving using [18F]FDG-PET. Ann Nucl Med 20: 623–628
13. Kudo Y, Okamura N, Furumoto S et al (2007) 2-(2-[2-Dimethylaminothiazol-5-yl]ethenyl)-6-(2-[fluoro]ethoxy)benzoxazole: a novel PET agent for in vivo detection of dense amyloid plaques in Alzheimer's disease patients. J Nucl Med 48: 553–561
14. Okamura N, Funaki Y, Tashiro M et al In vivo visualization of donepezil binding in the brain of patients with Alzheimer's disease. Br J Clin Pharmacol (in press)
15. Tashiro M, Itoh M, Fujimoto T et al (2008) Application of positron emission tomography to neuroimaging in sports sciences. Methods 45: 300–306
16. Fujimoto T, Itoh M, Kumano H et al (1996) Whole-body metabolic map with positron emission tomography of a man after running. Lancet 348: 266
17. Iemitsu M, Itoh M, Fujimoto T et al (2001) Regional cardiac glucose metabolism during running measured by 3D positron emission tomography in humans. Advances in Exercise and Sports Physiology 7: 53–58
18. Reivich M, Sokoloff L, Shapiro H et al (1974) Validation of an autoradiographic method for the determination of the rates of local cerebral glucose utilization. Trans Am Neurol Assoc 99: 238–240
19. Schwartzman RJ, Greenberg J, Revich M et al (1981) Functional metabolic mapping of a conditioned motor task in primates utilizing 2-[14C]deoxyglucose. Exp Neurol 72: 153–163

20. Sharp FR (1976) Relative cerebral glucose uptake of neuronal perikarya and neuropil determined with 2-deoxyglucose in resting and swimming rat. Brain Res 110: 127–139
21. Herholz K, Buskies W, Rist M et al (1987) Regional cerebral blood flow in man at rest and during exercise. J Neurol 234: 9–13
22. Fink GR, Adams L, Watson JD et al (1995) Hyperpnoea during and immediately after exercise in man: evidence of motor cortical involvement. J Physiol 489: 663–675
23. Mishina M, Senda M, Ishii K et al (1999) Cerebellar activation during ataxic gait in olivopontocerebellar atrophy: a PET study. Acta Neurol Scand 100: 369–376
24. Tashiro M, Itoh M, Fujimoto T et al (2001) 18F-FDG PET mapping of regional brain activity in runners. J Sports Med Phys Fitness 41: 11–17
25. Kemppainen J, Aalto S, Fujimoto T et al (2005) High intensity exercise decreases global brain glucose uptake in humans. J Physiol 568: 323–332
26. Baxter LR (1990) Brain imaging as a tool in establishing a theory of brain pathology in obsessive compulsive disorder. J Clin Psychiatry 51 (Suppl):22–5; discussion 26
27. Perani D, Colombo C, Bressi S et al (1995) [18F]FDG PET study in obsessive-compulsive disorder. A clinical/metabolic correlation study after treatment. Br J Psychiatry 166: 244–250
28. Boecker H, Sprenger T, Spilker ME et al (2008) The runner's high: opioidergic mechanisms in the human brain. Cereb Cortex 18: 2523–2531
29. Dietrich A, Sparling PB (2004) Endurance exercise selectively impairs prefrontal-dependent cognition. Brain Cogn 55: 516–524
30. Dietrich A (2006) Transient hypofrontality as a mechanism for the psychological effects of exercise. Psychiatry Res 145: 79–83
31. Tashiro M, Kubota K, Itoh M et al (1999) Hypometabolism in the limbic system of cancer patients observed by positron emission tomography. Psychooncology 8: 283–286
32. Tashiro M, Juengling FD, Reinhardt MJ et al (2000) Reproducibility of PET brain mapping of cancer patients. Psychooncology 9: 157–163
33. Tashiro M, Juengling FD, Reinhardt MJ et al (2001) Depressive state and regional cerebral activity in cancer patients—a preliminary study. Med Sci Monit 7: 687–695
34. Matsuda T, Takayama T, Tashiro M et al (2005) Mild cognitive impairment after adjuvant chemotherapy in breast cancer patients—evaluation of appropriate research design and methodology to measure symptoms. Breast Cancer 12: 279–287
35. Yanai K, Tashiro M (2007) The physiological and pathophysiological roles of neuronal histamine: an insight from human positron emission tomography studies. Pharmacol Ther 113: 1–15
36. Iwabuchi K, Ito C, Tashiro M et al (2005) Histamine H1 receptors in schizophrenic patients measured by positron emission tomography. Eur Neuropsychopharmacol 15: 185–191
37. Kano M, Fukudo S, Tashiro A et al (2004) Decreased histamine H1 receptor binding in the brain of depressed patients. Eur J Neurosci 20: 803–810
38. Yoshizawa M, Tashiro M, Fukudo S et al (2008) Increased brain histamine H1 receptor binding in patients with anorexia nervosa. Biol Psychiatry
39. Higuchi M, Yanai K, Okamura N et al (2000) Histamine H(1) receptors in patients with Alzheimer's disease assessed by positron emission tomography. Neuroscience 99: 721–729
40. Tashiro M, Duan X, Kato M et al (2008) Brain histamine H1 receptor occupancy of orally administered antihistamines, bepotastine and diphenhydramine, measured by PET with 11C-doxepin. Br J Clin Pharmacol 65: 811–821
41. Tashiro M, Sakurada Y, Iwabuchi K et al (2004) Central effects of fexofenadine and cetirizine: measurement of psychomotor performance, subjective sleepiness, and brain histamine H1-receptor occupancy using 11C-doxepin positron emission tomography. J Clin Pharmacol 44: 890–900
42. Tashiro M, Sakurada Y, Mochizuki H et al (2008) Effects of a sedative antihistamine, D-chlorpheniramine, on regional cerebral perfusion and performance during simulated car driving. Hum Psychopharmacol 23: 139–150
43. Goldman WP, Price JL, Storandt M et al (2001) Absence of cognitive impairment or decline in preclinical Alzheimer's disease. Neurology 56: 361–367

44. Price JL, Morris JC (1999) Tangles and plaques in nondemented aging and "preclinical" Alzheimer's disease. Ann Neurol 45: 358–368
45. Aisen PS (2005) The development of anti-amyloid therapy for Alzheimer's disease: from secretase modulators to polymerisation inhibitors. CNS Drugs 19: 989–996
46. Klunk WE, Engler H, Nordberg A et al (2004) Imaging brain amyloid in Alzheimer's disease with Pittsburgh Compound-B. Ann Neurol 55: 306–319
47. Price JC, Klunk WE, Lopresti BJ et al (2005) Kinetic modeling of amyloid binding in humans using PET imaging and Pittsburgh Compound-B. J Cereb Blood Flow Metab 25: 1528–1547
48. Lopresti BJ, Klunk WE, Mathis CA et al (2005) Simplified quantification of Pittsburgh Compound B amyloid imaging PET studies: a comparative analysis. J Nucl Med 46: 1959–1972
49. Okamura N, Suemoto T, Shimadzu H et al. (2004) Styrylbenzoxazole derivatives for in vivo imaging of amyloid plaques in the brain. J Neurosci 24: 2535–2541
50. Okamura N, Suemoto T, Shiomitsu T et al (2004) A novel imaging probe for in vivo detection of neuritic and diffuse amyloid plaques in the brain. J Mol Neurosci 24: 247–255

Part IV
Recent Trends in Drug Developments

The Sea as a Source of New Drugs

Fernando Albericio[1,2,3], Mercedes Álvarez[1,2,4], Carmen Cuevas[5], Andres Francesch[5], Daniel Pla[1,2], and Judit Tulla-Puche[1,2]

Summary

The role of natural products in drug discovery has suffered ups and downs during the past few years. Recently, and as the number of classical drugs has not been increasing greatly, it seems that pharmaceutical companies are looking again towards nature as an inspiration for new drugs. Especially interesting has been the fact that during the past few years the first drugs based on natural marine products have been introduced in the market, showing the enormous potential of the sea as a source of drugs. One of the main problems associated with the use of natural products as therapeutics is their poor pharmacokinetic properties. A strategy for improving these properties is through their chemical modification, which is exemplified herein by work on such structural edition of two natural products (thiocoraline and lamellarin) currently being carried out at our laboratories.

Key words Drug discovery · Heterocyclic compounds · Marine products · Natural products · Peptides · Solid-phase synthesis

Drug Discovery, Some Remarks

The roles of the pharmaceutical industry and medicinal chemists are currently changing [1]. Thus, the industry is moving forward to the so-called "extended enterprise," which is defined as a dynamic networked organization, based on out-

[1]Institute for Research in Biomedicine, Barcelona Science Park, University of Barcelona, Baldiri Reixac 10, E-08028 Barcelona, Spain
[2]CIBER-BBN Networking Centre on Bioengineering, Biomaterials and Nanomedicine, Barcelona Science Park, University of Barcelona, Baldiri Reixac 10, E-08028 Barcelona, Spain
[3]Department of Organic Chemistry, University of Barcelona, Martí Franqués 1, E-08028 Barcelona, Spain
[4]Laboratory of Organic Chemistry, Faculty of Pharmacy, University of Barcelona, Joan XXII s/n, E-08028 Barcelona, Spain
[5]Pharma Mar S. A., Avda de los Reyes 1, E-28770 Colmenar Viejo, Madrid, Spain

N. Tamaki, Y. Kuge (eds.)
Molecular Imaging for Integrated Medical Therapy and Drug Development.
DOI: 10.1007/978-4-431-98074-2_24, ©Springer 2010

sourcing (collaborations, contracts, and licenses) to other companies or academic groups [2]. This development is expected to facilitate the participation of academic groups in industrial programs and also to contribute to the collaboration/enrolment of Ph.D. students and post-doctoral fellows in such programs.

The role of the medicinal organic chemist in drug discovery has undergone many changes in the past 30 years and nowadays organic chemists are facing new challenges. Until the 1980s, the role of organic chemists was prominent because they were in charge of the design and the synthesis of the drug candidates for in vivo testing. Drugs in that era were discovered by random screening, serendipity, and the modification of known molecules. Many of these drugs were inspired by natural products.

At the beginning of the 1980s, with the advent of molecular and cell biology, which promoted improved knowledge of pathology and the set-up of in vitro screening methods, and the first era of X-ray and nuclear magnetic resonance (NMR) structural analysis and molecular design, pharmaceutical companies considered that new drugs should arise from a process starting only from an in silico-based decision, the synthesis of the proposed molecules, and in vitro testing. This approach was detrimental to natural product-based research.

In the 1990s, the decreased attention given by pharmaceutical companies to natural products was even more notable. Their efforts were oriented to the preparation of large libraries of, in many cases, simple organic compounds with moderate purity to feed the high-throughput screening systems developed from the new targets discovered during the "-omic" (e.g., genomic, proteomic.) era. Thus, the first large combinatorial libraries were synthesized, only to result in disappointingly low hit rates. These libraries were designed more on the basis of chemical accessibility and maximum achievable size than on biologically relevant chemical diversity [3]. In other words, it was believed, partially erroneously, that the possibility of discovering a new drug was proportional to the number of compounds, diversity, and number of biological essays used.

Given the long product-development cycles of the pharmaceutical industry, the downstream effects on the drug discovery process of the past two decades are still difficult to measure. However, a first analysis indicated that while 53 new chemical entities (NCEs) were introduced as drugs in 1996, only 17 were introduced in 2002. This number increased to 31 in 2004, but decreased again in 2005, with just 18 new drugs; there were 17 in 2007, and a slight rise, to 21, in 2008 [4, 5] (Fig. 1).

Analysis of the 21 drugs approved in 2008 has indicated that almost 50% of the new drugs can be considered "non-classical," in the sense that they are non-classical small molecules (Fig. 2).

Moreover, 11 of these drugs were developed and the New Drug Application (NDA) was submitted to the FDA by rather recent biotech companies; One NDA was from a "classical" biotech company. Four more drugs were developed by biotech companies, but the NDA was submitted by a Big Pharma company. Four drugs were developed by large/medium size companies and just one was developed by a Big Pharma company. These data reinforce the idea outlined at the beginning that any biotech companies and academic groups are able to participate in this fascinating world of drug discovery.

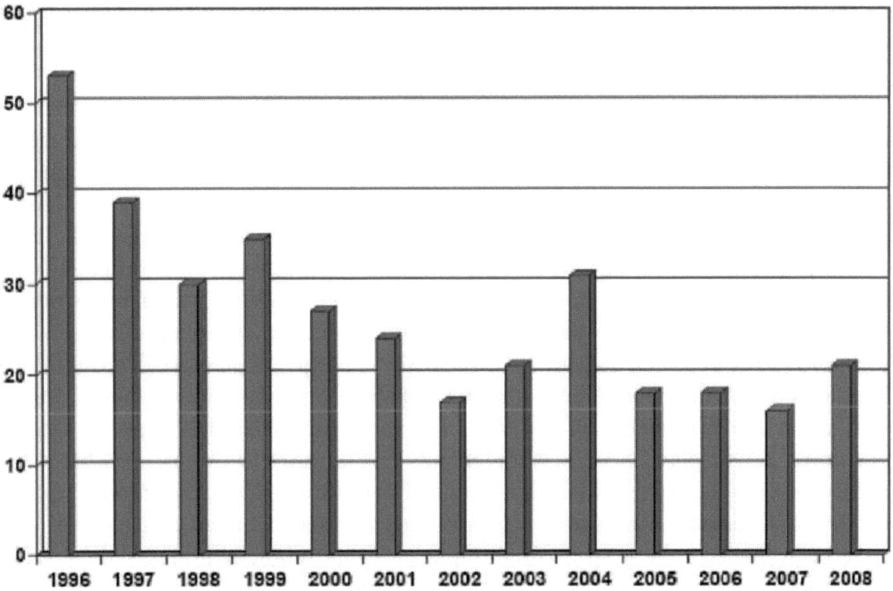

Fig. 1. New chemical entities approved by the FDA

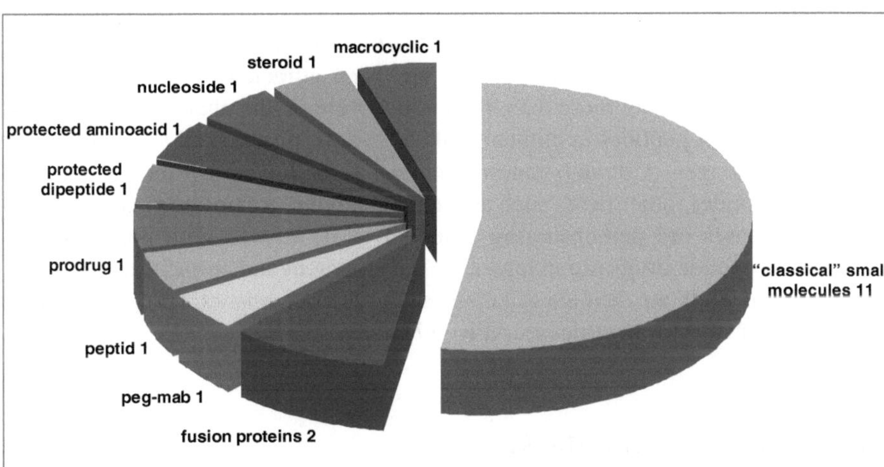

Fig. 2. Distribution by chemical structure of the new drugs approved by the FDA in 2008.

Several analysts have associated the problems encountered by the pharmaceutical industry with their diminished interest in natural products [6]. It is important to take into consideration that natural structures have a strong track record as pharmaceutical compounds. More than 60% of the 877 NCEs that reached the market from 1981 to 2002 had their origin in nature [7–9].

Nature is rich in antitumoral molecules and to a lesser extent in antiviral molecules, although less attention has been devoted to this field [10]. The immunomodulatory field is relatively unexplored and thus is propitious for discoveries. In cancer, natural molecules such as taxol, vincristine, doxorubicin, etoposide, and camptothecin are at the forefront of treatments.

Natural compounds very often contain complex structural features covering a more diverse chemical space than combinatorial chemistry libraries [11] and therefore offering great opportunity to discover novel biological activities [12]. Furthermore, a combination of natural products and combinatorial chemistry provides a potent approach for the discovery of new and safer therapeutic compounds [13].

Among the natural products, pharmaceutical industries have rekindled the interest in peptides, due to current novel technological accomplishments, strategy developments, advances in the areas of formulation, and enhanced drug delivery technology for peptides. Peptides as drugs show unique characteristics (high biological activity, high specificity, and low toxicity), thereby making them particularly attractive therapeutic agents [14]. The major obstacles that peptides face in becoming active pharmaceutical ingredients (APIs) are the lack of oral bioavailability and low stability under physiological conditions. Thus, the interest shown by the pharmaceutical industry in peptide research has declined and more efforts have been made in the field of small molecules that mimic peptides. Advances in the fields of formulation, drug delivery technology, and chemical modifications have rekindled the interest in peptides, thereby leading to highly successful peptide drugs such as the agonists of luteinizing hormone-releasing hormone (leuprolide and goserelin), which account for almost 25% ($US 1.8 billion) of the peptide drug market (2006 data) [15]. As a result, even the traditional pharmaceutical industry, which was focused on small molecules, has been attracted by the benefits that peptides confer. In 2006, more than 40 peptides were on the pharmaceutical world market, about 136 peptides in clinical trials of various phases, and a larger number in preclinical stages. Although cancer continues to be the largest therapeutic field (33%) for peptides, other fields such as metabolic, bone, gastrointestinal, and autoimmune diseases are demonstrating potential to be treated with peptides [16]. Today, new peptide drugs are in late-stage development and an increasing number of research projects are devoted to peptides, thereby demonstrating a continuously increasing demand for peptides as APIs [17].

Drug Discovery from the Sea

Marine ecosystems in particular have been demonstrated to be a wide source of biological and chemical biodiversity [18] (Fig. 3). Such chemical diversity will inevitably lead to new potential drug candidates [19, 20]. The reasons for this chemical diversity can be partially found in the following points [21]: (i) two-thirds of the planet is covered by the sea; (ii) life started in the sea 3.500 million years ago (on earth, only 600 million years ago); (iii) sea-water life conditions (pH, salinity, pressure, light, etc) create unique mechanisms of chemical defense; (iv) most marine organisms probably have not yet been described or have not been examined

Fig. 3. The sea as a source of new drugs

chemically; (v) natural products from marine invertebrates often have structures unlike those of compounds from terrestrial organisms; and (vi) more than 1% of marine organisms present pharmacological activity. However, we are probably far from understanding what these compounds do for the organisms that produce them, or what they might do for us, and this will require a whole new chemical vocabulary from the marine environment that has yet to be learned.

The past decade has witnessed a dramatic increase in the number of preclinical lead compounds of diverse marine life entering human clinical trials [22] It is pertinent to consider the evolution of the field of drug discovery of marine natural products in this context as it may help to identify future avenues which will be even more successful. However, it is encouraging to note that during recent months the first three marine products have arrived on the market, thereby demonstrating the "proof-of-concept" that the marine ecosystem is a source of new pharmaceutical compounds.

The Spanish company PharmaMar (Madrid, Spain) has launched Yondelis (trabectedin; ET-743), a three-fused tetrahydroisoquinoline ring member of the ecteinascidins isolated from the Caribbean tunicate *Ecteinascidia turbinata*, for the treatment of soft tissue sarcomas; Marinomed (Vienna, Austria), an Austrian biotech company, has received market authorization for a marine biopolymer, Carragelose, to treat the common cold; and the Israeli company Elan Pharmaceuticals (Dublin, Ireland) has received market authorization for Prialt (ziconotide) which, being a peptide, is a novel nonopioid, nonlocal analgesic, developed for the treatment of severe chronic pain (see Fig. 4 for chemical structures of these compounds).

PharmaMar is a world leading biopharmaceutical company committed to advancing the treatment of cancer through the discovery and development of new marine-derived medicines. PharmaMar has four novel compounds in clinical development, two of which are peptides. Kahalalide F is a depsipeptide isolated from the sea slug *Elysia rufescens*. The final elucidation of the structure and the synthetic

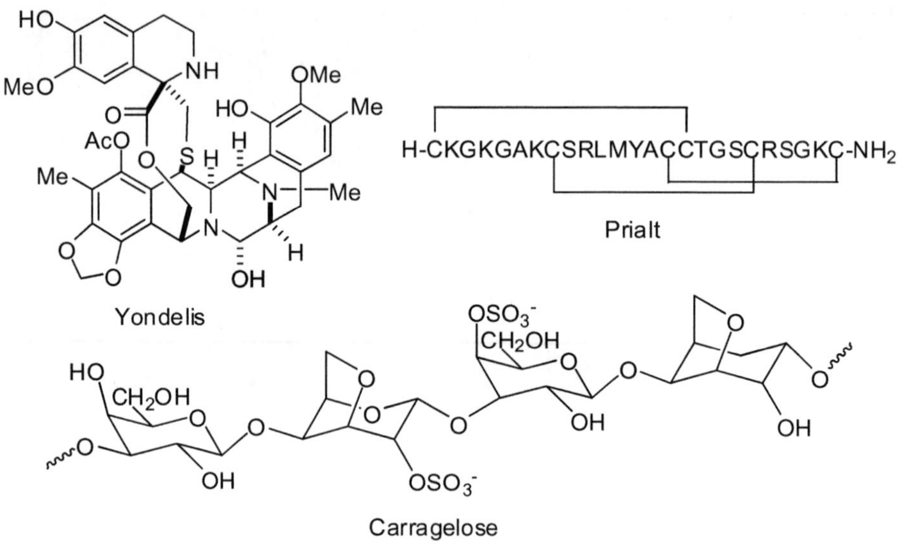

Fig. 4. Marine-based drugs: Yondelis (PharmaMar, Madrid, Spain); Prialt (Elan Pharmaceuticals, Dublin, Ireland); Carragelose (Marinomed, Vienna, Austria)

process of this compound were set up by our group in collaboration with PharmaMar [23]. Kahalalide F is a National Cancer Institute COMPARE-negative compound which induces oncosis and it has a 5- to 40-fold greater effect on cancer cells than on healthy cells. The mechanism of action of Kahalalide F does not affect the nuclear structure of the cancer cell, while the integrity of crucial organelles such as mitochondria, endoplasmic reticulum, and lysosomes is severely compromised. The alterations found indicate that the osmotic balance of the cell may be altered, possibly as a result of cellular membrane damage. Thus, it appears that Kahalalide F does not induce apoptosis, is caspase-independent, and is not affected by DNA expression. Patient recruitment has concluded for three phase II studies to assess Kahalalide F's efficacy and safety profile in patients with solid tumors (hepatocellular carcinoma, advanced malignant melanoma, and non-small-cell lung cancer). The results are currently being analyzed to decide on the strategy to be applied in the future. In parallel to the oncology studies, recruitment has been completed for a controlled pilot phase II trial to assess the efficacy and safety of Kahalalide F in patients with severe psoriasis. The data are currently being evaluated. Aplidin (plitidepsin) was originally isolated from the tunicate *Aplidium albicans* and is currently being obtained by total synthesis; this drug was also developed by our research group in collaboration with PharmaMar [24]. The mechanism of action of Aplidin is still under investigation. This compound induces apoptosis rapidly and persistently, blocks the secretion of vascular endothelial growth factor 1 (VEGF1), and blocks the cell cycle. Aplidin (designated as an orphan drug by the European Commission and the FDA), is currently in phase II clinical trials for the treatment of solid and hematological malignant neoplasias such as multiple myeloma, T-cell lymphoma, and myelofibrosis, as well as solid tumors such as melanoma. Likewise, it is in a phase I trial for the treatment of acute pediatric leukemia.

One of the main problems associated with the use of peptides and other natural products as therapeutics is their poor pharmacokinetic properties. A strategy for improving these properties is through their chemical modification. As an example, the ongoing work that is being carried out at our laboratories on thiocoraline and lamellarin is discussed.

Thiocoraline

Thiocoraline is one of several new potent antitumoral agents isolated from the marine organism *micromonospora sp* [25]. It has several motifs in common with a family of antitumoral peptide antibiotics, which includes BE-22179, triostin A, and echinomycin (Fig. 5).

This group of peptides shows: (a) a bicyclic structure; (b) C_2 symmetry; (c) an intercalation chromophore moiety; (d) an ester or thioester linkage at the terminal part of the peptide chain; (e) a disulfide or an analogue bridge in the middle of the peptide chain; (f) the presence of several *N*-methyl amino acids; and (g) non-natural amino acid of D configuration. Thus, the *N*-terminal amino function of thiocoraline is capped with 3-hydroxy-quinaldic acid, whose moiety acts as a intercalating chromophore group; the two peptide chains are bridged by thioester and disulfide linkages from Cys residues (these being the two that afford the *N*-methylated and D configuration disulfide bridge as well as the two Cys(Me) residues). All these characteristics confer on this family of peptides the capacity to bind with DNA by bisintercalation, and therefore to alter the vital cell cycle. Thiocoraline itself inhibits DNA elongation by DNA polymerase α at a concentration that inhibits cell-cycle progression and clonogenicity [26]. However, the main drawback for the clinical use of thiocoraline is its low solubility in all media used for delivery. An alternative approach is to prepare compounds that show a topology pattern close to that of thiocoraline but present a different solubility pattern.

Taking advantage of the arsenal of protecting groups, coupling reagents, and handles developed, our group have synthesized three analogues of natural

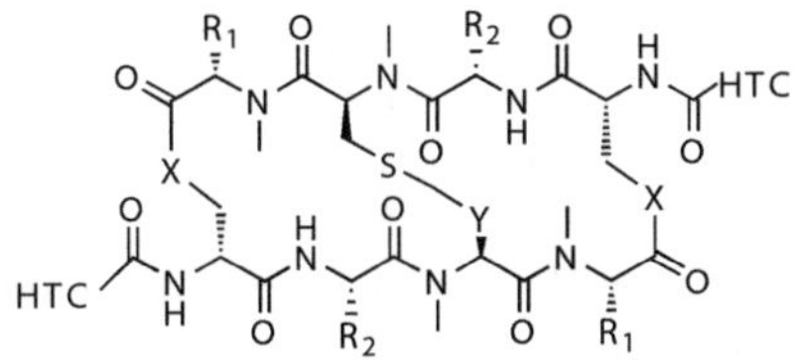

1, Triostin; X = O; Y = SCH_2; R_1 = $-CH(CH_3)_2$; R_2 = $-CH_3$; HTC = 2-quinoxaline carboxylic acid
2, Echinomycin; X = O; Y = $CH(SCH_3)$; R_1 = $-CH(CH_3)_2$ R_2 = CH_3; HTC = 2-quinoxaline carboxylic acid
3, Thiocoraline; X = S; Y = SCH_2; R_1 = $-CH_2S(Me)$; R_2 = H; HTC = 3-hydroxyquinaldic acid
4, BE-22179; X = S; Y = SCH_2; R_1 = CH_2; R_2 = H; HTC = 3-hydroxyquinaldic acid

Fig. 5. Common structure of antitumoral peptide antibiotics

Fig. 6. Structure of thiocoraline and analogues

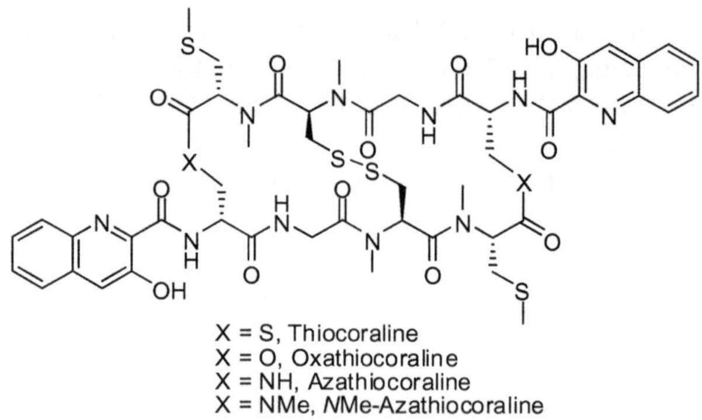

thiocoraline (X=S): the aza [27], the oxa [28], and the *N*Me-aza [29] derivatives (Fig. 6).

Although the synthesis of oxathiocoraline has been shown to be the most difficult, especially interesting has been the synthesis of *N*Me-azathiocoraline, because a hybrid approach of convergent solid-phase synthesis and solution synthesis has been carried out (Fig. 7).

The key steps of this strategy are [29]: (i) a [4 + 4] convergent solid-phase approach; (ii) the isostere of Cys, diaminopropionic acid (Dap) is used to construct the amide bridge; (iii) the use of chlorotrityl chloride (CTC) resin to minimize diketopiperazine (DKP) formation and racemization; (iv) the use of four amino protecting groups: Fmoc for the temporal protection of Gly and the β-amino of Dap, Alloc for the temporal protection of both Cys, because once the *N*Me-Cys(Me) has been introduced, Fmoc cannot be used because its use leads to a beta-elimination of the *N*Me-Cys(Me), Boc for the semipermanent protection of the α-amino of Dap, and *o*-nosyl for the protection/activation of the β-amino of Dap for facilitating methylation via a Mitsunobu reaction; (v) once the half molecule has been prepared in the solid phase, the resin is split into two portions and two-thirds are cleaved and the protected peptide is coupled on solid phase to the remaining one-third with the free amino function; (vi) disulfide formation is done on solid phase to overcome workup and polymerization; and (vii) three different coupling methods are used: the most energetic, HATU/HOAt/DIEA, for the elongation of the peptide chain, the phosphonium salts PyAOP or PyBOP/HOAt for the fragment coupling, and macrolactamization to avoid the guanidinium derivatives that can provoke the use of HATU, for the fragment coupling and macrolactamization to avoid the possible guanidinium side-products associated with the use of HATU; and the rather milder soluble carbodiimide and HOSu for avoiding over-acylation on the phenol of the 3-hydroxy-quinaldic acid.

*N*Me-azathiocoraline displays in vitro activity of the same order as the natural product [and superior to that of oxa (X=O) and aza (X=NH), in *N*Me-azathiocoraline order], and behavior similar to that of the natural product when interacting with DNA. Furthermore, this compound is significantly more stable than the natural product. Thus, we have demonstrated that the introduction of *N*Me-amides in bridges mimics the thioester bonds without imposing steric hindrance. Bridged-*N*Me amides allow conservation of the hydrogen bonding map of the natural

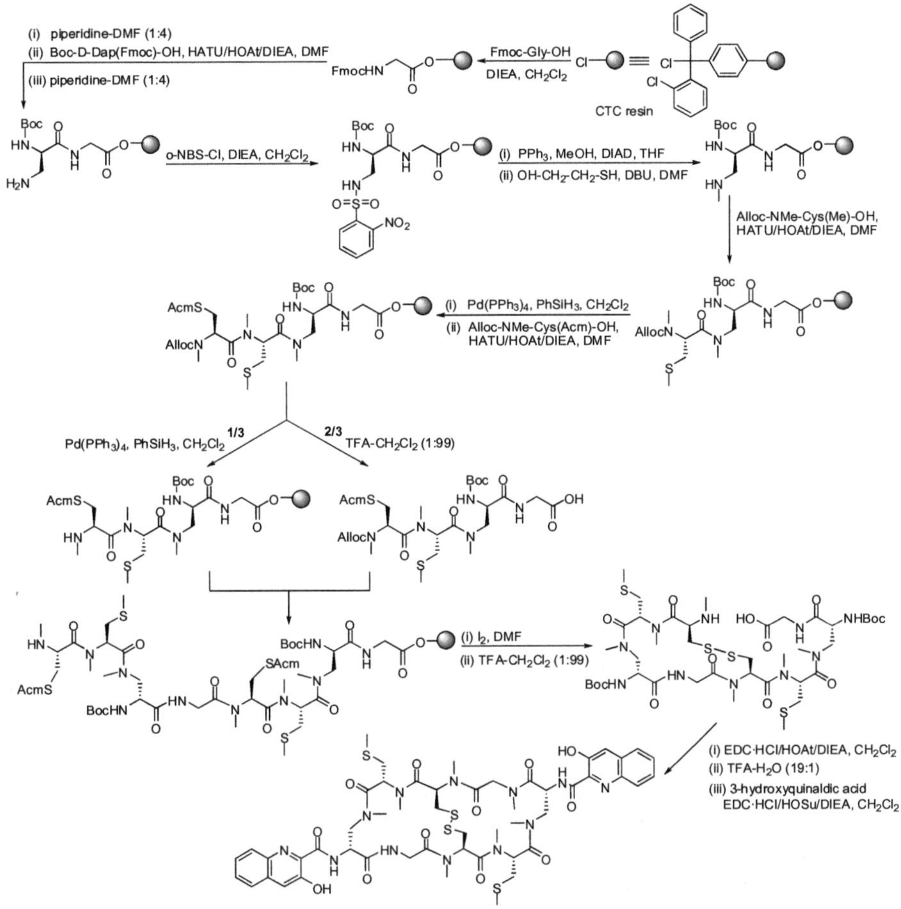

Fig. 7. Synthetic pathway followed for the synthesis of *N*Me-azathiocoraline. Alloc: allyloxycarbonyl; Acm: acetamidomethyl; Boc: *tert*-butoxycarbonyl; DBU: 1, 8-diazabicyclo [5.4.0] under-7-ene; DDQ: 2,3-dichloro-5, 6-dicyano-p-benzoquinone; DIAD: diisopropyl azodicarboxylate; DMF: *N,N*-dimethylformamide; Fmoc: fluorenylmethyloxycarbonyl; Im: imidazole; o-NBS-C1: 4-nitro-benzenesulfonyl chloride; TFA: trifluoroacetic acid

product. This approach could be used to enhance stability in other depsipeptides and side-chain to side-chain cyclic peptides with similar problems.

Lamellarins

Lamellarins constitute a family of marine alkaloids of more than 40 members that show important bioactivities [30]. Lamellarin D (Fig. 8), which is the most important member of the family, is a cytotoxic agent against various tumor cells, an inhibitor of topoisomerase I [31], and a potent pro-apoptotic agent [32].

Fig. 8. Structure of lamellarin D

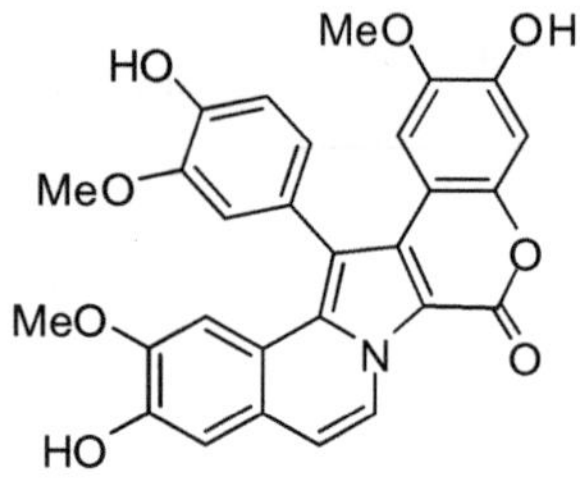

We recently described the total synthesis of lamellarin D [33] and the preparation of a library of more than 40 analogues and their corresponding structure-activity relationship (SAR) studies [34]. Preparation of poly(ethyleneglycol) (PEG)-conjugates as well as bioconjugates with a nuclear localization signal peptide improved the solubility. These conjugates were subjected to cell-penetration studies, and the apoptotic mechanism of cell death they induced has been recently described [35, 36]. As an example, the preparation of a lamellarin D (Lam-D) conjugate with poly(ethyleneglycol) is discussed (Fig. 9).

This Lam-D conjugate was initially prepared following the modular synthesis of Lam-D [33]. The scaffold was methyl 8-isopropoxy-9-methoxy-5,6-dihydropyrrolo[2,1-*a*]isoquinoline-3-carboxylate. The borolane building blocks were protected with Bn/*i*Pr. Regioselective and sequential introduction of aryl groups using Pd(0)-catalyzed cross-coupling reactions, aromatization, and phenol-deprotection provided useful Lam-D precursors for regioselective conjugation in its free phenolic groups. The PEG-group was introduced in the appropriate precursor by esterification of the free phenol, using N-(3-dimethylaminopropyl)-N′-ethylcarbodiimide (EDC)·HCl with a catalytic amount of 4-dimethylaminopyridine (DMAP) in CH_2Cl_2. Final O*i*Pr deprotection with $AlCl_3$ provided the proper conjugate [35].

Furthermore, another approach, based on the simultaneous formation of pyrrole and lactone rings using a [2 + 3] cycloaddition reaction, was used for the construction of the pentacyclic structure in Lam-D. This procedure was performed bearing three different orthogonal protecting groups: OMOM, BnO, and TBDPSO.

Due to the lability properties of the conjugates, a second generation of semiprotected precursors was designed, taking into consideration mild reaction deprotection conditions in order to optimize yields. Thus, a di-TBDPSO Lam-D precursor bearing the free phenol in position 3 was synthesized. Further attachment of the PEG-COOH by condensation under the same conditions as in the previous step, and final simultaneous deprotection of *N*-Boc and TBDPSO with HF at low temperature, rendered a conjugate with improved yields [36].

In summary, two synthetic strategies were employed for the synthesis of Lam-D conjugates; these strategies provide high versatility and robust methodologies for the preparation of pH-labile ester bond conjugates of the natural product. The application of these methodologies was useful for the preparation of 11 Lam-D conjugates. The conjugates were tested in a panel of three different human tumor cell lines

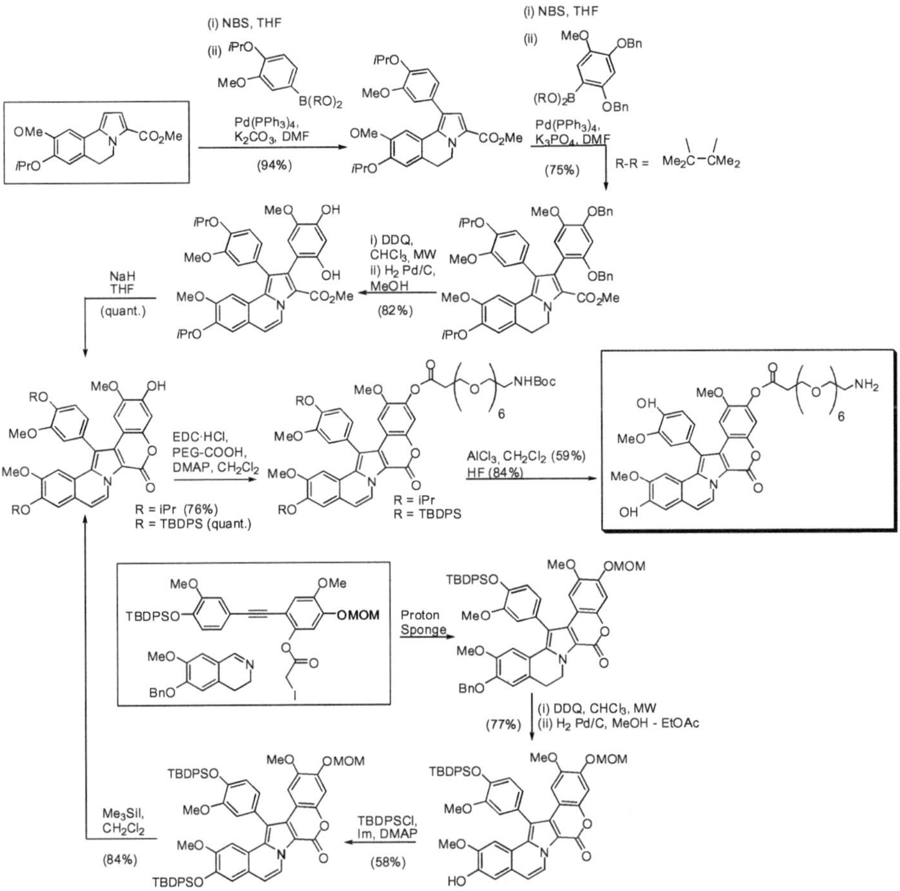

Fig. 9. Synthetic pathway followed for the synthesis of PEG conjugated lamellarin D

(MDA-MB-231 breast, A-549 lung, and HT-29 colon) to evaluate their cytotoxicity. Several compounds exhibited enhanced cellular internalization, and more than 85% of the derivatives showed better 50 percent growth inhibition (GI_{50}) (of one order of magnitude) as compared with lamellarin D. Furthermore, cell-cycle arrest at G2 phase and apoptotic cell-death pathways were determined for lamellarin D and these derivatives.

Conclusions

The new pharmaceutical industry arena where more "non-classical" drugs are introduced in the market, in many cases by biotech companies, should facilitate the participation of academic groups in industrial programs. Pharmaceutical companies are looking again towards nature as the inspiration for new drugs. In the case of marine natural products, the "proof-of-concept" that the sea can become a great source of new drugs has been reached with the first three marine products that have arrived on the market. Chemical modification of natural products should be the strategy for overcoming one of the main drawbacks associated with natural products as drugs—their poor pharmacokinetic properties.

Acknowledgments This study was partially supported by Comisión Interministerial de Investigaciones Científicas (CICYT) (CTQ2006-03794/BQU), the Instituto de Salud Carlos III (ISCIII) (Centro de Investigación Biomédica en Red (CIBER), nanomedicine), the Institute for Research in Biomedicine, and the Barcelona Science Park. J. T.-P. is a Juan de la Cierva fellow Ministerio de Ciencia e Innovatión (MICINN).

References

1. Gray NS (2006) Drug discovery through industry-academic partnerships. Nature Chem Biol 2: 643–649
2. Cavalla D (2003) The extended pharmaceutical enterprise. Drug Discov Today 8: 267–274
3. Martin YC, Critchlow RE (1999) Beyond mere diversity: tailoring combinatorial libraries for drug discovery. J Comb Chem 1: 32–45
4. Burrill & Company (2008) Analysis for pharmaceutical research and manufacturers of America
5. Hughes B (2009) 2008 FDA Drug approvals. Nat Rev Drug Discov 8: 93–96
6. Ganesan A (2008) The impact of natural products upon modern drug discovery. Curr Opin Chem Biol 12: 306–317
7. Newman DJ, Cragg GM, Snader KM (2003) Natural products as sources of new drugs over the period 1981–2002. J Nat Prod 66: 1022–1037
8. Newman DJ, Cragg GM (2005) The discovery of anticancer drugs from natural sources. In: Zhang L, Demain AL (eds) Natural products. Humana, Totowa, NJ, pp 129–168
9. Newman DJ, Cragg GM (2007) Natural products as sources of new drugs over the last 25 Years. J Nat Prod 70: 461–477
10. Butler MS, Newman DJ (2008) Mother nature's gifts to diseases of man: the impact of natural products on anti-infective, anticholestemics and anticancer drug discovery. Prog Drug Res 65: 3–44
11. Feher M, Schmidt JM (2003) Property distributions: differences between drugs, natural products, and molecules from combinatorial chemistry. J Chem Inf Comput Sci 43: 218–227
12. Newman DJ (2008) Natural products as leads to potential drugs: an old process or the new hope for drug discovery? J Med Chem 51: 2589–2599
13. Ganesan A (2004) Natural products as a hunting ground for combinatorial chemistry. Curr Opin Biotechnol 15: 584–590
14. Bruckdorfer T, Marder O, Albericio F (2004) From production of peptides in milligram amounts for research to multi-tons quantities for drugs of the future. Curr Pharm Biotechnol 5: 29–43
15. Sehgal A (2006) New applications in discovery, manufacturing, and therapeutics. In: Research and markets. Dublin, Ireland
16. Ayoub M, Scheidegger D (2006) Peptide drugs, overcoming the challenges, a growing business. Chem Today 24: 46–48
17. Zompra AZ, Galanis AS, Werbitzky O et al (2009) Manufacturing peptides as active pharmaceutical ingredients (API). Future Med Chem 1: 361–377

18. Hill RA (2008) Marine natural products. Ann Rep Prog Chem, Section B: Organic Chem, 104: 127–141
19. Simmons TL, Andrianasolo E, McPhail K et al (2005) Marine natural products as anticancer drugs. Mol Cancer Ther 4: 333–342
20. Bowling JJ, Kochanowska AJ, Kasanah N et al (2007) Nature's bounty—drug discovery from the sea. Exp Op Drug Disc 2: 1505–1522
21. Fusetami M (ed) (2000) Drugs from the sea. S Karger, Berlin
22. Molinski TF, Dalisay DS, Lievens SL et al (2009) Drug development from marine natural products. Nat Rev Drug Discov 8: 69–85
23. López-Macià A, Jiménez JC, Royo M et al (2001) Synthesis and structural determination of Kahalalide F. J Am Chem Soc 123: 11398–11401
24. Jou G, González I, Albericio F et al (1997) New syntheses of the didemnins: total synthesis of dehydrodidemnin B based upon two novel routes to didemnin A. Use of uronium and phosphonium salt coupling reagent in peptide synthesis in solution. J Org Chem 62: 354–366
25. Romero F, Espliego F, Perez Baz J et al (1997) Thiocoraline, a new depsipeptide with antitumor activity produced by a marine Micromonospora. I. Taxonomy, fermentation, isolation, and biological activities. J Antibiot 50: 734–737
26. Erba E, Bergamaschi D, Ronzoni S et al (1999) Mode of action of thiocoraline, a natural marine compound with anti-tumor activity. Br J Cancer 80: 971–980
27. Bayó N, Fernández A, Riego E, et al (2006) Solid-phase syntheses of azathiocoralines. Chem Eur J 12: 9001–9009
28. Tulla-Puche J, Bayó-Puxan N, Moreno JA et al (2007) Solid-phase synthesis of oxathiocoraline by a key intermolecular disulfide dimer. J Am Chem Soc 129: 5322–5323
29. Tulla-Puche J, Marcucci E, Bayó-Puxan N et al (2009) *N*-Me amide as synthetic surrogate for the thioester moiety in thiocoraline. J Med Chem 52: 834–839
30. Pla D, Albericio F, Álvarez M (2008) Recent advances in lamellarin alkaloids: isolation, synthesis and activity. Anti-Cancer Agents Med Chem 8: 746–760
31. Facompré M, Tardy C, Bal-Mahieu C et al (2003) Lamellarin D: a novel potent inhibitor of topoisomerase I. Cancer Res 63: 7392–7399
32. Vanhuyse M, Kluza J, Tardy C et al (2005) Lamellarin D: a novel pro-apoptotic agent from marine origin insensitive to P-glycoprotein-mediated drug efflux. Cancer Lett 221: 165–175
33. Pla D, Marchal A, Olsen CA et al (2005) Modular total synthesis of lamellarin D. J Org Chem 70: 8231–8234
34. Pla D, Marchal A, Olsen CA et al (2006) Synthesis and structure-activity relationship study of potent cytotoxic analogues of the marine alkaloid lamellarin D. J Med Chem 49: 3257–3268
35. Pla D, Francesch A, Calvo P et al (2009) Lamellarin D Bioconjugates I: Synthesis and cellular internalization of PEG-derivatives. Bioconjugate Chem 20: 1100–1111
36. Pla D, Martí M, Farrera-Sinfreu J et al (2009) Lamellarin D bioconjugates II: Synthesis and cellular internalization of dendrimer and nuclear location signal derivatives. Bioconjug Chem 20: 1112–1121

PET/AMS Applications in Drug Development

Tomio Inoue[1], Akiko Suzuki[2], and Ryogo Minamimoto[1]

Summary

On June 3, 2008, the Ministry of Health, Labor and Welfare (MHLW) of Japan published a document: "Japanese guidance for conducting microdose clinical trials" as a guideline for exploratory investigational new drug (eIND) studies that employ three highly sensitive measurements: positron emission tomography (PET); accelerator mass spectrometry (AMS), using a ^{14}C-labeled drug; and liquid chromatography/mass spectrometry/mass spectrometry(LC/MS/MS).

PET has three major roles in drug development: (1) ^{11}C-labeled or ^{18}F-labeled ligands with the same chemical structure as new drug candidates are used in eIND studies, including microdose testing, to evaluate pharmacodynamics (PD). (2) Receptor imaging with PET is used for determining the appropriate dose of the IND by measuring the receptor occupancy of the target organ and the blood concentration of the administered drug. (3) PET can be employed as a surrogate endpoint to evaluate the drug effect. These roles of PET may give us some information that will enable us to individualize drug therapy. Based on this concept, we are now conducting human PET imaging with ^{18}F-labeled 5-fluorouracil (FU) as a clinical study in healthy volunteers and in head/neck cancer patients to survey the different drug distribution in target and nontarget lesions.

At the same time, AMS is another highly sensitive measuring instrument for evaluating pharmacokinetics (PK) and metabolic data quantitatively over a long period of time, and it may provide more precise evaluation of PET imaging data. As baseline research with small animals, we investigated the correlation between PET imaging and AMS quantification data using ^{18}F fluorodeoxyglucose (FDG) labeled with both ^{18}F (^{18}F-FDG) and ^{14}C (^{14}C-FDG). Both radioactive FDG ligands were simultaneously injected to mice; the drug concentration in blood and the drug

[1]Department of Radiology, Yokohama City University School of Medicine, 3-9 Fukuura, Kanazawa-ku, Yokohama 236-0004, Japan
[2]Department of Radiology, Yokohama City University Hospital, 3-9 Fukuura, Kanazawa-ku, Yokohama 236-0004, Japan

N. Tamaki, Y. Kuge (eds.)
Molecular Imaging for Integrated Medical Therapy and Drug Development.
DOI: 10.1007/978-4-431-98074-2_25, ©Springer 2010

distribution to major organs were analyzed by three methods: ^{18}F FDG was measured with a gamma counter and with micro-PET imaging, while ^{14}C-FDG was analyzed by AMS. The results of organ distribution measured with the gamma counter and AMS showed a good correlation, and the PET imaging data well supported the PK data. The combination of PET and AMS, which provides drug distribution as well as PK data, may be used as a technology that will lead to individualized medicine in the future.

Key words Microdose · 5-FU · AMS · PET · eIND

Introduction

On June 3, 2008, the Ministry of Health, Labor and Welfare (MHLW) of Japan published a document: "Japanese guidance for conducting microdose clinical trials" as a guideline for exploratory investigational new drug (eIND) studies in which positron emission tomography (PET), accelerator mass spectrometry (AMS) using a ^{14}C-labeled drug, and liquid chromatography/mass spectrometry/mass spectrometry (LC/MS/MS) are defined as highly sensitive devices for measurements of drug concentrations. These content of the guidance suggest that we have opened a new era of PET as an effective measuring tool for developing new drugs and supporting drug treatment.

AMS is a modality that shows extremely high sensitivity for analyzing pharmacokinetics, by using very small amounts of ^{14}C-labeled drugs in blood, urine, and tissue. Although AMS itself does not detect radioactivity, AMS technology should be included in the field of nuclear medicine from the viewpoint of radionuclide utilization. Applications of PET and AMS in drug development are new fields in nuclear medicine. In this chapter, we review the role of PET/AMS in drug development and briefly introduce our preliminary data on PET/AMS applications in analyzing pharmacokinetics.

Roles of PET in Clinical Trials for Drug Development

The potential roles of PET in drug development are classified into three categories. Two of these are performed in exploratory clinical trials for screening candidate drugs and the other one is the application of receptor PET imaging for setting appropriate clinical doses of the candidate drug.

The first role of PET, screening of candidate drugs, is in exploratory clinical studies, which are intended to expedite drug development and to be conducted early in phase 1; they involve limited human exposure, have no therapeutic or diagnostic intent, and are not intended to examine the maximum tolerated dose. Microdose studies are classified as exploratory clinical studies in which the administered dose of the candidate drug is not more than 100 μg and not more than 1/10th of no observed adverse effect level (NOAEL). According to the draft of the document: "Guidance on Nonclinical Safety Studies for the Conduct of Human Clinical Trials and Marketing Authorization for Pharmaceuticals" released for consultation at the International Conference on Harmonization of Technical Requirements for Registration of Pharmaceuticals for Human Use (ICH) process, in a PET study target receptor binding or tissue distribution can be applied as a microdose study in which not more than a total dose of 100 μg can be divided among up to five doses in any subject. In the field of nuclear medicine imaging, there is the concept of a tracer dose; that is, nuclear medicine imaging technology can provide extremely sensitive measurement and it requires only tracer amounts; namely, an amount far less than the pharmacological dose. Theoretically, the range of tracer doses in nuclear imaging meets the definition of the microdose level.

In PET imaging, radionuclides for PET, such as ^{11}C, ^{13}N, or ^{18}F, are generated by an in-hospital cyclotron and transferred to an autosynthetic device for the labeling of radionuclides attached to the new candidate drug in a "hot" laboratory. After the quality control (QC) test, we inject the PET ligands to volunteers or patients to obtain biodistribution images using PET or PET/computed tomography (PET/CT). For a preclinical test, microPET/CT is also useful to confirm the proof of concept (POC). PET with ^{11}C, ^{13}N, or ^{18}F-labeled ligands can keep the same chemical structure as that of new candidate drugs that are employed in eIND studies, including microdose tests, to evaluate pharmacodynamics (PD) and screen the candidate drugs. However, the limitation of this pharmacodynamic analysis is that only the initial in vivo distribution of the drug is obtained because of the short radioactive half life.

In drugs used for treating psychiatric disease, the drug must permeate through the blood brain barrier (BBB), and an image showing only venous blood pool in the brain may suggest a low probability of success in IND studies, while an image showing permeability through the BBB suggests the potential to go forward into a regular phase I trial. An earlier decision may save costs and time in drug development (Fig. 1).

The effectiveness of the microdose concept in drug development is still controversial among pharmacologists, and there have been few studies of living human in vivo distribution of drugs. Radiolabeled PET drugs, which have the same chemical structures as candidate drugs, have a limitation in that they provide initial distribution only, because of their short radioactive half life. Theoretically C-11 is the most appropriate radionuclide for the screening of candidate drugs.

As a radiolabeled drug with a positron emitter, ^{18}F 5-fluorouracil (5-FU) was employed in a clinical study of drug development in 1973 [1], when PET/CT did not exist. Tumor uptake of ^{18}F 5-FU and tumor response to 5-FU was demonstrated in an animal experiment [2] in 1977. There were some reports of human clinical studies about pharmacokinetic analysis with ^{18}F 5-FU and PET between 1993 and 2000 [3–5]. The metabolism of 5-FU in normal tissue and tumor cells seemed to be complex, as shown in Fig. 2. More than 30 years after the first report of ^{18}F 5-FU, quantification of 5-FU with PET is still in the experimental stage. At

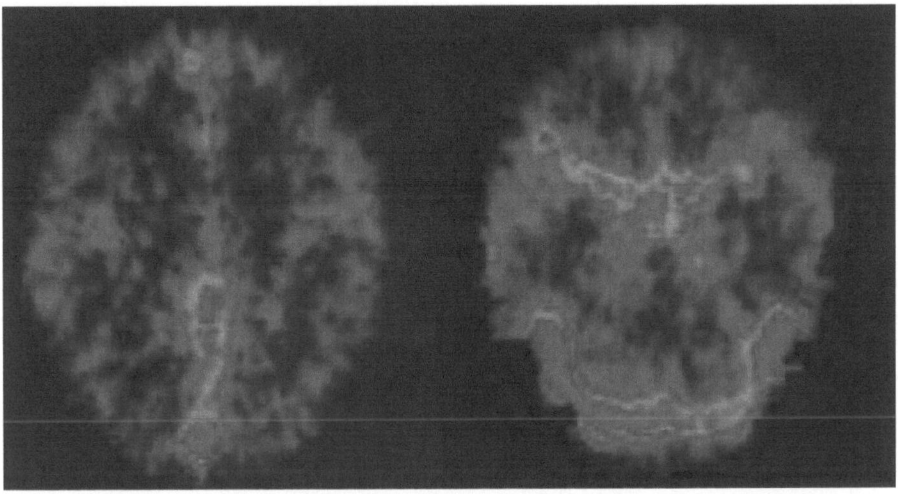

Not permeate through blood-brain barrier = Stop

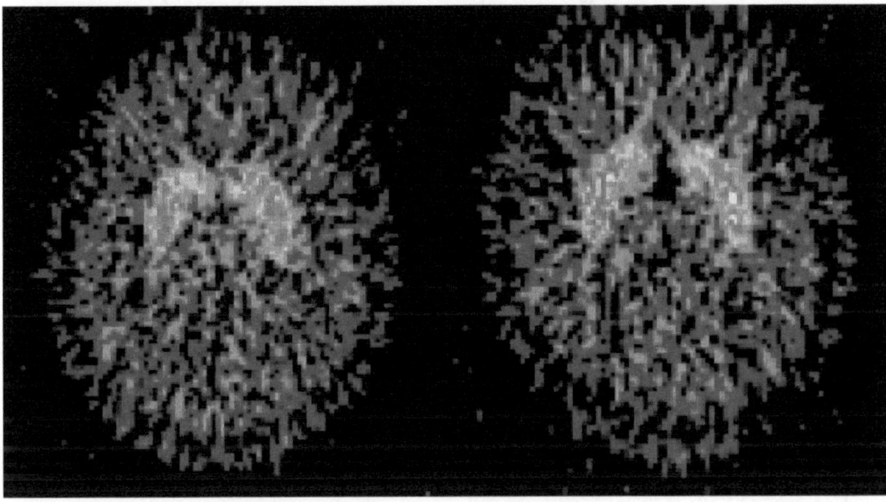

Permeate through blood-brain barrier = Go

Fig. 1. The role of PET under the notion of microdose studies

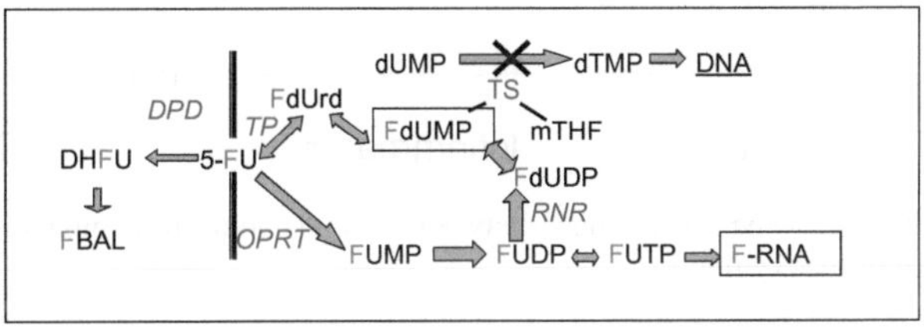

Fig. 2. Complex process of 5-fluorouracil (*5-FU*) metabolism in vivo

Yokohama City University, ^{18}F 5-FU PET/CT human clinical studies were initiated in 2008 after small-animal experiments with normal mice and a mouse disease model [6]. A comparative study of ^{18}F fluorodeoxy D-glucose (FDG) and ^{18}F 5-FU uptake in lesions showed a difference of retention times between cancer and inflammatory lesions for the two PET ligands, and this suggested that ^{18}F 5-FU may be useful for the differential diagnosis of cancer and inflammation (Fig. 3). ^{18}F 5-FU PET images showed liver uptake and excretion of 5-FU in urine (Fig. 4).

The second expected role of PET in drug development is to determine the clinical dose by measuring drug receptor occupancy. In dopamine D2 receptor-binding drugs, brain PET images with a dopamine D2 receptor PET ligand and the displacement of cold D2 receptor-binding drug candidates allow us to calculate the receptor occupancy. The function of the blood concentration of a candidate drug and the

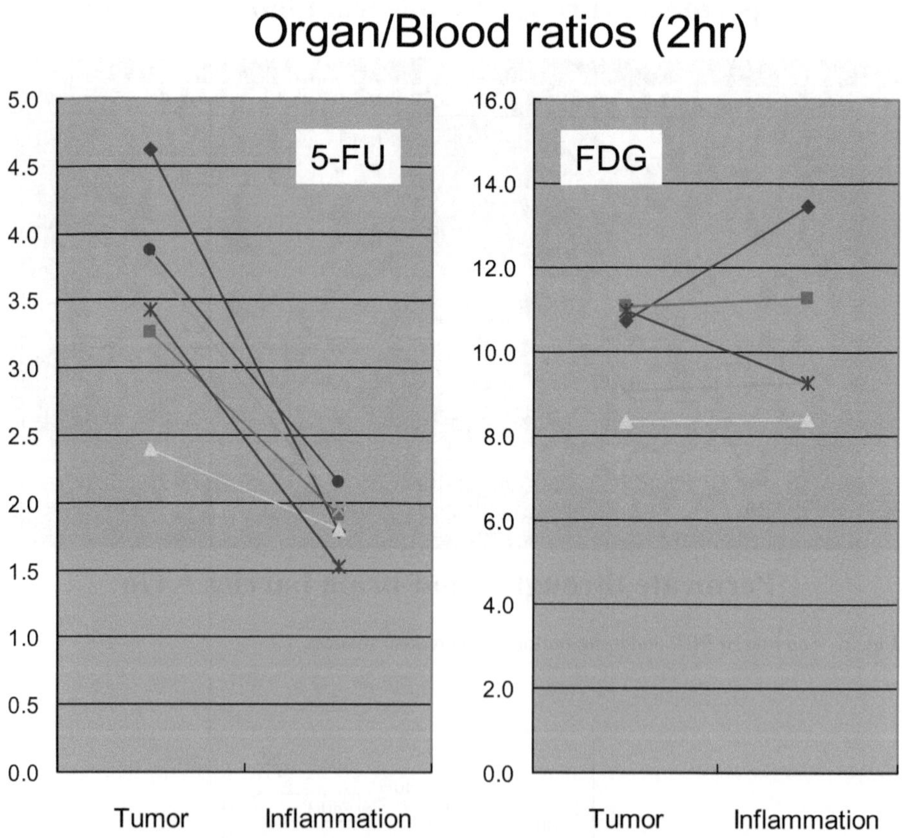

Fig. 3. The difference between tumor and inflammation uptake of ^{18}F-5-FU. *FDG*, fluorodeoxyglucose

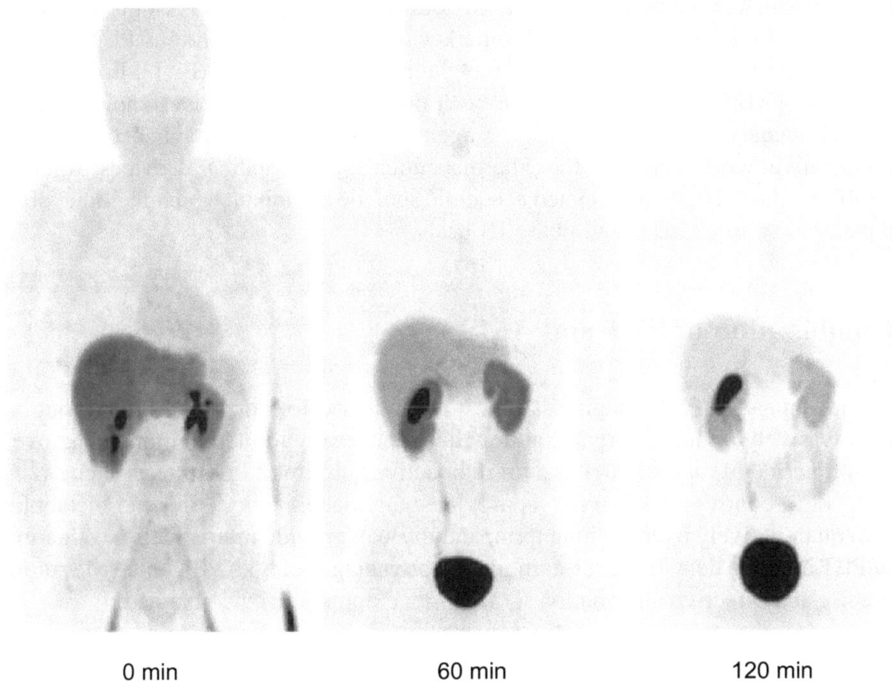

Fig. 4. Sequential images of ^{18}F 5-FU biodistribution in healthy male volunteer after intravenous injection

receptor occupancy can provide scientific information on the appropriate clinical intake dose to use in clinical trials.

The third role of PET in drug development is to be employed as a surrogate endpoint, thus reducing the duration of clinical phase II and phase III trials and supporting drug approval. Traditionally, response evaluation criteria in solid tumors (RECIST) are the criteria used to determine the effectiveness of anticancer drugs, based on changes in tumor diameter. PET may be useful as an alternative diagnostic tool for the evaluation of the therapeutic effects of anticancer drugs. ^{18}F-fluorodeoxy D-glucose (^{18}F-FDG) is the most widely used PET ligand in oncology, and it has proven to be useful for the evaluation and prediction of the effects of chemotherapeutic agents, including molecular targeting agents. The Oncology Biomarker Qualification Initiative (OBQI) announced, in February 2006, an agreement between the National Cancer Institute (NCI), the United States Food and Drug Administration (FDA), and the Center for Medicare and Medicaid Services (CMS) to develop a clinical trial design in which FDG PET/CT was used as a surrogate endpoint. The Critical Path Opportunities Report published in 2006 by the FDA suggested that biomarker and imaging were key technologies for facilitating clinical trials in drug development, and the term "biomarker imaging" was employed in the report [7]. PET has potential for biomarker imaging because various PET ligands expressing the feature of tumor such as fluorodeoxythymidine (FLT), fluoromisonidazole (FMISO), and others, have been developed. The relatively new concept of PET ligands as biomarker imaging agents will have to be established based on cooperative work between the Pharmaceuticals and Medical Devices Agency (PMDA), the MHLW, and related academic societies in Japan, which facilitate drug development in phase II and phase III trials.

Combination of PET and AMS

An advantage of PET imaging is that it can be used to monitor drug distribution noninvasively in human organs, although it is generally limited to monitoring over a short period of time due to the short radioactive half life of PET tracers such as ^{11}C and ^{18}F. In contrast, AMS analysis provides pharmacokinetics (PK) and metabolic data quantitatively over the long term, and this can provide more precise evaluation of PET imaging data. If we can combine the advantages of both PET and AMS, more precise and effective information for drug development can be obtained.

Outline of AMS

AMS was developed in the 1970s and has been applied to the fields of archeology, geology, and environmental and earth sciences. The measurement principle of AMS is based on the separation and direct measurement of ^{14}C and ^{12}C atoms without any disintegration time requirements. The AMS system is a highly sensitive analysis method that can be used for measuring radiocarbon concentration ($^{14}C/^{12}C$ ratio). The sensitivity of AMS is more than 1000 times higher than that of conventional beta-ray counting methods such as the liquid scintillation counter. Because the required amount of radioactivity for AMS is far less than that required for a conventional test with ^{14}C, AMS can be employed without any concerns about radiation exposure in volunteers in clinical trials. The American Research Group at the National Laboratory is a pioneer of AMS work within the biochemical community [7, 8]. This technology is now being applied in microdose studies for new drug development because ^{14}C is stably incorporated into many organic drug molecules and ^{14}C-labeled drugs are available for studies designed to determine absorption, distribution, metabolism, and excretion (ADME).

After sample preparation and graphitization were performed for AMS measurements, approximately one-half of the graphite was packed into an aluminum cathode. The cathodes of the test samples were randomly placed into a 133-position AMS sample wheel with standards and controls. The sample wheel was inserted into the ion source of the tandem electrostatic accelerator. It took about 300 s (3000 cycles) for one measurement, and the ratio of $^{14}C/^{12}C$ content was calculated based on three measurements. Because AMS provides a relative measurement, the ratio of the $^{14}C/^{12}C$ content of a sample was normalized based on the ratio of the $^{14}C/^{12}C$ content of the simultaneously measured standard sample. The AMS results were expressed as the percentage of modern carbon (pMC), where 100 pMC was defined as: 100 pMC = 13.56 dpm/gC (gC, g of carbon). Therefore, A (pMC) × 0.1356 = B (dpm ^{14}C/gC).

The ^{14}C-concentration (dpm/g) in the sample was calculated by using the following equation:

$$^{14}\text{C-concentration (dpm/g)} = \text{B (dpm } ^{14}\text{C/gC)} \times \text{carbon content (gC/g)}$$

where the carbon content (gC/g) in the sample was calculated from the amount of CO_2 gas measured with a capacitance gauge.

Initial Experience of AMS Analysis

Because the application of AMS to the drug kinetics starts but it is not long yet, we carried out two primitive experiments as the future next steps in AMS application.

First, we measured background ^{14}C concentration in human blood to obtain a reference for further microdosing studies in Japan [9].

Blood samples of five healthy Japanese volunteers (three males, two females; median age, 40.4 ± 9.8 years) were collected around the same time, just prior to when the subjects ate a meal. Centrifugal separation of blood that was allowed to clot and the plasma was performed at 3000 rpm/min for 2 min at 4°C. The background ^{14}C concentration of each of the samples was measured using an AMS system. The Institute of Accelerator Analysis (IAA), which is the first contract research organization in Japan that is capable of providing AMS analysis services for carbon dating and bioanalysis work, performed the AMS analysis. The mean ^{14}C-concentration in blood was 1.613 ± 0.125 dpm/ml. The ^{14}C-concentrations in the clot and blood were higher than those found for plasma. The ^{14}C-concentrations in the blood and plasma were slightly different among individuals as compared to the values for the clot, although the differences were quite small, with a coefficient of variation (CV) value under 7.8%. Even though the ^{14}C-concentration differed only slightly between individuals, the ^{14}C-concentrations in the clot and blood were higher than those seen for the plasma. Therefore, the variation and difference of the background data for blood and plasma might be of use as a reference for microdosing test evaluations [10].

As the second primitive experiment, we investigated the possibility of combined data analysis of PET and AMS. In this study, the correlation between PET imaging and AMS quantification data was investigated using FDG, which is one of the most popular substances used in PET imaging. FDG labeled with both ^{18}F (^{18}F-FDG) and ^{14}C (^{14}C-FDG) was simultaneously injected to mice, and the drug concentra-

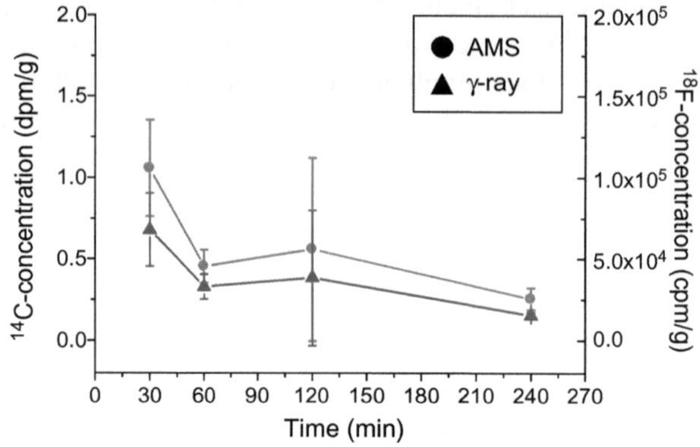

Fig. 5. Comparison of ^{14}C concentration, measured by accelerator mass spectrometry (*AMS*), and ^{18}F concentration, measured with a γ-counter

tions in blood and the drug distribution to major organs (brain, heart, liver, kidney, and blood) were analyzed by three methods: ^{18}F-concentration was measured with a gamma counter and micro-PET imaging, while ^{14}C-concentration was analyzed by AMS. The results of organ distribution analyzed with the gamma counter and AMS had a good correlation (Fig. 5), and the PET imaging data well supported the PK data.

Our primitive experiments have shown that AMS analysis of blood samples is reliable, and the PET/AMS combined analysis has the potential for providing more detailed PK/PD data in human clinical trials.

References

1. Fowler JS, Fin RD, Lambrecht RM et al (1973) The synthesis of ^{18}F-5-fluorouracil. J Nucl Med 14: 63–64
2. Shani J, Wolf W (1977) A model for prediction of chemotherapy response to 5-fluorouracil based on the differential distribution of 5-[^{18}F]fluorouracil in sensitive versus resistant lymphocytic leukemia in mice. Cancer Res 37: 2306–2308
3. Saleem A, Yap J, Osman S et al (2000) Modulation of fluorouracil tissue pharmacokinetics by eniluracil: in vivo imaging of drug action. Lancet 355: 2125–2131
4. Dimitrakopoulou A, Strauss LG, Clorius JH et al (1993) Studies with positron emission tomography after systemic administration of fluorine-18-uracil in patients with liver metastases from colorectal carcinoma. J Nucl Med 34: 1075–1081
5. Hohenberger P, Strauss LG, Lehner B et al (1993) Perfusion of colorectal liver metastases and uptake of fluorouracil assessed by H2(15)O and [^{18}F] uracil positron emission tomography (PET). Eur J Cancer 29A:1682–1686
6. Sugae S, Suzuki A, Takahashi N et al (2008) Fluorine-18-labeled 5-fluorouracil is a useful radiotracer for differentiation of malignant tumors from inflammatory lesions. Ann Nucl Med 22: 65–72
7. US Department of Health and Human Services Food and Drug Administration (2006) Critical Path Opportunities Report March 2006. Washington, DC

8. Vogel JS, Turteltaub KW, Finkel R et al (1995) Accelerator mass spectrometry. Anal Chem 67: 353A–359A
9. Kaye B, Garner RC, Mauthe RJ et al (1997) A preliminary evaluation of accelerator mass spectrometry in the biochemical field. J Pharm Biomed Anal 16: 541–543
10. Minamimoto R, Hamabe Y, Miyaoka T et al (2008) Accelerator mass spectrometry analysis of background 14C-concentrations in human blood; aiming at reference data for further microdosing studies. Ann Nucl Med 22: 883–889

Functional and Structural Analysis Reveals Dual Function on C-Terminal α Helix of Alg13 Protein

Xiao-Dong Gao, Satoru Moriyama, Nobuaki Miura, and Shin-Ichiro Nishimura

Summary

As one of the most frequent and common post-translational modifications, protein asparagine *N*-glycosylation is initiated with the biosynthesis of a highly conserved dolichol-linked oligosaccharide (LLO) $Glc_3Man_9GlcNAc_2$ on the cytoplasmic face of the endoplasmic reticulum (ER) and ends within the lumen. The second step of this process is catalyzed by a UDP-*N*-acetylglucosamine transferase that is comprised of two subunits, Alg13 and Alg14. The interaction between Alg13 and 14 is crucial for UDP-GlcNAc transferase activity, so formation of the Alg13/14 complex is likely to play a key role in the regulation of *N*-glycosylation. Using a combination of bioinformatics and molecular biological methods, we have undertaken a functional and structural analysis of yeast Alg13 and Alg14 proteins to elucidate the mechanism of their interaction. Mutational studies have demonstrated that a short C-terminal α helix of Alg13 is required for interaction with Alg14 and enzyme activity. Furthermore, the highly conserved hydrophobic amino acids in this α helix are required for the membrane-associated localization of Alg13. Electrostatic surface views of the modeled Alg13/14 complex suggest the presence of a hydrophobic cleft in Alg14 that provides a pocket for the Alg13 C-terminal α helix. Co-immunoprecipitation assays have confirmed that the C-terminal three amino acids of Alg14 are required for maintaining the integrity of the Alg13/Alg14 complex and this depends on their hydrophobicity. Interestingly, a protein stability assay indicated that deletion or destruction of the C-terminal α helix stabilized an inactive free Alg13, indicating the importance of this structure in the degradation of cytosolic Alg13. These results demonstrate that: (1) the formation of this hetero-oligomeric complex is mediated by a short C-terminal α helix of Alg13 in cooperation with the last three amino acids of Alg14 and (2) the terminal α helix also functions as a degradation signal for controlling the protein level of free Alg13 in cytosol.

Graduate School of Advanced Life Science, Frontier Research Center for Post-Genomic Science and Technology, Hokkaido University, N21, W11, Kita-ku, Sapporo 001-0021, Japan

N. Tamaki, Y. Kuge (eds.)
Molecular Imaging for Integrated Medical Therapy and Drug Development.
DOI: 10.1007/978-4-431-98074-2_26, ©Springer 2010

Key words Alg13p · Alg14p · UDP-GlcNAc glycosyltransferase · Protein complex · Proteasomal degradation

Introduction

Protein asparagine *N*-glycosylation is one of the most frequent and common protein modifications; the process is important because it is required for the structure and function of glycoproteins. The process of *N*-glycosylation is initiated with the biosynthesis of a highly conserved dolichol-linked oligosaccharide (LLO) $Glc_3Man_9GlcNAc_2$ that begins on the cytoplasmic face of the endoplasmic reticulum (ER) and ends within the lumen [1–3]. In this process, the formation of a 14-sugar oligosaccharide is sequentially catalyzed by conserved ER glycosyltransferases [4]. *N*-glycosylation is essential for viability and mutations that affect glycosyltransferases that synthesize early steps of LLO are lethal [5–7].

The second step of LLO synthesis produces $GlcNAc_2$PP-Dol by transferring an *N*-acetylglucosamine (GlcNAc) from UDP-GlcNAc to GlcNAc-PP-Dol. This reaction is catalyzed by an unusual eukaryotic glycosyltransferase that is comprised of two separate polypeptides, Alg13 and Alg14 [6, 8, 9]. In yeast, both Alg13 and Alg14 subunits are essential for cell viability. Alg14 is a membrane protein that recruits the cytosolic Alg13 protein to the ER to form a hetero-oligomeric complex that catalyzes the biosynthesis of $GlcNAc_2$-PP-Dol [6]. Alg13 contains the catalytic domain of the UDP-GlcNAc transferase, but cytosolic Alg13 is not active unless bound to Alg14 at the ER membrane [9], suggesting that the formation of the Alg13/14 complex is crucial for UDP-GlcNAc transferase activity. The formation of the Alg13/14 complex has also been suggested to be a target for the regulation of *N*-linked glycosylation. Unassembled excess cytosolic Alg13p inhibits *N*-linked glycosylation, and is prevented to accumulate intracellularly by proteasomal degradation [10], providing further evidence for the importance of Alg13/14 complex formation.

Alg13 and Alg14 were first identified in silico by their structural homology to the bacterial MurG protein [8]. MurG is a UDP-GlcNAc undecaprenyl-PP-MurNAc pentapeptide: *N*-acetylglucosaminyl transferase, that plays an essential role in peptidoglycan biosynthesis in *Escherichia coli* [11]. MurG belongs to the GT-B glycosyltransferase (GTase) superfamily [12] and is one of the few family members whose crystal structure has been solved. MurG contains two distinct domains, each of which contains several Rossmann folds that are postulated to be involved in nucleotide-sugar binding and transfer, as well as lipid-acceptor recognition. The C-terminal domain contains the UDP-GlcNAc binding and catalytic domains where the enzymatic reaction occurs . . . ?, while the N-terminal domain contains the putative lipid-acceptor recognition domain [13, 14]. Alg13 protein is predicted to contain the conserved catalytic domain, found in the C-terminal domain of MurG, while Alg14 contains the predicted lipid acceptor- and membrane-associating domain, found in the N-terminal domain [8]. Recently, a NMR structure of yeast soluble Alg13 (2jzc) has been described. Although the N-terminal half of Alg13 contains an irregular mixed α/β domain rather than the predicted motif, this MRI analysis still suggests the presence of a highly conserved Rossmann fold in the C-terminal half of Alg13 [15], consistent with the predicted model [8]. These structural similarities between MurG and the Alg13/14 complex lead to the idea that eukaryotic LLO synthesis evolved from the bacterial glycosylation process, but the questions of why the Alg13/14 UDP-GlcNAc transferase split into two subunits and what regulates their interaction remain. In this study, we report that complex formation of Alg13 and Alg14 requires only the C-terminal α helix of Alg13 and the last three amino acids of Alg14. As the second function of the C-terminal α helix of Alg13, we also present the evidence that this structure can serves as an autonomous degradation signal required for regulating the protein level of cytosolic Alg13.

Experimental Procedures

Homology Modeling of the Alg13/Alg14 Complex

A refined three-dimensional (3D) homology model of the yeast Alg13/14 complex was constructed based on the crystal structure of *E. coli* MurG (PDB code, 1NLM Chain A) using the Modeller 9.1 [16, 17]. The sequence/template alignment between the MurG and Alg13/14 subunits was used for the modeling [8]. Divisions between the Alg13 and Alg14 sequences were described by using the character "/" and gaps were inserted. The model was constructed in the following three steps: (i) 200 models of the first construction were generated using the Automodel class. The model with the best objective score was selected and advanced to the next step. (ii) The geometry of its loop regions was roughly optimized using Loopmodel, and the best model in 100 candidates was selected for the next calculation. (iii) To further optimize a large loop in the yeast Alg13 subunit (from 53 to 79 residues), the best model selected in the second step was again applied to the Loopmodel class. The final 3D model of the Alg13/14 complex was then selected from 50 candidates that had the best objective score. The DOPE scores calculated for the model (yeast Alg13/14 complex) and MurG were −38 927.46 and −41 430.95, respectively. To represent the molecular surface of the Alg14 and the C-terminal region of Alg13 shown in Fig. 4 the SYBYL 7.1 program package (Tripos Inc, St. Louis, MO, USA) was used [18].

Plasmids, Yeast Strains and Media

Plasmids used in this work and their important features are listed in Table 1. Standard molecular biology techniques were used for all plasmid constructions [19]. The correct sequence of all polymerase chain reaction (PCR)-amplified products

Table 1. Plasmids used in this study

Plasmid	Description	Source (reference number)
pRS305	*LEU2/CEN6* yeast integration vector	29
pRS306	*URA3/CEN6* yeast integration vector	29
pRS304	*TRP1/CEN6* yeast integration vector	29
pXG202	*3HA-ALG14* expressed from the *ALG14* promoter in pRS306	6
pXG208	*ALG13-3FLAG* expressed from the *ALG13* promoter in pRS306	22
pXG211	*3FLAG-ALG13* expressed from the *ALG13* promoter in pRS305	22
pSA1	*alg13-NΔ6-3FLAG* expressed from the *ALG13* promoter in pRS306	22
pSA2	*alg13-NΔ10-3FLAG* expressed from the *ALG13* promoter in pRS306	22
pSA4	*3FLAG- alg13-CΔ3* expressed from the *ALG13* promoter in pRS305	22
pSA6	*3FLAG- alg13-CΔ15* expressed from the *ALG13* promoter in pRS305	22
pSA7	*3HA- alg14-CΔ3* expressed from the *ALG14* promoter in pRS306	22
pSA8	*3HA- alg14-CΔ12* expressed from the *ALG14* promoter in pRS306	22
pSA13	*3HA-alg14-L236G,V237G* expressed from the *ALG14* promoter in pRS306	22
pSA14	*3HA-alg14- I235G, L236G, V237G* expressed from the *ALG14* promoter in pRS306	22
pSA15	*3FLAG- alg13-F192G* expressed from the *ALG13* promoter in pRS306	This study
pSA16	*3FLAG- alg13-L196G* expressed from the *ALG13* promoter in pRS306	This study
pSA17	*3FLAG- alg13-I200G* expressed from the *ALG13* promoter in pRS306	This study
pSA18	*3FLAG- alg13-F192, L196G* expressed from the *ALG13* promoter in pRS306	This study
pSA19	*3FLAG- ALG13* expressed from the *GAL1/10* promoter in pRS304	This study
pSA20	*3FLAG- Alg13* expressed from the *GAL1/10* promoter in pRS304	This study
pSA21	*3FLAG- alg13-F192, L196, I200, Y201A* expressed from the *GAL1/10* promoter in pRS304	This study
pSA22	*3FLAG- alg13-F192, L196, I200, Y201G* expressed from the *GAL1/10* promoter in pRS304	This study
pSA23	*3FLAG- alg13-CΔ15* expressed from the *GAL1/10* promoter in pRS304	This study

was verified by DNA sequencing. The sequences of primers used in this study are available upon request. W303a (*MATa ade2-1 ura3-1 his3-11 trp1-1 leu2-3, 112 can1-100*) is the parental strain for all the strains used in the work. XGY151 and XGY154, which contain replacements of the *ALG14* and *ALG13* promoters with the glucose-repressible *GAL1/10* promoter, respectively [6], were used for testing the activity of truncated and mutated Alg13 or Alg14 proteins by monitoring complementation of the lethality associated with the loss of *ALG13* or *ALG14* function. XGY155 contains a C-terminal triple FLAG-tagged *ALG13* allele, marked by the *S. pombe his5*$^+$ gene [6]. Standard yeast media, growth conditions, and genetic techniques were used [20].

Preparation of Cell-Free Lysates and Western Analysis

Exponentially growing yeast cells were harvested at an OD_{600} of 1–3 and converted to spheroplasts with lyticase. To prepare detergent-solubilized extracts, spheroplasts (6–7 OD units) were resuspended in 500 μl of ice-cold lysis buffer (150 mM NaCl, 10 mM [HEPES]-KOH (pH 7.5), 5 mM $MgCl_2$) with protease inhibitors and 1% Triton X-100 as described previously [6]. To prepare an ER enriched membrane fraction, 6–7 OD units of exponentially growing yeast cells were lysed by glass bead beating in 500 μl of ice-cold lysis buffer. The lysate was centrifuged at 3000 *g* for 5 min to remove the unbroken cells and wall debris. The collected post-3000 *g* supernatant was centrifuged at 20 000 *g* for 30 min in a TOMY MX-301 centrifuge. The pellet (P20) was re-suspended in 500 μl of lysis buffer followed by the determination of its protein concentration. This P20 fraction was used as the ER membrane fraction.

To make the membrane fraction that was used in the experiments described in Fig. 6, exponentially growing yeast cells (10 OD_{600}) were homogenized by vortexing with glass beads in ice-cold lysis buffer and centrifuged at 3 kg for 10 min to remove unbroken cells and wall debris. The 3 kg supernatant was directly centrifuged at 100 kg for 30 min in a Beckman Optima TL ultracentrifuge (Beckman, Fullerton, CA, USA). The soluble (cytosolic) protein-containing supernatant (S100) was collected, and the membrane-containing pellet (P100) was re-suspended in 100 μl lysis buffer. The protein concentrations of both fractions were determined by BCA Protein Assay (Pierce Chemical, Rockford, IL, USA).

Equivalent amounts (about 80 μg) of protein in each fraction were separated by 12% sodium dodecylsulfate (SDS)-polyacrylamide gel electrophoresis (PAGE) and transferred to Immobilon-PVDF membranes (Millipore, Billerica, MA, USA). The membrane was blotted with anti-FLAG M2 monoclonal antibody conjugated to alkaline phosphatase (Sigma-Aldrich, St. Louis, MO, USA) or a mouse anti-HA monoclonal antibody conjugated to peroxidase (Anti-[HA]-Peroxidase; Roche, Basel, Switzerland) and detected by chemiluminescence [CDP-*Star* Detection Reagent; Roche or ECL, GE (Healthcare UK, Buckinghamshire, UK)].

Co-Immunoprecipitation

Epitope-tagged proteins in 1% Triton X-100 extracts were immunoprecipitated with anti-HA Affinity Matrix (Sigma-Aldrich) as described [6]. After separating proteins by 12% SDS-PAGE, immunoprecipitates were transferred to Immobilon-PVDF membranes (Millipore). To detect the FLAG-tagged proteins, the membrane was blotted with a rabbit anti-FLAG polyclonal antibody (Rockland, Gilbertsville, PA, USA) followed by a secondary anti-rabbit antibody conjugated to alkaline phosphatase (Chemicon International, Temecula, CA, USA).

Results

Stereo-Structure Suggests the Importance of Termini for Formation of the Alg13/Alg14 Complex

To gain more information about the Alg13/14 complex, its stereo-structure was modeled, based on the sequence alignments and secondary structure predictions of Alg13 and Alg14 with those of MurG (see "Experimental Procedures") [8]. The structure represented by the ribbon in Fig. 1 shows the Alg13/Alg14 complex (left panel) adopting a GT-B fold with a two-domain architecture that shows remarkable similarities to MurG (right panel). Like the C- and N-terminal domains of MurG, Alg13 and Alg14 both contain classical Rossman folds with central seven-stranded parallel β sheets (Fig. 1, drawn in yellow) flanked by α helices (Fig. 1, drawn in red). Like MurG, Alg13 possesses an extended C-terminal α helix (Alg13C) that is predicted to closely appose Alg14 (Fig. 1, left panel, drawn in blue). This extended α helix of MurG is considered to be the major feature of the GT-B fold (Fig. 1, right panel, drawn in blue). Notable structural differences between the predicted Alg13/Alg14 complex and MurG were also found. Alg14 possesses a transmembrane domain at its N-terminus that is absent in MurG (Fig. 1, left panel, indicated by a solid arrow; real structure extends out of the panel). Another major difference is the absence of the linker peptide that connects the two Rossman fold domains of MurG (Fig. 1, right panel, drawn in green). According to our predicted model, a consequence of the absence of this linker is that the N-terminus of Alg13 and the C-terminus of Alg14 are exposed at the surface of the complex (Fig. 1, left panel). Especially, the exposed C-terminus of Alg14 positions is near the end of the extended C-terminal α helix of Alg13, suggesting the possibility that these terminal regions of Alg13 and Alg14 mediate their interaction. To test this model, we performed a series of molecular biological experiments using a yeast cell system to examine the structural and functional importance of these terminal regions.

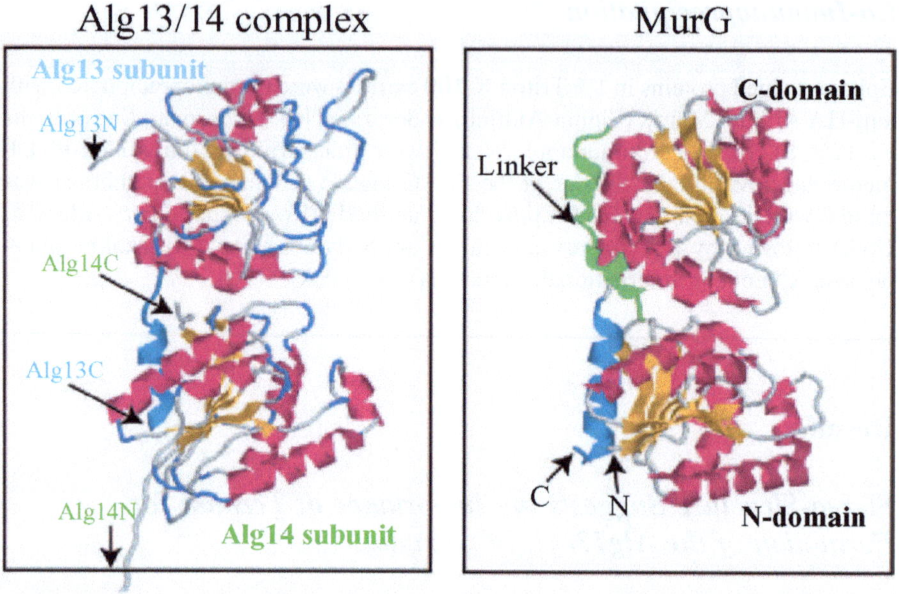

Fig. 1. Theoretical structure of the Alg13/14 complex is structurally similar to MurG. The *ribbon representation* shows the predicted Alg13/14 complex (*left panel*) adopting the GT-B fold with a structure very similar to that of MurG (*right panel*). (See "Experimental Procedures" for details of how these models were generated). For both structures, the β strands in the two Rossmann fold subdomains are drawn in *yellow*, and the flanking α helices are in *red*. The linker peptide that connects the two Rossmann fold domains in MurG but that is absent from the Alg13/14 complex is drawn in *green*. The extended C-terminal α helixes of MurG and Alg13 are drawn in *blue*. The termini of all proteins are indicated by *arrows*

The Predicted C-Terminal α Helix of Alg13 Is Important for Its Interaction with Alg14

To test if the C-terminus of Alg13 is required for enzyme activity and/or complex formation, mutant alleles encoding Alg13 proteins deleted for 3 (*alg13-CΔ3*) or 15 (*alg13-CΔ15*) C-terminal amino acids were constructed. Truncated proteins were tagged with an N-terminal triple FLAG tag. The C-terminus of yeast Alg13 contains hydrophobic amino acids (-FXXLLVXXIX-) and is predicted to adopt an α helical structure (Fig. 2a). Deletion of 15 amino acids completely disrupts this α helix. To test their functionality, mutant *alg13* alleles were introduced into a P_{GAL1}-*ALG13* strain (XGY154) containing *ALG13* under the control of the glucose-repressible *GAL1* promoter. This strain grows normally in galactose, but fails to grow when *ALG13* gene expression is repressed by glucose. These complementation experiments demonstrated that while the deletion of 3 amino acids had no affect on Alg13, the deletion of 15 C-terminal amino acids completely impaired Alg13 function,

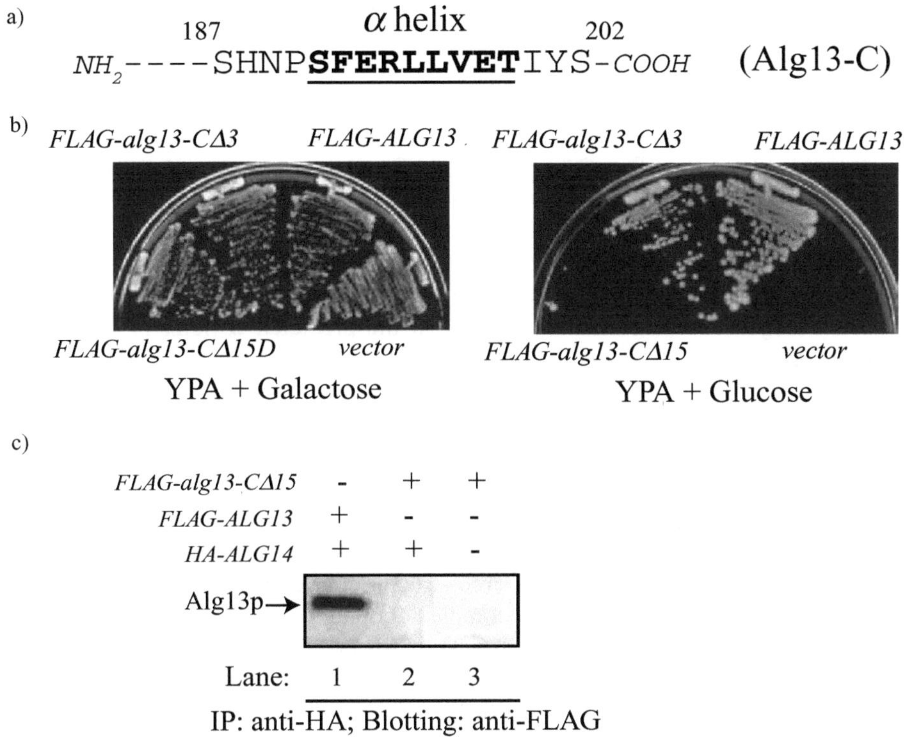

Fig. 2a–c. The C-terminal α helix of yeast Alg13 is required for viability and for interaction with Alg14. **a** The C-terminal sequence of yeast Alg13 protein. The predicted α helix is shown in *boldface and underlined.* **b** Yeast strain (XGY154) with *ALG13* under control of the glucose repressible *GAL1* promoter (P_{GAL1}) was transformed with plasmids containing *FLAG-ALG13* (pXG211), *FLAG-alg13-CΔ3* (pSA4), or *FLAG-alg13-CΔ15* (pSA6), and streaked onto YPA plates supplemented with galactose (*left panel*) or glucose (*right panel*). Cells were incubated for 2 days at 30° C. **c** Whole-cell detergent extracts were prepared from a wild-type strain (W303a) that co-expresses *HA-ALG14* (pXG202) and *FLAG-alg13-CΔ15* (pSA6) or *FLAG-ALG13* (pXG211). Samples were immunoprecipitated with anti-HA affinity matrix by Western blotting. Proteins or extracts were separated by 12% SDS-PAGE, immunoblotted with rabbit anti-FLAG antibodies, and detected by chemiluminescence as described in "Experimental Procedures"

because *FLAG-alg13-CΔ15* failed to complement the growth of P_{GAL1}-*ALG13* strain XGY154 cells in glucose (Fig. 2b). These results suggested that the C-terminal α helix domain of Alg13 is essential for its function.

Our structural analysis suggested that the C-terminus of Alg13 was involved in its interaction with Alg14. To test whether this idea is correct, a co-immunoprecipitation assay was performed. Yeast strains were constructed that co-express N-terminally HA-tagged Alg14 with N-terminally FLAG-tagged Alg13-CΔ15p or wild-type Alg13p as a positive control (Fig. 2c). Detergent extracts were prepared from these strains and clarified by centrifugation at 100 000 *g* to remove any nonspecific protein aggregates (see "Experimental Procedures"). Alg14 proteins were immunoprecipitated with an anti-HA affinity matrix. The immunoprecipitates were separated by SDS-PAGE, and blotted with anti-FLAG rabbit antibodies to determine whether the truncated alg13-CΔ15p bound to Alg14. Unlike wild-type Alg13p, we found that HA-tagged Alg14 failed to bind FLAG-tagged Alg13-CΔ15p (Fig. 2c, lanes 1 and 2). Our failure to detect any bound Alg13-CΔ15p was not due to lowered intracellular levels of this protein, because Alg13 lacking its C-terminal 15 amino acids is as stable and as abundant as wild-type Alg13 protein in detergent extracts (data not shown). These results demonstrated that the C-terminal 15 amino acids of Alg13 are required for the interaction between Alg14 and Alg13 and suggest that the predicted α helix mediates this interaction.

The C-Terminus of Alg14 Is Required for Its Interaction with Alg13

Our computational analysis of the stereo-structure of yeast Alg14 implied that the C-terminal region was involved in complex formation (Fig. 1). Figure 3a shows the C-terminal amino acid sequence of yeast Alg14 protein that includes the predicted β strand. To test whether or not this conserved region of Alg14 protein is important for its interaction with Alg13, mutant alleles were constructed that encoded truncated Alg14, lacking 3 (*HA-alg14-CΔ3*) or 12 (*HA-alg14-CΔ12*) C-terminal amino acids. These truncated Alg14 proteins were tagged with an N-terminal triple HA. These truncated Alg14 proteins were expressed in a yeast strain containing *ALG14* under the control of the glucose-repressible *GAL1* promoter (XGY151) and tested by complementation of loss of Alg14 function in the presence of glucose. Surprisingly, neither *HA-alg14-CΔ3* nor *HA-alg14-CΔ12* complemented the growth of the XGY151 strain in the presence of glucose (Fig. 3b), suggesting that deletion of just three amino acids from the C-terminus of Alg14 protein completely impaired Alg14 function.

To study how these C-terminal residues affect Alg14 function, the stability of these truncated Alg14 proteins and their interaction with Alg13 were tested. Strains were constructed that co-expressed these HA-tagged truncated Alg14 proteins and FLAG-tagged Alg13 proteins. The steady-state levels of these mutant proteins (HA-Alg14-CΔ3p or HA-Alg14-CΔ12p) in ER membranes, prepared by differential centrifugation, were compared quantitatively to wild-type HA-Alg14 protein by Western blotting with anti-HA antibody. This experiment demonstrated that both the normal and truncated Alg14 proteins were present at relatively similar levels in the ER (Fig. 3c, lanes 2–4). These results demonstrated that the deletion of up to 12 C-terminal amino acids had no significant effect on Alg14 protein stability.

To determine if these residues are required for an interaction with Alg13, a co-immunoprecipitation assay was used. To measure the relative amount of FLAG-

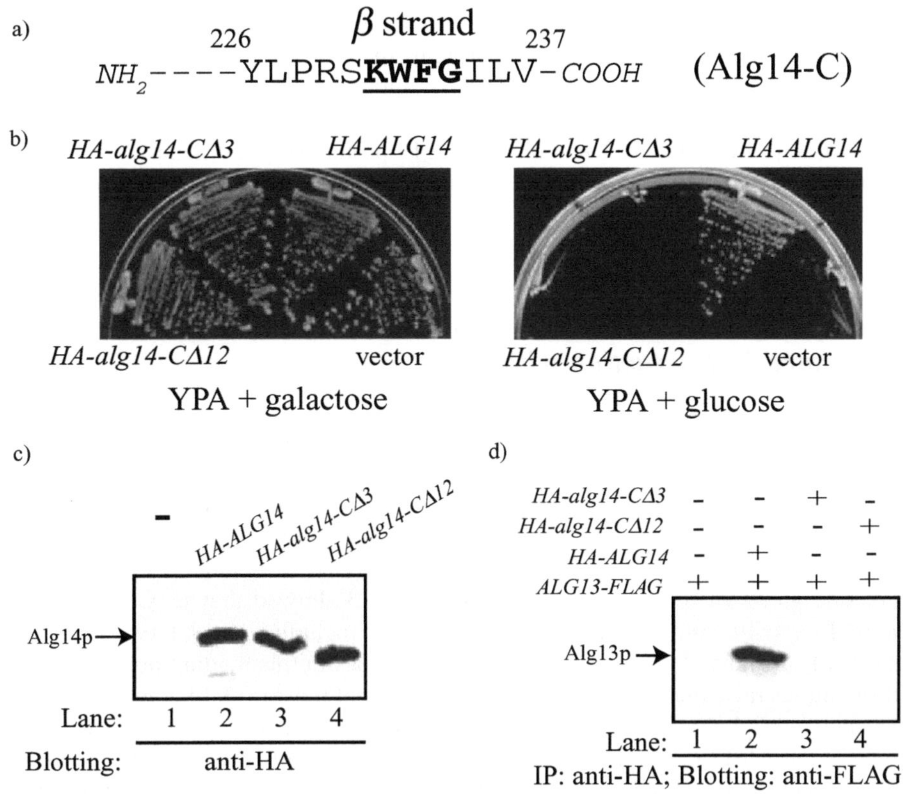

Fig. 3a–d. The C-terminal amino acids of Alg14 are required for viability and interaction with Alg13. **a** The C-terminal sequence of yeast Alg14 protein. The predicted β strand region is *shown in boldface and underlined.* **b** Yeast strain (XGY151) with *ALG14* under control of the glucose repressible *GAL1* promoter (P_{GAL1}) was transformed with plasmids containing *HA-ALG14* (pXG202), *HA-alg14-CΔ3* (pSA7) or *HA-alg14-CΔ12* (pSA8), and streaked onto YPA plates supplemented with galactose (*left panel*) or glucose (*right panel*). Cells were incubated for 2 days at 30° C. **c** A strain that contains a C-terminal triple FLAG-tagged *ALG13* at the chromosomal ALG13 locus (XGY155) was transformed with plasmids encoding *HA-ALG14* (pXG202) or *HA-alg14-CΔ3* (pSA7) or *HA-alg14-CΔ12* (pSA8). Endoplasmic reticulum (ER)-enriched membrane fractions (P20) from the transformed cells were prepared as described in "Experimental Procedures." Protein (80 μg) from each sample was separated by 12% SDS-PAGE and quantitatively analyzed for HA-tagged Alg14 proteins by immunoblotting with a mouse anti-HA monoclonal antibody conjugated to peroxidase. **d** Whole-cell detergent extracts were prepared from the same strains as those used in the Western blotting assay described above. Samples were immunoprecipitated with anti-HA affinity matrix. Immunoprecipitated proteins or extracts were separated by 12% SDS-PAGE, immunoblotted with rabbit anti-FLAG antibodies, and detected by chemiluminescence, as described in "Experimental Procedures"

tagged Alg13 that associated with the HA-tagged mutant or wild-type Alg14, detergent extracts were prepared, and proteins that precipitated with anti-HA antibody were separated by SDS-PAGE and immunoblotted with anti-FLAG rabbit antibodies (Fig. 3d). Unlike wild-type Alg14p, neither HA-Alg14-CΔ12p nor HA-Alg14-CΔ3p precipitated FLAG-tagged Alg13 protein (Fig. 3d, lanes 2–4). These results demonstrated that just the last three amino acids of Alg14 protein are required for its interaction with Alg13.

The Hydrophobic Face in C-Terminal α Helix of Alg13 Is Implicated in Interaction with Alg14

A close-up electrostatic surface view of the predicted yeast Alg13/14 complex from our computational analyses suggested that the C-terminal α helix of Alg13 (Fig. 4a, b. shown in the cartoon) is inserted in a cleft formed in the Alg14 subunit (Fig. 4a, b. colored in brown inside the white circle). This model suggests that the interaction between Alg13 and Alg14 takes place within a hydrophobic pocket. The top and side views of the C-terminal region of Alg13 showed that its C-terminal α helix (Fig. 4c in window had its hydrophobic face including F190, L196, and I200 (Fig. 4d) surrounded by Alg14 hydrophobic residues in this binding pocket. These observations raise the possibility that the assembly of the Alg13/14 complex occurs through hydrophobic interactions between these two domains. The sequence alignment of the Alg13 C-terminal α helix domain also supports this idea. As shown in Fig. 4, three hydrophobic residues, F190, L196, and I200, which constitute the hydrophobic face in yeast Alg13, were highly conserved in all eukaryotic Alg13 homologues. Especially, the F190 and L196 residues were contained at exactly the same position in all Alg13 homologues that we have searched, from yeast to human (Fig. 5).

To test this idea and understand the involvement of the hydrophobic face of the Alg13 C-terminal α helix in interaction with Alg14, these three hydrophobic amino acids of yeast Alg13 were mutagenized. Plasmids expressing triple FLAG-tagged Alg13 proteins with F^{190}, L^{196}, I^{200} or both F^{190} and L^{196} replaced by G, respectively, were constructed and introduced into the P_{GAL1}-*ALG13* strain (XGY154) to test for complementation of *alg13*. As shown in Fig. 6a, replacement of both F^{190} and L^{196} to G completely impaired FLAG-Alg13 function, because P_{GAL1}-*ALG13* cells expressing the F192G, L196G allele failed to grow in the presence of glucose. But Alg13-F192G, Alg13-L196G, or Alg13-I200G proteins supported the growth of the P_{GAL1}-*ALG13* cells in the presence of glucose, indicating that mutants with only a single mutation on the hydrophobic face have no serious effect on FLAG-Alg13 activity. To further confirm the effect of these hydrophobic residues on Alg13 protein, the localization of mutant proteins with a single mutation was analyzed. Cell lysates were first prepared from strains that had been used for the complementation test shown in Fig. 6a, and separated into membrane and soluble (cytosolic) proteins by differential centrifugation (see "Experimental Procedures" for details).

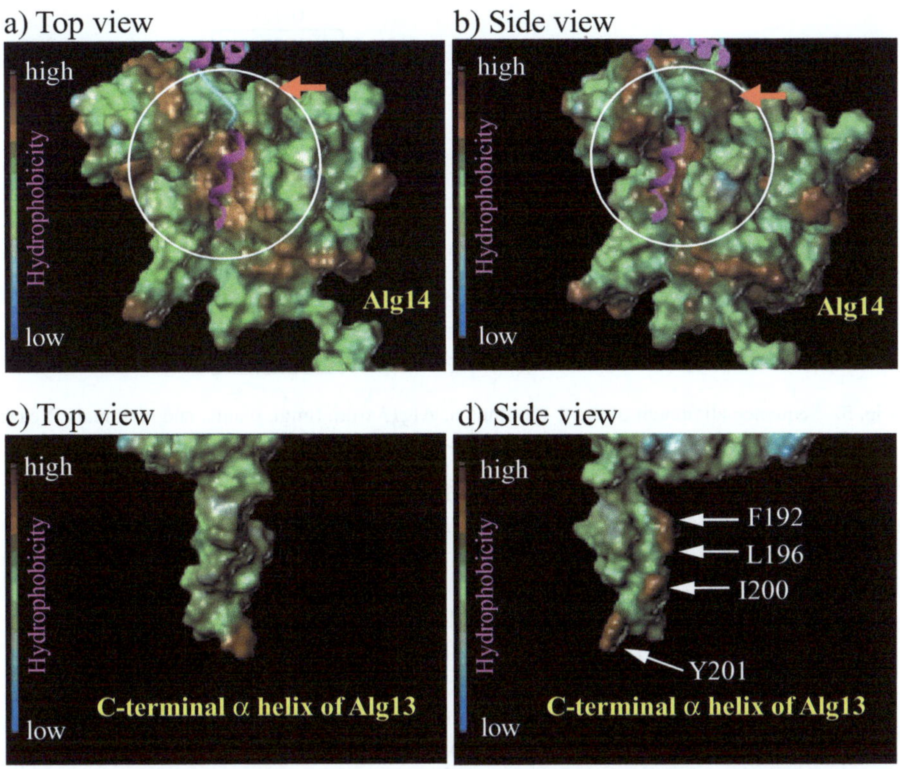

Fig. 4a–d. Molecular surface view of Alg14 suggests a hydrophobic cleft that serves as a binding pocket for the C-terminal α helix of Alg13. The molecular surface of yeast Alg13 and 14 proteins is colored according to the residue hydrophobicity in which the high-to-low lipophilicity scale corresponds to the color ramp from *blue to brown*. **a** Shows top view and **b** shows side view of Alg14. The *areas within the white circles* indicate the putative binding pocket for the C-terminal α helix of Alg13. The *arrows in red* indicate the last three hydrophobic residues of Alg14. The structure of the C-terminal region of Alg13, including the last α helix, is shown in the cartoon. **c** Shows top view and **d** shows side view of the C-terminal region of Alg13, including the proposed α helix that interacts with Alg14 at the binding pocket. The hydrophobic amino acids *shown with arrows* on the side face of the C-terminal region are predicted to be involved in the interaction with the Alg14 in the binding pocket

The amount of various triple FLAG-tagged Alg13 proteins in each fraction was then assayed by immunoblotting with anti-FLAG antibodies (Fig. 6b, c). Wild-type Alg13 protein was observed in both the membrane and the soluble (cytosolic) fractions, but at about a 5-to-1 ratio (Fig. 6b lanes 1 and 2; and Fig. 6c), demonstrating its predominant localization in the membrane fraction. In marked contrast, mutant proteins with a single mutation in conserved hydrophobic residues at the C-terminal domain were found mainly in the cytosolic fraction at amounts three to four times higher than those in the membrane fraction (Fig. 6b, lanes 3–8; and Fig. 6c). Y201, an unconserved hydrophobic residue found in the C-terminal domain of yeast

		α-helix	
Fungi	*S. cerevisiae*	---VSHNPSFERLLVETIYS---*COOH*	202
	A. gossypii	---PAYKQGFAELLQDVAHR---*COOH*	203
	K. lactis	----TYNKQFEKLLYNVAID---*COOH*	197
	E. gossypii	---PAYKQGFAELLQDVAHR---*COOH*	203
Fish& insects	*S. salar*	-FLPGQPKHFANFLDRALGVQ--*COOH*	164
	C. quinquefasciatus	---PGSVADFISYLDAFMGF---*COOH*	170
	P. humanus corporis	---RVNPNIFPIYLNNLMGFQ--*COOH*	168
Mammals	*H. sapiens*	---PGQPEKFSAFLDKVVGLQK-*COOH*	165
	R. norvegicus	---PGQPEKFSAFLDKVVGLQK-*COOH*	165
	M. musculus	---PGQPEKFSAFLDKVVGLQK-*COOH*	165
	B. taurus	---PGQPEKFSAFLDKVVGLQK-*COOH*	165

Fig. 5. Sequence alignment of the C-terminus of Alg13 from fungi, plants, and mammals. Yeast Alg13 protein was used as the query for PSI-BLAST [28] analysis against the UniRef90 protein dataset at UniProt. The last 17 to 20 C-terminal amino acids of Alg13 orthologues selected from the PSI-BLAST results were manually aligned. The predicted α helix is *boxed*, and hydrophobic residues that are conserved in all Alg14 orthologues are shaded in *gray*

Alg13 (Fig. 4d) was replaced to G, and its localization was also tested. Unlike the Alg13-F192G, Alg13-L196G, and Alg13-I200G proteins, Alg13-Y201G was observed in both membrane and cytosolic fractions with a 1-to-1.4 ratio, indicating less effect of this hydrophobic residue on the localization of Alg13. Taken together, our results demonstrate that hydrophobic residues highly conserved in the C-terminal α helix of Alg13 are implicated in its localization on the ER membrane.

The Hydrophobicity of the Last Three Amino Acids of Alg14 Is Also Important for the Formation of the Alg13/14 Complex

The molecular surface view also supports the idea that the last three amino acids of Alg14 are involved in Alg13/14 complex formation. These three hydrophobic amino acids (Fig. 3a) are found at the entrance of the hydrophobic pocket (Fig. 4a, b, indicated by red arrow), where we hypothesize they could function as a key to lock the C-terminal α helix of Alg13 into the hydrophobic pocket. To further understand the involvement of C-terminal residues of Alg14 in complex formation, this domain of Alg14 was compared in other species (Fig. 7a). As shown in Fig. 7a, Alg14 proteins in most eukaryotes contain a similar C-terminus that includes an invariant glycine residue (Fig. 7a; marked with asterisks). These amino acids are predicted to fold into a β strand that is followed by a tail consisting of the last three amino acids (Fig. 7a). The hydrophobicity of this "tail" has been conserved; fungal Alg14 proteins contain hydrophobic amino acids at their last three positions, while plants and animals contain two. This conservation suggested that the hydrophobicity of this tail might be important for function.

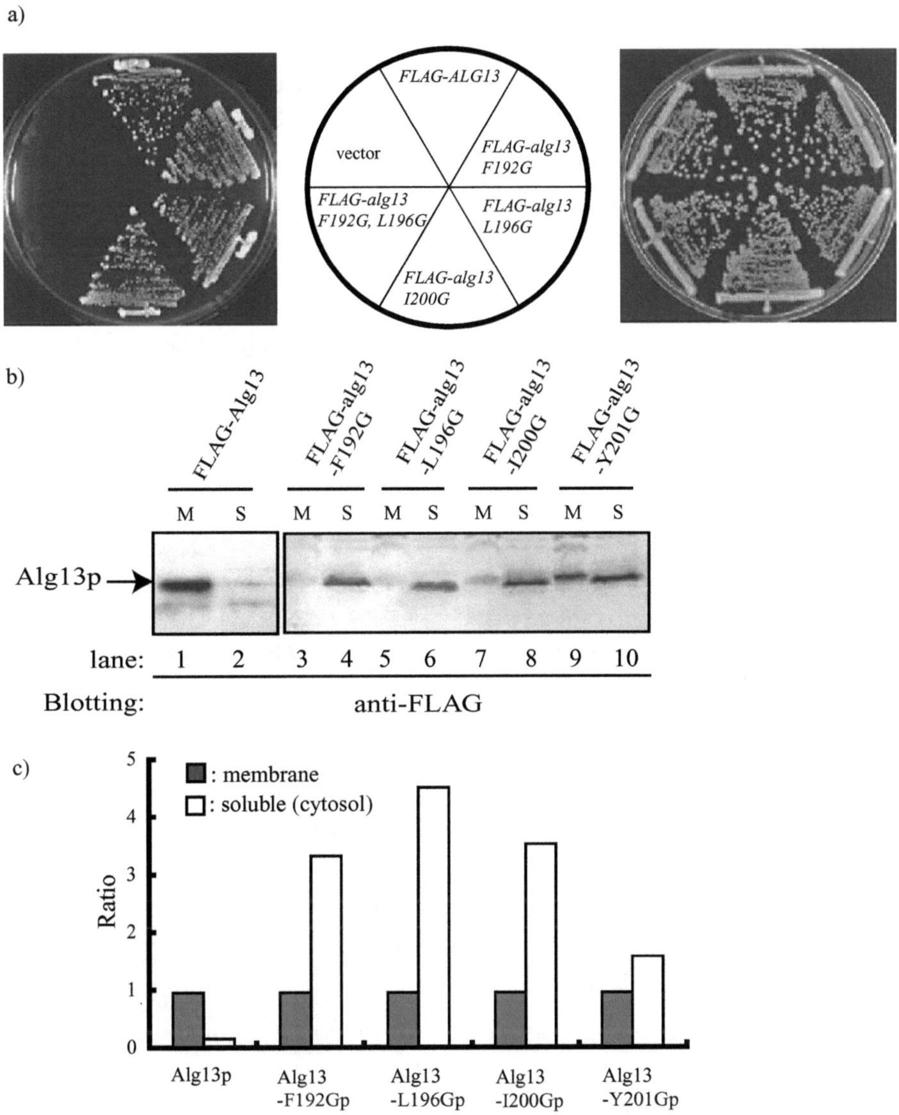

Fig. 6a–c. Hydrophobic residues conserved in C-terminal α helix of Alg13 are required for enzyme activity and complex formation. **a** Alg13 proteins with highly conserved hydrophobic residues in the C-terminal α helix replaced by glycine were tested for activity. The P_{GAL1}-*ALG13* strain (XGY154) was transformed with plasmids containing *FLAG-ALG13* (pXG211), *FLAG-alg13-F192G* (pSA15), *FLAG-alg13-L196G* (pSA16), *FLAG-alg13-I200G* (pSA17), or *FLAG-alg13-F192. L196G* (pSA18) was streaked onto YPA plates supplemented with galactose (*right panel*) or glucose (*left panel*). Cells were incubated for 2 days at 30°C. **b** The soluble (cytosolic) and membrane fractions from the strains shown in **a** were prepared as described in "Experimental Procedures." Protein (80 μg) from each sample was separated by 12% SDS-PAGE and quantitatively analyzed for FLAG-tagged Alg13 proteins by immunoblotting with a mouse anti-FLAG monoclonal antibody conjugated to horseradish peroxidase. **c** The relative amounts of triple-FLAG-tagged Alg13 proteins in each sample for immunoblotting were quantitated by densitometry using the LAS4000 System (Fujifilm, Tokyo, Japan). The values are shown graphically, with the density of each membrane-associated Alg13 protein set as 1; the values of the cytosolic proteins are displayed as ratios compared to the membrane proteins

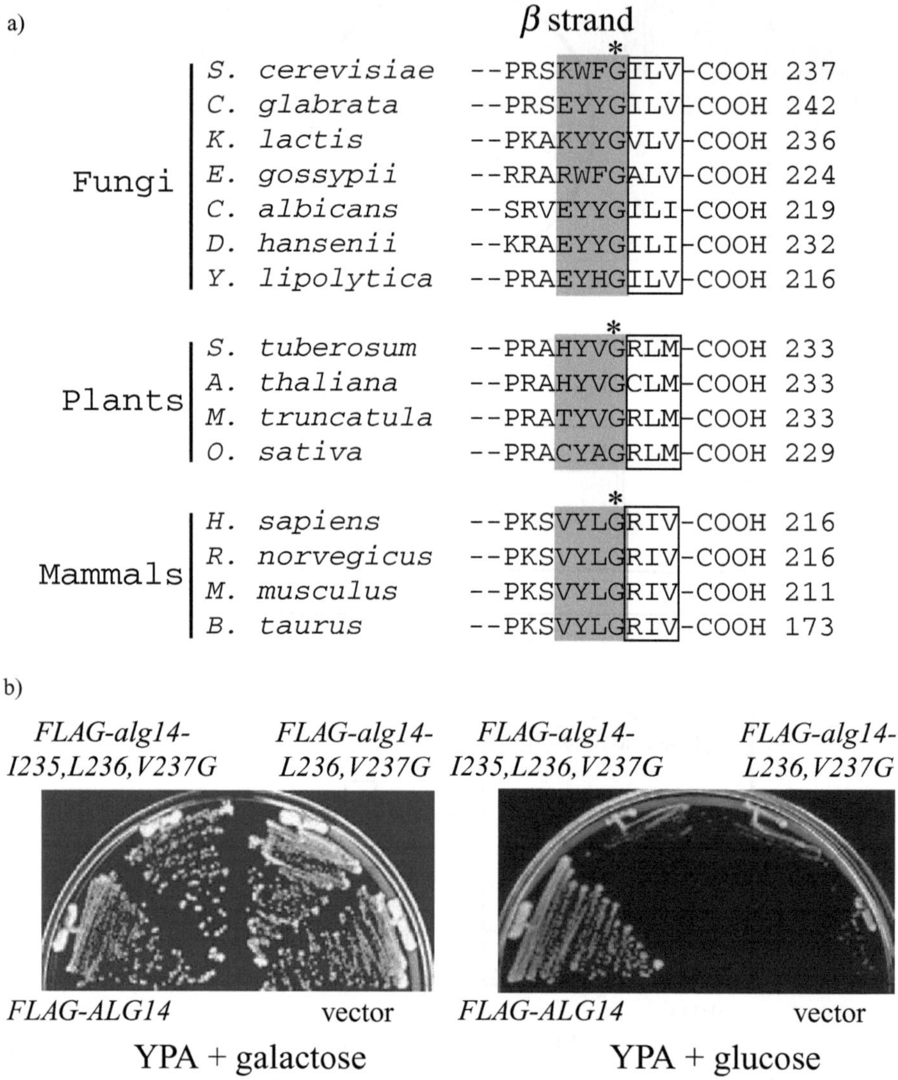

Fig. 7a,b. The hydrophobicity of Alg14 C-terminal amino acids is required for the formation of an activity Alg13/14 complex. **a** Yeast Alg14 protein was used as the query for PSI-BLAST [28] analysis against the UniRef90 protein dataset at UniProt. The last ten C-terminal amino acids of Alg14 orthologues selected from the PSI-BLAST results were manually aligned. The predicted β strands are *shaded in gray*, and the glycine residue that is conserved in all Alg14 orthologues is marked with *asterisks*. The last three amino acids are *boxed*. **b** Alg14 proteins with the last two or three residues that were replaced by glycine were tested for activity. The P_{GAL1}-*ALG14* strain (XGY151) was transformed with plasmids containing *HA-ALG14* (pXG202), *HA-alg14-L236G, V237G* (pSA13), or *HA-alg14-I235G, L236G, V237G* (pSA14), and streaked onto YPA plates supplemented with glucose (*right panel*) or galactose (*left panel*). Cells were incubated for 2 days at 30°C

To test this idea, the last three (ILV) or two (LV) amino acids of yeast Alg14 were altered. Plasmids expressing Alg14p with the last two or three residues replaced by G were constructed and introduced into the P_{GAL1}-*ALG14* strain (XGY151) to test for complementation of *alg14*. Replacement of the ILV or LV residues to G led to severe Alg14 dysfunction, because the P_{GAL1}-*ALG14* cells expressing Alg14 ILV or LV to G mutant proteins grew very slowly, even after incubation at 30° C for 3 days on YPD medium supplemented with glucose (Fig. 7b, right panel). Each of these proteins was also FLAG-tagged at the N-terminus. We confirmed, by Western blotting using anti-FLAG antibody, that these mutations did not affect protein production or stability, because each of these mutant proteins was produced at levels comparable to wild-type Alg14p (data not shown). These results demonstrate that the hydrophobicity of the last three or two C-terminal amino acids of Alg14 protein plays an important role in Alg13/14 function.

Degradation of Cytosolic Free Alg13 Requires Its C-Terminal α Helix

It has been demonstrated that excess cytosolic free Alg13 is regulated intracellularly by proteasomal degradation, and the C-terminal domain including 100 amino acids serves as an autonomous degradation signal [10]. Furthermore, NMR analysis reveals a dangling structure of the terminal α helix in unassembled Alg13, suggesting a possibility that this structure may also be recognized by other cytosolic factors [15]. Based on this biological and structural evidence, we decided to test whether this α helix was involved in the degradation of cytosolic Alg13. First, mutant alleles encoding triple FLAG-tagged yeast Alg13 proteins with F^{190}, L^{196}, I^{200}, and Y^{201}, replaced, respectively, by G (*alg13-F192, L196, I200, Y201G*) or A (*alg13-F192, L196, I200, Y201A*) were constructed and placed under the control of galactose-driven *GAL1* promoter. As a control for the disruption of the last α helix, plasmids overexpressing triple FLAG-tagged Alg13 protein deleted for 15 (*alg13-CΔ15*) C-terminal amino acids were also prepared. The C-terminal sequences of the constructed Alg13 mutants were applied to a DSSP program for predicting their secondary structures [21]. As shown in Fig. 8a, replacement of four residues with glycine (alg13-F192, L196, I200, Y201Gp) resulted in the disruption of the α helix, while alg13-F192, L196, I200, Y201A protein (replacement with alanine) still maintained the α helix structure (Fig. 8a). It was confirmed that both mutant proteins failed to complement the growth of P_{GAL1}-*ALG13* strain XGY154 cells in glucose (data not shown). Whole-cell detergent extracts were prepared from wild-type strain (W303a) harboring plasmids that encode mutant Alg13 proteins (see Table 1), which were grown in YPAD medium supplemented with galactose for overexpressing mutant proteins. Then, the steady-state levels of these mutant proteins were compared quantitatively to wild-type 3FLAG-Alg13p by Western blotting with anti-FLAG antibody (Fig. 8b). Although both alg13-F192, L196, I200, Y201Gp and alg13-F192, L196, I200, Y201Ap are inactive proteins, the stability

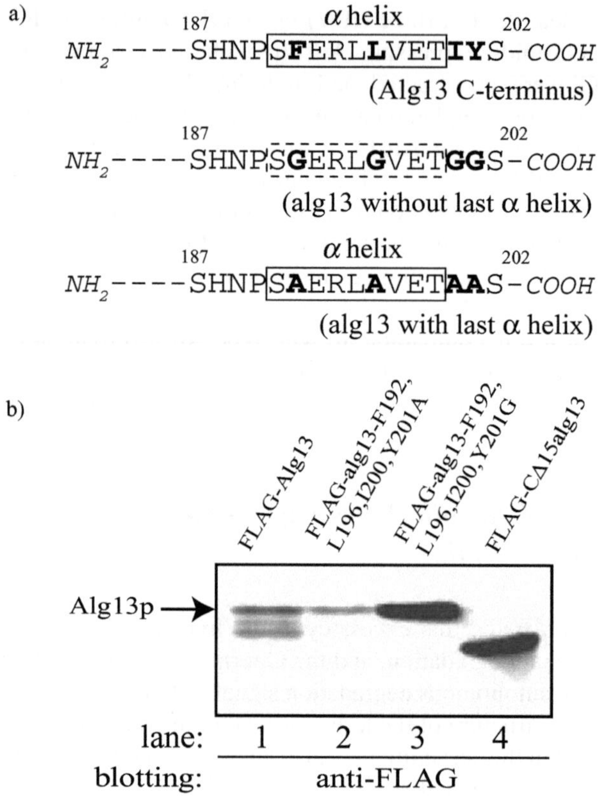

Fig. 8a,b. Degradation of cytosolic free Alg13 requires its C-terminal α helix. **a** The predicted secondary structures possessed by the C-termini of yeast Alg13 protein and its mutant proteins are shown. The α helix is *boxed with a continuous line*, and the perturbed α helix is *boxed with a dashed line*. The amino acids that were mutagenized in the experiment are shown in *boldface*. **b** Wild-type strain (W303a) transformed with plasmids encoding *3FLAG-alg13-F192, L196, I200, Y201G; 3FLAG-alg13-F192, L196, I200, Y201A*; *3FLAG-alg13-CΔ15*, or *3FLAG-Alg13* under the galactose-driven *GAL1* promoter was grown in YPA medium supplemented with galactose. Whole-cell detergent extracts were prepared, and samples were quantitatively analyzed for the presence of triple-FLAG-tagged Alg13p by Western blotting. Extracts were separated by 12% SDS-PAGE, immunoblotted with rabbit anti-FLAG antibodies, and detected by chemiluminescence, as described in "Experimental Procedures"

of their intracellular overexpressed products was markedly different. While the intracellular level of alg13-F192, L196, I200, Y201Ap was relatively similar to the wild-type Alg13, alg13-F192, L196, I200, Y201Gp showed a similar level to the C-terminal truncated protein alg13-CΔ15p with tenfold more abundance than wild-type protein (Fig. 8b). This experiment demonstrated that mutant proteins with a perturbed last α helix (alg13-F192, L196, I200, Y201G, and alg13-CΔ15 proteins) are more stable than wild-type Alg13 protein and mutant protein with a maintained last α helix (alg13-F192, L196, I200, and Y201A protein), indicating that the

C-terminal α helix is required for the proteasomal degradation of excess cytosolic Alg13 protein.

Discussion

The assembly of the Alg13 and Alg14 subunits is critical for the activity of UDP-*N*-acetylglucosamine transferase and contributes to the regulation of *N*-linked glycosylation in eukaryotic cells. Despite the importance of the formation of the Alg13/14 complex, little is known about the molecular mechanism that mediates its assembly. Using computational and molecular biological analyses, we report the functional importance of the C-termini of Alg13 and Alg14 in mediating the formation of the complex and the degradation of excess cytosolic Alg13 protein.

Molecular Mechanism of the Interaction Between Alg13 and Alg14

Despite the low sequence similarity, we found that Alg13 and Alg14 could be modeled on the MurG structure as a threading template. Like most GT-B structures, MurG has two distinct domains that are connected directly by a linker peptide (Fig. 1; right panel, shown in green). Although the MRI structure of Alg13 indicates its unconventional N-terminal Rossmann motif, Alg13 in the modeled Alg13/14 complex still possesses structural similarity to the C-terminal domain of MurG (Fig. 1) [15, 22]. Compared to Alg13, Alg14 in the modeled complex presents a classical Rossmann fold resembling the N-terminal domain of MurG (Fig. 1) [22]. An extended C-terminal α helix of MurG (Fig. 1, shown in blue) has been proposed to function as a connector of the two domains, and in the formation of the architecture of the two Rossmann folds [23]. The Alg13/14 complex shares much of this conserved architecture but lacks the linker peptide to connect the subunits (Fig. 1). Our computational modeling studies of the Alg13/14 complex suggested a potential conformational role of the terminal regions of both Alg13 and Alg14 in mediating their association. This structural information suggests the involvement of the C-terminal α helix of Alg13 in the formation of the Alg13/14 complex because it is the only structure capable of tying the Alg13 and Alg14 subunits together (Fig. 1). Our mutational analyses of Alg13 and Alg14 proteins deleted in these domains demonstrated that these termini were essential for the viability of the cells (Figs. 2, 3), and for the physical interaction between Alg13 and Alg14 (Figs. 2, 3). In particular, mutations that affect the C-terminal α helix of Alg13 completely blocked the interaction between Alg13 and Alg14 subunits (Fig. 2). Furthermore, replacement of just the last three amino acids of Alg14 revealed their involvement in the formation of the Alg13/14 complex (Fig. 7). Using the molecular surface view of the Alg13/14 complex (Fig. 4) to guide our mutational analyses, our results regarding the molecular requirements for the interaction between Alg13 and Alg14 can be explained as follows. (i) The C-terminal α helix of Alg13 interacts with Alg14 at a hydrophobic binding pocket. It is reasonable to think that Alg13 and Alg14 subunits assemble through hydrophobic interaction, because the role of hydrophobic residues in protein-protein recognition and in the formation of multimeric protein assemblies has long been recognized [24, 25]. Mutational analyses that exchanged the highly conserved hydrophobic residues on the surface of the C-terminal α helix of Alg13 (Figs. 5, 6) have given clear answers on the explanation of this interaction. (ii) The hydrophobic character of the last three residues (ILV) of Alg14 is important for the formation of the complex. From the position in the predicted model (Fig. 4a, b, indicated by red arrows) and their mutant phenotypes (Fig. 7), we propose that these three residues of Alg14 work as a key that locks the C-terminal α helix of Alg13 into the Alg14 hydrophobic pocket to stabilize the connection. These hydrophobic interactions could provide the driving force to lock Alg13 in, while the conserved G residue found at the fourth last position of all Alg14 C-terminal domains [22] (Fig. 7a; shown by asterisks) could conceivably provide the flexibility for the movement of the last three residues.

Function of the C-Terminal α Helix in the Degradation of Excess Cytosolic Alg13

Despite the importance of the terminal α helix of Alg13 protein in complex formation with Alg14, our mutational analysis revealed that the terminal α helix of Alg13 protein also plays important roles in its free cytosolic form. We found that deletion or perturbation of the C-terminal α helix inversely stabilized Alg13 protein (Fig. 8b). It has been clearly demonstrated that the C-terminal domain of Alg13 (including 100 amino acids) can serve as an autonomous degradation signal required for the degradation of cytosolic Alg13 [10]. Our current studies are consistent with this previous finding and demonstrate that this α helix (11 amino acids) at the Alg13 C-terminus plays a central role in the degradation of cytosolic, free Alg13 protein. While speculative, one idea is that the C-terminal α helix of Alg13 functions differently depending on the status of the Alg13 protein. In the Alg13/14 complex, the Alg13 protein is embedded in the hydrophobic pocket of Alg14 and functions as a connecter for the formation of the complex. In the unassembled, cytosolic, Alg13 protein, this α helix may signal degradation by the proteasome. A more detailed investigation of the role of the C-terminal α helix during the degradation of cytosolic Alg13 will be required to confirm this idea. A two-hybrid screening search for cytosolic factors that recognize the C-terminal α helix and tie unassembled cytosolic Alg13 to the proteasome system is currently in progress.

In conclusion, our structural analysis of the formation of the Alg13/14 complex contributes not only to the fundamental research of eukaryotic *N*-linked glycosylation but also provides useful information for the development of antibacterial drugs. The bacterial cell wall is an essential organelle. As a consequence of this essentiality, MurG has emerged as an attractive target for the development of new antibiotics that target cell-wall biosynthesis [26, 27]. It will be extremely important to unravel the mechanism of Alg13/14 complex formation, because the similarity between MurG and the Alg13/14 complex, which catalyzes an essential step of *N*-linked glycosylation, requires the rational design of drugs that specifically inhibit the bacterial enzyme without affecting the human one.

References

1. Helenius A, Aebi M (2004) Roles of N-linked glycans in the endoplasmic reticulum. Annu Rev Biochem 73: 1019–1049
2. Kelleher DJ, Gilmore R (2006) An evolving view of the eukaryotic oligosaccharyltransferase. Glycobiology 16: 47R–62R
3. Lehle L, Strahl S, Tanner W (2006) Protein glycosylation, conserved from yeast to man: a model organism helps elucidate congenital human diseases. Angew Chem Int Ed Engl 45: 6802–6818
4. Weerapana E, Imperiali B (2006) Asparagine-linked protein glycosylation: from eukaryotic to prokaryotic systems. Glycobiology 16: 91R–101R
5. Kukuruzinska MA, Robbins PW (1987) Protein glycosylation in yeast: transcript heterogeneity of the ALG7 gene. Proc Natl Acad Sci USA 84: 2145–2149
6. Gao XD, Tachikawa H, Sato T et al (2005) Alg14 recruits Alg13 to the cytoplasmic face of the endoplasmic reticulum to form a novel bipartite UDP-N-acetylglucosamine transferase required for the second step of N-linked glycosylation. J Biol Chem 280: 36254–36262
7. Albright CF, Robbins RW (1990) The sequence and transcript heterogeneity of the yeast gene ALG1, an essential mannosyltransferase involved in N-glycosylation. J Biol Chem 265: 7042–7049
8. Chantret I, Dancourt J, Barbat A, Moore SE (2005) Two proteins homologous to the N- and C-terminal domains of the bacterial glycosyltransferase Murg are required for the second step of dolichyl-linked oligosaccharide synthesis in Saccharomyces cerevisiae. J Biol Chem 280: 9236–9242
9. Bickel T, Lehle L, Schwarz M et al (2005) Biosynthesis of lipid-linked oligosaccharides in Saccharomyces cerevisiae: Alg13p and Alg14p form a complex required for the formation of GlcNAc(2)-PP-dolichol. J Biol Chem 280: 34500–34506
10. Averbeck N, Gao XD, Nishimura S, Dean N (2008) Alg13p, the Catalytic Subunit of the Endoplasmic Reticulum UDP-GlcNAc Glycosyltransferase, Is a Target for Proteasomal Degradation. Mol Biol Cell 19: 2169–2178
11. Mengin-Lecreulx D, Texier L, Rousseau M, van Heijenoort J (1991) The murG gene of Escherichia coli codes for the UDP-N-acetylglucosamine: N-acetylmuramyl-(pentapeptide) pyrophosphoryl-undecaprenol N-acetylglucosamine transferase involved in the membrane steps of peptidoglycan synthesis. J Bacteriol 173: 4625–4636
12. Bourne Y, Henrissat B (2001) Glycoside hydrolases and glycosyltransferases: families and functional modules. Curr Opin Struct Biol 11: 593–600
13. Ha S, Walker D, Shi Y, Walker S (2000) The 1.9 A crystal structure of Escherichia coli MurG, a membrane-associated glycosyltransferase involved in peptidoglycan biosynthesis. Protein Sci 9: 1045–1052
14. Hu Y, Chen L, Ha S et al (2003) Crystal structure of the MurG:UDP-GlcNAc complex reveals common structural principles of a superfamily of glycosyltransferases. Proc Natl Acad Sci USA 100: 845–849
15. Wang X, Weldeghiorghis T, Zhang G et al (2008) Solution structure of Alg13: the sugar donor subunit of a yeast N-acetylglucosamine transferase. Structure 16: 965–975

16. Sali A, Blundell TL (1993) Comparative protein modelling by satisfaction of spatial restraints. J Mol Biol 234: 779–815
17. Fiser A, Do RK, Sali A (2000) Modeling of loops in protein structures. Protein Sci 9: 1753–1773
18. Ghose AK, Viswanadhan VN, W JJ (1998) Prediction of Hydrophobic (Lipophilic) Properties of Small Organic Molecules Using Fragmental Methods: An Analysis of ALOGP and CLOGP Methods. J Phys Chem A 102: 3762–3772
19. Sambrook J, Fristsch EF, Maniatis T (1998) Molecular Cloning: A Laboratory Manual, Cold Spring Harbor Laboratory, Cold Spring Harbor, NY
20. Guthrie C, Fink GR (1991) Methods Enzymol 194: 3–24
21. Aleman C, Roca R, Luque FJ, Orozco M (1997) Helical preferences of alanine, glycine, and aminoisobutyric homopeptides. Proteins 28: 83–93
22. Gao XD, Moriyama S, Miura N et al (2008) Interaction between the C Termini of Alg13 and Alg14 Mediates Formation of the Active UDP-N-acetylglucosamine Transferase Complex. J Biol Chem 283: 32534–32541
23. Grizot S, Salem M, Vongsouthi V et al (2006) Structure of the Escherichia coli heptosyltransferase WaaC: binary complexes with ADP and ADP-2-deoxy-2-fluoro heptose. J Mol Biol 363: 383–394
24. Guharoy M, Chakrabarti P (2005) Conservation and relative importance of residues across protein-protein interfaces. Proc Natl Acad Sci U S A 102: 15447–15452
25. Chothia C, Janin J (1975) Principles of protein-protein recognition. Nature 256: 705–708
26. Zawadzke LE, Wu P, Cook L et al (2003) Targeting the MraY and MurG bacterial enzymes for antimicrobial therapeutic intervention. Anal Biochem 314: 243–252
27. Ha S, Gross B, Walker S (2001) E. Coli MurG: a paradigm for a superfamily of glycosyltransferases. Curr Drug Targets Infect Disord 1: 201–213
28. Altschul SF, Madden TL, Schaffer AA et al (1997) Gapped BLAST and PSI-BLAST: a new generation of protein database search programs. Nucleic Acids Res 25: 3389–3402
29. Sikorski RS, Hieter P (1989) A system of shuttle vectors and yeast host strains designed for efficient manipulation of DNA in Saccharomyces cerevisiae. Genetics 122: 19–27

Discovery of a First-in-Class Drug, a Prostaglandin D_2 Antagonist, for the Treatment of Allergic Diseases

Akinori Arimura

Summary

Prostaglandin D_2 (PGD_2) is a major lipid mediator produced by mast cells in response to IgE-dependent stimuli. In patients with allergic diseases such as allergic rhinitis and asthma, the level of PGD_2 is known to increase in nasal and bronchial lavage fluids after allergen challenge. Although PGD_2 has been reported to exert a variety of inflammatory effects such as increases in nasal airway resistance and vascular permeability and eosinophil infiltration, there are few reports on the efficacy of PGD_2 receptor (DP1) antagonists, either in subjects with allergic diseases or in animal models. In 1994 we started a PGD_2 project to discover DP1 antagonists for the treatment of allergic diseases and, by carrying out structure-activity relationship studies, we eventually established S-5751 as a candidate first-in-class drug. In vivo pharmacology studies demonstrated that it dramatically inhibited antigen-induced nasal congestion and inflammatory cell migration in allergic rhinitis models as well as dramatically suppressing bronchial hyper-responsiveness and lung inflammation in asthma models. In 2000, the world's first phase 2 clinical study with a DP1 antagonist was carried out in patients with allergic rhinitis. However, no significant efficacy was shown, although some favorable findings were seen in a subanalysis. Based on pharmacokinetics (PK)/pharmacodynamics (PD) analysis in the human subjects and animals used for pharmacology studies, we speculated that the failure of the phase 2 study was due to insufficient exposure in humans compared to animal models rather than being due to a minor role of PGD_2 in the pathogenesis of allergic diseases. Subsequently, a structure-activity relationship study was conducted again, using alternative lead compounds and in 2007 we found a back-up compound, S-555739, in which the PK profile and DP1 antagonistic activity was markedly improved. Phase 1 studies have demonstrated that S-555739 is well tolerated and shows a good PK profile with once-a-day dosing, and now a phase 2 study is being planned. Using S-555739, not only the potential of a DP1 antagonist as a first-in-class drug in the treatment of allergic diseases but also the role of PGD_2 in the pathogenesis of allergic diseases will be clarified.

Strategic Development Department, Pharmaceutical Development Division, Shionogi & Co., Ltd, 5-12-4 Sagisu, Fukushima-ku, Osaka 553-0002, Japan

N. Tamaki, Y. Kuge (eds.)
Molecular Imaging for Integrated Medical Therapy and Drug Development.
DOI: 10.1007/978-4-431-98074-2_27, ©Springer 2010

Key words Prostaglandin D_2 · DP1 antagonist · Allergy · S-5751 · S-555739

Introduction

Prostaglandin D_2 (PGD_2) is a lipid mediator synthesized from arachidonic acid through the catalytic activities of cyclooxygenases and PGD synthases in mast cells, alveolar macrophages, airway epithelium, and other cellular sources. Because PGD_2 is predominantly generated from mast cells and is the major cyclooxygenase metabolite produced by immunoglobulin E (IgE) dependently activated mast cells [1], PGD_2 has been measured as a marker of mast-cell activation. In patients with allergic rhinitis, PGD_2 levels were increased in nasal lavage fluid after nasal antigen provocation [2]. Also, local allergen challenge leads to the rapid production of PGD_2, which can be detected in the bronchoalveolar lavage fluid of patients with asthma and also in the skin of patients with atopic dermatitis.

Recently, PGD_2 has been shown to produce a variety of inflammatory actions through two types of transmembrane receptors, DP1 and CRTH2 [3]. DP1 is a classic PGD_2 receptor expressed ubiquitously and it mediates vasodilation and inflammatory cell migration; CRTH2, a chemoattractant receptor-homologous molecule expressed on T-helper type-2 (Th2) cells, is expressed on Th2 lymphocytes, eosinophils, and basophils and may contribute to the activation of Th2 lymphocytes and eosinophils. Exogenously injected PGD_2 has been shown to cause erythema by itself and also to enhance plasma exudation caused by other inflammatory mediators in the skin and conjunctiva. This action is considered to be due to an increased blood flow to the inflamed lesion through DP1. In DP1-deficient mice, inflammatory cell infiltration into lung tissue following repeated antigen challenge was reduced compared with that in wild-type mice in an asthma model [4]. In addition, bronchial hyper-responsiveness and Th2 cytokine response were also attenuated. These findings suggest that PGD_2 plays an important role in the onset of allergic inflammation through DP1.

In healthy volunteers, nasal challenge with PGD_2 has been shown to produce nasal congestion, and it is more effective in this respect than either histamine or bradykinin [5]. The PGD_2-induced nasal congestion could not be attenuated by treatment with ramatroban (Bay u 3405), an inhibitor of both CRTH2 and thromboxane A_2 receptor (TP) but not DP1 [6]. These findings suggest that PGD_2 produced by activated mast cells in nasal mucosa contributes to the vascular dilation of nasal capacitance vessels through DP1, leading to the nasal congestion. Allergic rhinitis is characterized by three major symptoms: sneezing, rhinorrhea, and nasal congestion, which are considered to be triggered by multiple mediators released from mast cells and other inflammatory cells. Histamine is thought to be the main mediator, and antihistamines are the most commonly prescribed pharmacotherapeutic agents for allergic rhinitis. However, they are not efficacious against nasal congestion, which is a key symptom influencing the quality of life, suggesting that mediators other than histamine, probably PGD_2, may be involved in the nasal congestion in allergic rhinitis. However, there have been no data available on the activity of DP1 antagonists either in patients with allergic diseases or in animal models. Therefore we started a PGD_2 program at Shionogi Research Laboratories to discover DP1 antagonists suitable for humans and to clarify their potential as a new class of drug for the treatment of allergic diseases.

Discovery of S-5751

An orally active DP1 selective antagonist, S-5751 ((Z)-7[1R,2R,3S,5S]-2-(5-hydroxybenzo[b]thiophen-3-ylcarbonylamino)-10-norpinan-3-yl)hept-5-enoic acid), was established at Shionogi Research Laboratories by structure-activity relationship studies, using a lead compound possessing a prostanoid skeleton that was found by screening a compound library for the inhibitory effects on specific PGD_2 binding to human platelet membranes [7–9]. The specific binding of $[^3H]$-PGD_2 to human platelet membrane was suppressed by S-5751, with a *Ki* value of 1.6 nM. In a functional second-messenger assay, S-5751 strongly inhibited the cyclic adenosine monophosphate (cAMP) elevation elicited by PGD_2 (0.1 μM) in human washed platelets, with an IC_{50} value of 0.9 nM. However, using human platelet-rich plasma, the inhibition of cAMP production induced by PGD_2 (0.3 μM) by S-5751 was reduced due to high protein binding, the IC_{50} value being 840 nM. S-5751 showed weak antagonism for TP and EP2 (a PGE_2 receptor subtype) in an in vitro assay, but no significant antagonism in an in vivo study even at the highest dose used for pharmacology studies, indicating that S-5751 could act as a selective DP1 antagonist in an in vivo model. Toxicology studies in animals conducted under Good Laboratory Practice (GLP) demonstrated that there was no safety issue to prevent the conduct of clinical studies with S-5751.

Efficacy of S-5751 in Allergic Diseases in Animal Models

First of all, we evaluated the effect of S-5751 on PGD_2-induced increase in intranasal pressure in a guinea-pig model. Similar to reports in humans, in guinea pigs nasal airway resistance was immediately increased by PGD_2 application to the nose, and this was almost completely suppressed by orally administered S-5751 at 1 mg/kg, suggesting that PGD_2-induced nasal blockage is mediated via DP1 [10]. To clarify the role of PGD_2 in the onset of allergic rhinitis and the potential of a DP1 antagonist to be developed as a drug for treating allergic rhinitis, the efficacy of S-5751 was evaluated in an allergic rhinitis model and compared with that of an antihistamine, terfenadine [9]. In patients with allergic rhinitis, nasal antigen challenge causes not only early-phase responses characterized by sneezing, rhinorrhea, and nasal blockage but also late-phase responses characterized by nasal blockage occurring several hours later in approximately 50% of the patients, with accompanying accumulation of inflammatory cells such as eosinophils in the nasal cavity [11]. Therefore we evaluated the effect of S-5751 and terfenadine on both early and late nasal responses in guinea pig models of allergic rhinitis that were sensitized by repeated antigen challenge. Orally administered S-5751 at 10 mg/kg almost completely suppressed the increase in intranasal pressure that occurred immediately after intranasal antigen challenge, while terfenadine showed only partial inhibition, even at a dose of 10 mg/kg which is sufficient to exert antihistamine activity [9]. Interestingly, not only terfenadine but also S-5751 inhibited sneezing that occurred for 30 min after antigen challenge in the animal model. However, sneezing is considered to be caused mainly by histamine in patients with allergic rhinitis, and PGD_2 by itself is not an elicitor of sneezing in healthy volunteers [5]. Thus, PGD_2 may contribute to the onset of allergen-induced sneezing via some indirect actions. The late nasal response, assessed by plasma exudation into nasal mucosa, was also suppressed by S-5751 but not by terfenadine, and, interestingly, the increase in the number of inflammatory cells such as eosinophils and macrophages in nasal lavage fluid at 5 h after the antigen challenge was significantly reduced by S-5751, suggesting that PGD_2 may trigger the migration of inflammatory cells via DP1, leading to late nasal responses.

In a guinea pig model of allergic conjunctivitis, S-5751 almost completely suppressed the allergen-induced increase in microvascular permeability in the conjunctiva [9]. Most patients with allergic rhinitis show allergic symptoms in the eyes, such as conjunctivitis and itching. Thus it is considered that S-5751, as a drug for the treatment of allergic rhinitis, can relieve both nasal and eye symptoms.

Lastly, in order to evaluate the anti-asthma activity of S-5751, we assessed the effect of S-5751 on early and late asthmatic responses and bronchial hyperresponsiveness in a sheep asthma model, in collaboration with Dr. Abraham at Miami University. Oral administration of S-5751 at 30 mg/kg partially but significantly inhibited the early-phase asthmatic response by 41% and the late-phase response by 81%, and also almost completely suppressed the onset of bronchial hyper-responsiveness [12]. These inhibitions are almost comparable to those of the leukotriene receptor antagonist montelukast, which is being used as a controller for asthma, suggesting that S-5751 could also be expected to be a useful drug in the treatment of allergic asthma.

In conclusion, based on pharmacological studies with S-5751 using various allergic disease models, PGD_2, through DP1, is considered to play a critical role in the development of allergic inflammation, resulting in various allergic symptoms. A DP1 antagonist would thus be useful for the treatment of various allergic diseases.

Clinical Studies of S-5751

A proof-of-concept (POC) study (phase 2 study) of a first-in-class drug, S-5751, was carried out in patients with allergic rhinitis, because PGD_2, through DP1, has been reported to be a potent elicitor of nasal congestions in humans, and S-5751, a DP1 antagonist showed excellent efficacy in an allergic rhinitis model. To evaluate the efficacy of this drug for treating allergic rhinitis, a double-blind, placebo-controlled phase 2 study was conducted in more than 600 patients with seasonal allergic rhinitis. However, clear efficacy was not seen in either the total nasal symptom score or other secondary endpoints such as nasal congestion and sneezing. It was considered that one of the reasons why efficacy was not shown may have been because the efficacy was masked by a high placebo effect due to a much lower pollen count in the year of the study than in a typical year. To minimize the placebo effect, we conducted a subanalysis of the clinical study by limiting the study population to those patients with a high score for nasal symptoms at the baseline; this included approximately one-fourth of the total number of patients. As a result, a tendency toward reduction not only in the total nasal symptom score but also in scores for other symptoms such as sneezing, congestion, and rhinorrhea was seen in patients treated with S-5751 at doses of 100 and 400 mg twice daily compared to those treated with placebo.

To confirm whether a sufficient DP1 antagonistic action was produced by S-5751 in humans, a PGD_2 provocation study was conducted in healthy volunteers. The dose-response curve for nasal airway resistance induced by increasing doses of PGD_2 was partially but significantly shifted to the right by treatment with S-5751 at 400 mg twice daily. However, MK-0524, a DP1 antagonist newly developed by Merck, demonstrated complete suppression of PGD_2-induced nasal congestion [13], indicating that S-5751 was insufficient to produce a maximum pharmacological effect in subjects even at the highest dose. Therefore, further development of S-5751 was stopped and we made the decision to switch to a back-up compound even though no back-up compound had been established yet at that time.

Development of Second-Generation DP1 Antagonist

Based on nonclinical and clinical studies with S-5751, we concluded that the failure to obtain clear efficacy with S-5751 was due to its inadequate DP1 antagonistic action, particularly in the presence of albumin, and insufficient plasma concentration to show a maximum effect in humans even at the highest dose used in clinical studies. To overcome the shortcomings of S-5751, a second-generation DP1 antagonist, S-555739, was successfully produced at Shionogi Research Laboratories through structure-activity relationship studies, using several lead compounds with nonprostanoid skeletons. Preclinical studies have demonstrated that S-555739 shows more potent and selective DP1 antagonistic activity, with a much better pharmacokinetics (PK) profile than S-5751. Pharmacological studies in several allergic disease models have shown that S-555739 is far superior to S-5751 in suppressing allergic symptoms such as nasal congestion and inflammatory cell migration. In addition, GLP toxicology studies demonstrated that there was no concern about safety to prevent the conduct of phase 1 studies. Now phase 1 studies are ongoing, and, according to PK data obtained from a phase 1 single-dose study in healthy volunteers, S-555739 is anticipated to be effective when given once daily and it is anticipated that it will reach a sufficient plasma concentration at a reasonable dosage. Phase 2 studies with S-555739 will be conducted in the near future.

Conclusion

A DP1 antagonist, S-5751, was developed as a first-in-class drug for the treatment of allergic rhinitis, but it was not proven to be clearly more effective than placebo in a phase 2 study. However, clinical studies as well as nonclinical studies using S-5751 provided us with much valuable information, which prompted us to try again with a back-up compound, S-555739. Using S-555739, which has a more potent DP1 antagonistic action and a much better PK profile than S-5751 in humans, we will be able to clarify the role of PGD_2 in the onset of allergic diseases and the potential of a DP1 antagonist to be developed as an innovative drug for the treatment of these diseases.

References

1. Lewis RA, Soter NA, Diamond PT et al (1982) Prostaglandin D_2 generation after activation of rat and human mast cells with anti-IgE. J Immunol 129: 1627–1631
2. Naclerio RM, Meier HL, Kagey-Sobotka A et al (1983) Mediator release after nasal airway challenge with allergen. Am Rev Respir Dis 128: 597–602
3. Hirai H, Tanaka K, Yoshie O et al (2001) Prostaglandin D_2 selectively induces chemotaxis in T helper type 2 cells, eosinophils, and basophils via seven-transmembrane receptor CRTH2. J Exp Med 193: 255–261
4. Matsuoka T, Hirata M, Tanaka H et al (2000) Prostaglandin D_2 as a mediator of allergic asthma. Science 287: 2013–2017
5. Doyle WJ, Boehm S, Skoner DP (1990) Physiologic responses to intranasal dose-response challenges with histamine, methacholine, bradykinin, and prostaglandin in adult volunteers with and without nasal allergy. J Allergy Clin Immunol 86: 924–935
6. Johnston SL, Smith S, Harrison J et al (1993) The effect of BAY u 3405, a thromboxane receptor antagonist, on prostaglandin D_2-induced nasal blockage. J Allergy Clin Immunol 91: 903–909
7. Tsuri T, Honma T, Hiramatzu Y et al (1997) Bicyclo[2.2.1]heptane and 6,6-dimethylbicyclo [3.1.1]heptane derivatives: orally active, potent, and selective prostaglandin D_2 receptor antagonists. J Med Chem 40: 3504–3507
8. Mitzumori S, Tsuri T, Honma T et al (2003) Synthesis and biological activity of various derivatives of a novel class of potent, selective, and orally active prostaglandin D_2 receptor antagonists. 2. 6, 6-Dimethylbicyclo[3.1.1]heptane derivatives. J Med Chem 46: 2446–2455
9. Arimura A, Yasui K, Kishino J et al (2001) Prevention of allergic inflammation by a novel prostaglandin receptor antagonist, S-5751. J Pharmacol Exp Ther 298: 411–419.
10. Yasui K, Asanuma F, Hirano Y et al (2007) Contribution of prostaglandin D_2 via prostanoid DP receptor to nasal hyperresponsiveness in guinea pigs repeatedly exposed to antigen. Eur J Pharmacol 578: 286–291

11. Naclerio RM (1991) Allergic rhinitis. N Engl J Med 325: 860–869
12. Abraham WM, Shichijo M, Deguchi M et al (2007) A prostaglandin D_2 receptor antagonist modifies experimental asthma in sheep. Am J Respir Crit Care Med 175:A560
13. Van-Hecken A, Depre M, De Lepeleire I et al (2007) The effect of MK-0524, a prostaglandin D_2 receptor antagonist, on prostaglandin D_2-induced nasal airway obstruction in healthy volunteers. Eur J Clin Pharmacol 63: 135–141

Towards Developing a Golgi Simulator: Microfluidic Device Enabling Synthesis of a Tetrasaccharide

Yasunari Ono[1], Shusaku Daikoku[1], Yasuko Hasegawa[2], Toshiyuki Sato[1], Mariko Kobayashi[1], Katsuhiko Suzuki[1], and Osamu Kanie[1]

Summary

Despite the important roles of glycans and glycoconjugates in many biological processes, the regulatory mechanism underlying the synthesis of these molecular structures in the Golgi apparatus is not understood. Thus, investigation of the synthetic process is quite important. However, ordinary methods cannot allow us to visualize such a process. Therefore, an alternative method has to be developed. A micro total analysis system (μTAS) that enables analysis of sequential glycosyltransfer reactions was developed. This microchip-based analysis platform, by mimicking Golgi function, may become a valuable tool not only in the synthesis of compounds but also in situations in which direct observation is quite difficult.

Key words Microchip · LC-MS · Golgi · Oligosaccharide · Glycosyltransferase

Introduction

Glycoconjugates play important roles in many biological processes, including cell adhesion, viral infection, tumor growth, and metastasis [1, 2]. Thus, synthetic oligosaccharides and glycoconjugates have been recognized as indispensable tools for probing biological functions. Despite the importance of glycoconjugates, however, very little is known of the regulatory mechanism of glycan synthesis in the Golgi apparatus. Recent advances in time-resolved imaging techniques that visualize detailed cellular organelles have allowed us to perform spatio-temporal investigation of certain molecules such as proteins fused with green fluorescence proteins [3]. However, such investigations do not provide information regarding the molecular structures being synthesized in the Golgi apparatus. In order to investigate the

[1]Mitsubisi Kagaku Institute of Life Sciences (MITILS), 11 Minami-ooya, Machida-shi, Tokyo 194-8511, Japan
[2]Tokyo Institute of Technology, Graduate School of Bioscience and Biotechnology, 4529 Nagatsuta-cho, Midori-ku, Yokohama 226-0018, Japan

N. Tamaki, Y. Kuge (eds.)
Molecular Imaging for Integrated Medical Therapy and Drug Development.
DOI: 10.1007/978-4-431-98074-2_28, ©Springer 2010

structures, in addition to the spatio-temporal resolution, we considered that an indirect method mimicking the process taking place in the Golgi apparatus would be useful to understand the regulatory mechanism of glycan synthesis, by simulating the sequential glycosyltransfer reactions taking place in the Golgi apparatus. In this study, we introduce a challenge aiming at "visualizing the Golgi mechanism" where we use a single microchip, consisting of a microfluidic channel with three reaction chambers that allow three sequential glycosyltransfer reactions [4].

Analysis Platform

In order to analyze the complex mixture formed by a series of enzymatic reactions, a system consisting of a liquid chromatograph (LC), coupled with either an ultraviolet (UV) or fluorescence (FL) detector for peak separation and quantification, and a mass spectrometer (MS) for structural analysis is required. Thus, an analytical system was constructed; it consisted of a sample loop, an LC pumping unit (LC-20A; Shimadzu, Kyoto, Japan), an analytical column (Pegasil ODS; 3-μm particles, 1 mm × 100 mm; Senshu Scientific, Tokyo, Japan), a UV detector (SPD-20A; Shimadzu), and an MS (LCMS-IT-TOF; Shimadzu) with an electrospray ionization (ESI) interface. The sample loop was constructed to allow the coupling of the independent systems with different flow rates employed for the synthesis chip and LC system (Fig. 1a,b).

a

b

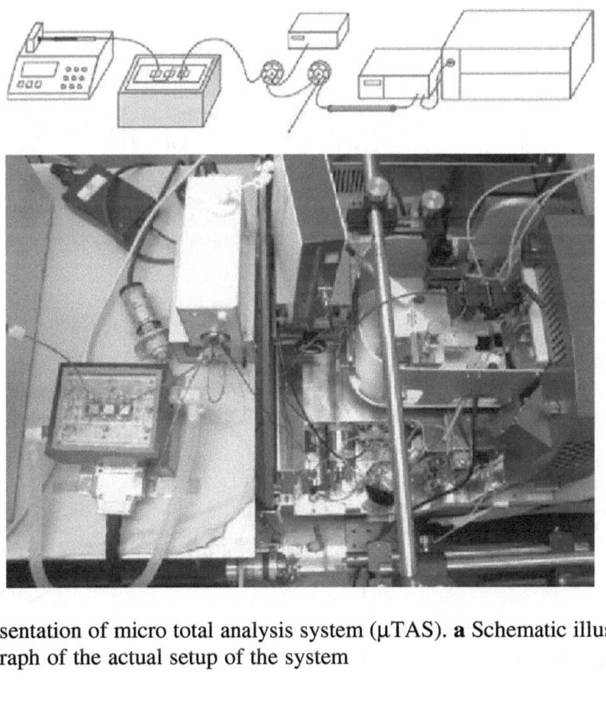

Fig. 1a,b. Representation of micro total analysis system (μTAS). **a** Schematic illustration of the system. **b** Photograph of the actual setup of the system

The setup can be upgraded to adopt nano LC for further scale-down, especially in proteomics and metabolomics research. Furthermore, such a setup could be applied effectively in the analysis of the glycoforms of a glycoprotein and glycolipids.

Target Compounds

Glycoprotein, proteoglycan, and glycolipid were the three major classes of glycoconjugates to be analyzed. Eventually all of these have to be addressed, but a tetrasaccharide derivative corresponding to the core part of the glycan structure in glycosaminoglycans (GAGs), called a linkage region tetrasaccharide; namely, β-D-GlcA-(1→4)-β-D-Gal-(1→3)-β-D-Gal-(1→4)-β-D-Xyl-PNP (compound **1**; Fig. 2) was chosen as a target compound for synthesis in our previous study [4]. A series of GAGs are formed by the action of various glycosyltransfer reactions, to form heparin, heparin sulfate, chondroitin sulfate, dermatan sulfate, and hyaluronan; however, the regulatory mechanism is not understood. These substances are involved in a variety of important biological functions [5, 6] and all of these GAGs are synthesized from the linkage region tetrasaccharide. Thus, the tetrasaccharide has a key role as the precursor of GAGs and is important as a primer molecule to investigate the synthetic process.

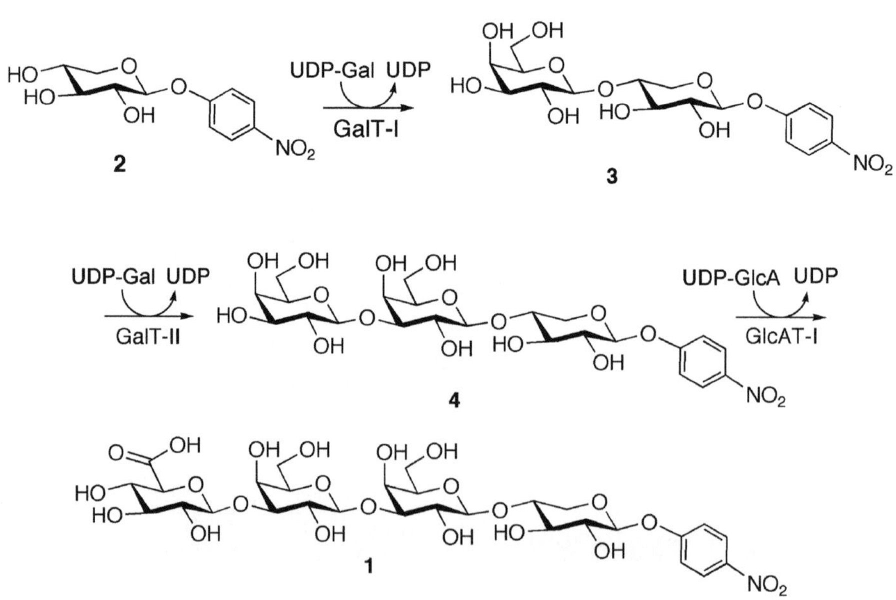

Fig. 2. Synthetic Scheme of compound **1**

The Microchip

The schematic design of the entire system, consisting of the microchip, temperature control system, LC, and MS, is shown in Fig. 1a. Two independent flow systems were connected to allow both the enzymatic reactions in the chip and the analysis. The chip has a single-channel pattern with three reaction chambers enabling three sequential enzymatic reactions (Fig. 3). The chip is composed of two Pyrex glass plates (30 mm × 70 mm) with the thicknesses of the top and bottom plates being 500 and 700 μm, respectively. Four access ports, three for media and beads inlets and one as an outlet, were drilled on the top plate. A microfluidic channel was formed on the bottom plate using standard photolithography and wet etching techniques. The channel and chamber size were 130 μm (depth) × 400 μm (width) and 130 μm (depth) × 550 μm (width), respectively. In order to retain the beads in the chamber, a series of dam structures was formed according to published methods, utilizing isotropy of wet glass etching [7, 8]. The glass plates were joined by fusion bonding at high temperature. The channel was cleaned with hydroxyethylpiperazine ethanesulfonic acid (HEPES)-KOH buffer (pH 7.4) prior to use, and then negative pressure was applied to the outlet of the chip to load the beads into the chamber.

The microchip was placed on a Peltier element-based temperature control device (Mitsubishi Chemical Engineering, Yokohama, Japan), which was designed and fabricated to control the temperature of three individual areas. A syringe pump (PHD2000; Harvard Apparatus, Hulliston, MA, USA) was used to introduce substrate solution into the microchip. *Drosophila* proteoglycan β1,4-galactosyltransferase I (GalT-I; 2.4.1.133) [9, 10], β1,3-galactosyltransferase II (GalT-II) [11, 12],

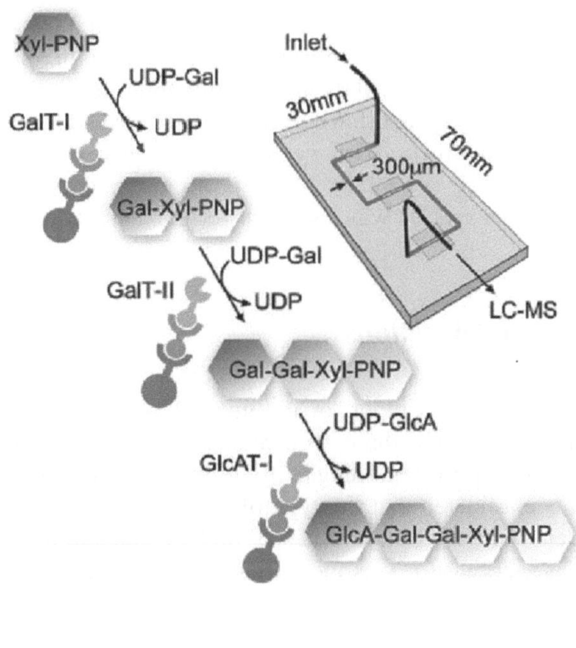

Fig. 3. Schematic representation of the microchip used in the system and the synthesis of tetrasaccharide. *LC*, liquid chromatograph; *MS*, mass spectrometer

and β1,3-glucuronic acid transferase (GlcAT-I) [13] were expressed as the secreted proteins fused with a FLAG peptide (DYKDDDDK) in insect cells. High Five cells were incubated with recombinant Vacuro virus at a multiplicity of infection of 5 and incubated for 72 h to yield conditioned media containing recombinant proteins fused with FLAG peptide. A 1-ml volume of culture medium was mixed with 1 μl of anti-FLAG M2 Agarose from mouse. The protein-gel mixture was washed twice with 50 mM Tris-buffered saline (50 mM Tris-HCl, pH 7.4, and 150 mM NaCl) containing 1 mM $MnCl_2$. The details of the expression of GlcAT-I will be reported elsewhere (Y. Ohmae et al., unpublished data).

Synthesis of Tetrasaccharide in Microchip

In order to perform three sequential glycosylation reactions in the microchip for the synthesis of the target tetrasaccharide **1** (Fig. 2;), GalT-I, GalT-II, and GlcAT-I were introduced into the designated reaction chambers. The microchip was held on the temperature control system (Fig. 1a, b). A substrate solution (3 mM Xyl-PNP (**2**), 2 mM uridine diphosphate (UDP)-Gal, and 1 mM UDP-GlcA in 14 mM HEPES-KOH containing 11 mM $MnCl_2$ and phosphatase; pH 7.4) was introduced into the microchannel at a flow rate of 0.1 μl min^{-1}. The reaction temperatures in the reaction chambers from the first to third reaction were controlled at 37°C, 37°C, and 25°C, respectively, based on the preliminary examination of these enzyme transformations. The phosphatase was added because the second and third enzymes also used UDP-sugars and, thus, accumulation of inhibitor was anticipated. The temperature applied was based on the reported conditions.

The mixture of enzymatic reactions was then separated, using the following LC, where compound **2** (*Rt* = 6.6 min), compound **3** (*Rt* = 10.9 min), and compound **4** (*Rt* = 6.2 min) were confirmed in the reaction mixture. Although individual peaks could not be assigned by the retention times because standard compounds were not available, the single-ion chromatogram based on mass spectrometry was successful in the assignment. Also, negative ion mode revealed the formation of tetrasaccharide **1** ($[\mathbf{1}\text{-H}]^-$: *m/z* 770). Mass spectrometry/mass spectrometry (MS/MS) analysis of the ion species was also used for the confirmation of these structures.

Future Directions

This is the first report of the microchip-based synthesis of tetrasaccharide. This process required three sequential enzymatic reactions, and until now there have been no reports regarding multiple enzymatic reactions in tetrasaccharide synthesis. It is considered that the type of microchip we have described might be useful in identifying drug metabolites as well. A simulation study is certainly the next challenge, because it was the initial objective. Nevertheless, it is considered that the microchip-based synthetic device we have described may be used not only in the synthesis of compounds but also in situations in which direct observation is quite difficult. The present investigation may be an important first step forward to such a goal.

Acknowledgments This work was supported in part by a Grant-in-Aid for Scientific Research on Priority Areas "Lifesurveyor" from the Ministry of Education, Culture, Sports, Science and Technology of Japan. The support of the Key Technology Research Promotion Program of the New Energy and Industrial Development Organization (NEDO) of the Ministry of Economy, Trade and Industry of Japan is also acknowledged.

References

1. Varki A (1993) Biological roles of oligosaccharides: all of the theories are correct. Glycobiology 3: 97–130
2. Ritchie GE, Moffatt BE, Sim RB et al (2002) Glycosylation and the complement system. Chem Rev 102: 305–320
3. Matsuura-Tokita K, Takeuchi M, Ichihara A et al (2006) Live imaging of yeast Golgi cisternal maturation. Nature 441: 1007–1010
4. Ono Y, Kitajima M, Daikoku S et al (2008) Sequential enzymatic glycosyltransfer reactions on a microfluidic device: synthesis of a glycosaminoglycan linkage region tetrasaccharide. Lab Chip 8: 2168–2173
5. Lander AD, Selleck SB (2000) The elusive functions of proteoglycans: in vivo veritas. J Cell Biol 148: 227–232
6. Tamura J (2001) Recent advances in the synthetic studies of glycosaminoglycans. Trends Glycosci Glycotechnol 13: 65–88
7. Sato K, Yamamoto M, Hagino T et al (2004) Microchip-based enzyme-linked immunosorbent assay (microELISA) system with thermal lens detection. Lab Chip 4: 570–575
8. Haes AJ, Terray A, Collins GE (2006) Bead-assisted displacement immunoassay for Staphylococcal enterotoxin B on a microchip. Anal Chem 78: 8412–8420
9. Nakamura Y, Haines N, Chen J et al (2002) Identification of a *Drosophila* gene encoding xylosylprotein β4-galactosyltransferase that is essential for the synthesis of glycosaminoglycans and for morphogenesis. J Biol Chem 277: 46280–46288
10. Takemae H, Ueda R, Okubo R et al (2003) Proteoglycan UDP-galactose: β-xylose β1,4galactosyltransferase I is essential for viability in *Drosophila melanogaster*. J Biol Chem 278: 15571–15578
11. Bai X, Zhou D, Brown JR et al (2001) A triple mutant of *Escherichia coli* lacking secondary acyl chains on lipid A. J Biol Chem 276: 48189–48195
12. Ueyama M, Takemae H, Ohmae Y et al (2008) Functional analysis of proteoglycan galactosyltransferase II RNAi mutant flies. J Biol Chem 283: 6076–6084
13. Kim BT, Tsuchida K, Lincecum J et al (2003) Identification and characterization of three *Drosophila melanogaster* glucuronyltransferases responsible for the synthesis of the conserved glycosaminoglycan-protein linkage region of proteoglycans: two novel homologs exhibit broad specificity toward oligosaccharides from proteoglycans, glycoproteins, and glycosphingolipids. J Biol Chem 278: 9116–9124

An Efficient Strategy for the Exploration of Specific Inhibitors of Sialyltransferases

Kensaku Hosoguchi[1], Takahiro Maeda[2], Jun-ichi Furukawa[2], Hiroshi Hinou[2], and Shin-Ichiro Nishimura[2]

Summary

The characteristic changes of glycan composition detected in the process of cell transformation are often caused by the alteration of sialyltransferase (SiaT) activities, as shown in, for example, human colorectal cancer and breast carcinoma.

To develop potent inhibitors of SiaTs, a novel and efficient approach was established by combining the use of two types of cytidine monophosphate *N*-acetylneuraminic acid (CMP-NANA)-based compound libraries and a matrix assisted laser desorption adsorption ionization-time of flight mass spectrometry (MALDI-TOF MS)-based high-throughput screening system. "Click chemistry" between two CMP-NANA-based precursors bearing an azide group at the C-9 or C-5 position and commercially available alkynes afforded more than 70 novel compounds quickly.

Aminooxy-functionalized peptides (N^{α}-((aminooxy)acetyl)tryptophanyl arginine methyl ester (aoWR) reagent) conjugated with an acceptor substrate of SiaTs enabled a high-throughput and quantitative inhibition assay of the compound libraries, using common MALDI-TOF MS equipment. It was clearly demonstrated, by using an α2,3- and an α2,6-SiaT, that the present protocol allowed for high-throughput screening, showing the inhibitory activities of newly synthesized compounds. Some compounds exhibited specific and strong inhibitory effects on α2,3-SiaT. On the other hand, interestingly, the inhibitors of α2,3-SiaT turned out to be donor substrates in the presence of α2,6-SiaT.

Key words Sialyltransferase · Inhibitor · Library · MALDI-TOF MS · High-throughput screening

[1]Graduate School of Life Science, Frontier Research Center for Post-Genomic Science and Technology, Hokkaido University, Kita 21 Nishi 11, Sapporo 001-0021, Japan
[2]Graduate School of Advanced Life Science, Frontier Research Center for Post-Genomic Science and Technology, Hokkaido University, Kita 21 Nishi 11, Sapporo 001-0021, Japan

N. Tamaki, Y. Kuge (eds.)
Molecular Imaging for Integrated Medical Therapy and Drug Development.
DOI: 10.1007/978-4-431-98074-2_29, ©Springer 2010

Introduction

It is well known that glycoproteins and glycolipids on the cell surface play an important role in several biological processes, such as cell interaction and infection by viruses and bacteria [1]. From the standpoint of medical biology, much research has been accomplished to elucidate the relationships between diseases and the alteration of glycoforms. Especially, sialylLewis^x, types of sialylated glycans that possess sialic acid at the nonreducing end, are reported to be massively expressed on the cell surfaces of lung, breast, and ovarian cancers in humans, and these compounds are greatly concerned in cancer metastasis as the ligands of selectins on epidermal cell surfaces [2, 3]. Thus, specific inhibitors which interrupt the synthesis of sialylated glycans have potential as a new type of therapeutic agent in the abovementioned diseases.

α2,3/6-Sialyltransferases (SiaTs) are members of the enzyme family that synthesizes sialylated glycans; α2,3/6-SiaTs transfer sialic acid residues from cytidine monophosphate *N*-acetylneuraminic acid (CMP-NANA), as a common donor substrate, to galactose residues at the nonreducing end of glycans, as acceptor substrates. To date, several inhibitors of these enzymes have been synthesized in order to regulate the biosynthetic process of sialylated glycans. However, because the structures of mammalian SiaTs have not yet been resolved, it is impossible to synthesize specific and strong inhibitors based on structure-based drug design (SBDD). Thus, it is necessary to find a novel method for the development of SiaT inhibitors.

In this study, we used a high-throughput screening system for exploring new inhibitors of SiaTs, and demonstrated the system to be efficient. The system was based on the combination of constructing a compound library and carrying out a rapid assay, using MALDI-TOF mass spectrometry (MS).

Methods and Materials

Materials

Recombinant rat α2,3-(*N*)-sialyltransferase (α2,3-SiaT; one unit of this enzyme is defined as the amount of enzyme that transfers 1.0 μmol of sialic acid from CMP-NANA to *N*-acetyllactosamine per min at 37°C, pH 7.4) and recombinant rat α2,6-(*N*)-sialyltransferase (α2,6-SiaT; one unit of this enzyme is defined as the amount of enzyme that transfers 1.0 μmol of sialic acid from CMP-NANA to *N*-acetyllactosamine per min at 37°C, pH 6.0) were purchased from Calbiochem (San Diego, CA, USA). Proton and carbon NMR were recorded with Bruker lambda 600 MHz (Bruker BioSpin, Rheinstten, Germany). Chemical shifts are given in ppm and referenced to CD_3OD (δ_H 3.31, δ_C 49.15), D_2O (δ_H 4.80), CH_3OH (δ_C 49.5 in D_2O).

MALDI-TOF Mass Spectrometry

All measurements were performed using an Ultraflex TOF/TOF mass spectrometer equipped with a reflector and controlled with the Flexcontrol 1.2 software package (Bruker Daltonics, Bremen, Germany). Ions generated by a pulsed UV laser beam (nitrogen laser; λ = 337 nm) were accelerated to a kinetic energy of 23.5 kV. External calibration of MALDI mass spectra was carried out using singly charged monoisotopic peaks of a mixture of human angiotensin II (mass/charge ratio 1046.542), bombesin (*m/z* 1619.823), adrenocorticotropic hormone (ACTH; *m/z* 2465.199), and somatostatin 28 (*m/z* 3147.472). The mixture of these peptides was measured on the central spot of a 3 × 3 square with external calibration. To achieve mass accuracy better than 60 ppm, internal calibration was carried out by doping the matrix solution with a mixture of the calibration peptides. Calibration of these mass spectra was performed automatically utilizing a customized macro command of the XMASS 5.1.2 NT software package (Bruker Daltonics, Bremen, Germany). The macro command was used for the calibration of the monoisotopic singly charged peaks of the above-mentioned peptides.

Conditions for 1,3-Dipolar Cycloaddition (Click Reaction)

Reaction solution containing 2 mM of compound **1** or **2**, 2.0 mM alkyne compound, 1 mM $CuSO_4$, 10 mM sodium ascorbate, 1 mM tris-(benzyltriazoylmethyl)amine (TBTA), 20 mM Tris-HCl (pH 7.6), 40% *t*-BuOH, 10% dimethylsulfoxide (DMSO), total volume 50 μl was incubated for 6 h at room temperature, and then was concentrated by evaporation (25 °C; 3 h). To the obtained residue was added 2 mM ethylenediamine tetraacetic acid (EDTA; 2 Na^+) to a total volume of 100 μl.

General Method of High-Throughput Screening

The reaction mixture containing 200 μM CMP-NANA, 200 μM inhibitor (calculated as the click reaction was completed), 1 mM Gal-$(GlcNAc)_3$-aminooxy (ao) tryptophanyl arginine methyl ester (WR), 12 mM sodium cacodylate buffer (pH 7.4), 0.15 mg/ml bovine serum albumin (BSA), 0.03% Triton CF-54, and 0.5 mU sialyltransferase, was adjusted to a total volume of 25 μl. The reaction mixture was incubated at 37 °C for 30 min. After incubation, 100 μl of CH_3CN was added to stop the reaction, and 125 μl of water was added to the mixture to have a total volume of 250 μl. To the 20 μl of 2.5-dihydroxy benzoic acid DHB solution (10 mg/ml DHB in 30% CH_3CN) was added 2 μl each of reaction mixture and deuterium-labeled internal standard, and then 0.5 μl of mixed solution was mounted on the MALDI MS target plate.

Preparation of Compound 3

(2R,4R,5R)-2-((((2R,3S,4R,5R)-5-(4-amino-2-oxopyrimidin-1(2H)-yl)-3,4-dihydroxytetrahydrofuran-2-yl)methoxy)oxidophosphoryloxy)-5-(2-(4-((8R,9S,13S,14S,17S)-3,17-dihydroxy-13-methyl-7,8,9,11,12,13,14,15,16,17-decahydro-6H-cyclopenta[a]phenanthren-17-yl)-1H-1,2,3-triazol-1-yl) acetamido)-4-hydroxy-6-((1R,2R)-1,2,3-trihydroxypropyl)tetrahydro-2H-pyran-2-carboxylate.

The reaction mixture containing 2 mM of compound **2**, 2.0 mM alkyne compound, 1 mM $CuSO_4$, 10 mM sodium ascorbate, 1 mM TBTA, 20 mM Tris-HCl (pH 7.6), 40% *t*-BuOH, and 10% DMSO; total volume 4.0 ml was incubated for 6 h at room temperature, and the mixture was concentrated by evaporation (25 °C; 6 h). The residue was dissolved with water, and filtered with a membrane filter. The mixture was purified with reverse-phase HPLC to give compound **3** (2.69 mg; 39%).^{1}H NMR δ (600 MHz, D_2O), 1.06 (s, 3H), 1.27–1.53 (m, 4H), 1.63–1.71 (m, 2H), 1.72 (td, 1H, H-3ax.), 1.86 (br, 1H), 1,95–2.00 (br, 2H), 2.16–2.20 (m, 2H), 2.47–2.52 (m, 1H), 2.56 (dd, 1H, H-3eq.), 2.80 (s, 2H), 3.70 (q, 1H, H-9b), 3.94 (d, 1H, H-9a), 4.00 (m, 1H, H-8), 4.06 (t, 1H, H-5), 4.21(td, 1H, H-4), 4.29(m, 4H, H-6, H-4 of ribose, H-5a of ribose, H-5b of ribose), 4.34 (t, 1H, H-2 of ribose), 4.39 (br, 1H, H-3 of ribose), 5.38 (s, 2H, methylene at acetamide), 6.03 (d, 1H, H-1 of ribose), 6.17 (d, 1H, H-5 of cytosine), 6.65 (s, 1H, aromatic), 6.69 (d, 1H, aromatic), 7.16 (d, 1H, aromatic), 8.00 (s, 1H, triazole), 8.02 (s, 1H, H-6 of cytosine); ^{13}C NMR δ (125 MHz, D_2O), 14.44, 23.67, 26.52, 27.57, 29.73, 33.30, 37.49, 39.84, 41.79, 43.55, 47.65, 48.72, 52.93, 63.68, 65.57, 67.42, 69.63, 70.05, 70.35, 72.25, 75.02, 83.23, 83.61, 89.75, 97.31, 125.79, 127.40, 139.49, 142.33, 154.05, 158.53, 166.92, 169.00, 174.94; HRMS (ESI) Anal. calcd for $C_{40}H_{53}N_7O_{18}P$ $[M-H]^-$ 950.31847; found 950.31749.

Determination of Km Value of Compound 3 for α2,6-Sialyltransferase

The kinetic parameter of the *K*m value of compound **3** for α2,6-sialyltransferase was determined at six different concentrations between 200 μM and 12.5 μM. The reaction solution containing compound **3**, 1 mM Gal-$(GlcNAc)_3$-aoWR, 40 mM sodium cacodylate buffer (pH 7.4), 0.5 mg/ml BSA, 0.1% Triton CF-54, 0.5 mU SiaT was adjusted to a total volume 25 μl. The reaction solution was incubated at 37 °C for 20 min. After incubation, 100 μl of CH_3CN was added to stop the reaction, and 125 μl of water was added to the mixture to have a total volume of 250 μl. To the 20 μl of DHB solution (10 mg/ml DHB in 30% CH_3CN) was added 2 μl each of reaction mixture and deuterium-labeled internal standard, and then 0.5 μl of the mixed solution was mounted on the MALDI MS target plate.

Determination of IC_{50} Value of Compound 3 for α2,3-Sialyltransferase

The kinetic parameter of the IC_{50} value of compound **3** for α2,3-sialyltransferase was determined at 11 different concentrations of inhibitor between 200 μM and 0.39 μM. The reaction solution containing 200 μM CMP-NANA, 1 mM Gal-$(GlcNAc)_3$-aoWR, 40 mM sodium cacodylate buffer (pH 7.4), 0.5 mg/ml BSA, 0.1% Triton CF-54, and 0.5 mU SiaT was adjusted to a total volume 25 μl. The reaction mixture was incubated at 25 °C for 15 min. After incubation, 100 μl of CH_3CN was added to stop the reaction, and 125 μl of water was added to the mixture to have a total volume of 250 μl. To the 20 μl of DHB solution (10 mg/ml DHB in 30% CH_3CN) was added 2 μl each of reaction mixture and deuterium-labeled internal standard, and then 0.5 μl of the mixed solution was mounted on the MALDI MS target plate.

Results and Discussion

Construction of CMP-NANA Derivative Library

In order to obtain specific and strong inhibitors of SiaTs, two libraries of inhibitor candidates, of which the chemical structures were based on CMP-NANA, were constructed rapidly. The 1,3-dipolar cycloaddition reaction (click reaction) [4, 5], which can be conducted in aqueous solution under neutral conditions, was adapted for the modification to avoid the decomposition of chemically unstable CMP-NANA.

As shown in Fig. 1, CMP-NANA derivatives that possess an azide moiety at the 9 position (compound **1**) or at the 5-Ac position (compound **2**) were synthesized as key intermediate compounds, then the click reaction was performed to join them with commercially available alkyne compounds. Alkyne compounds are classified as six groups; phenyl, carboxyl, amine, cholesterol, alkyl, and alcohol groups; hence, it is possible to know the relationship between the compound structure and inhibition activity. Thereby, the inhibitor candidate libraries could be obtained successfully and were used for the inhibition assay without purification.

Rapid Inhibition Assay by MALDI-TOF MS

Although radiochemical and photometric methods are used frequently for assays of glycosyltransferases [6], the throughput is limited by the associated separation steps. Therefore, a MALDI-TOF MS-based assay method using the acceptor oligosaccharide conjugated with N^{α}-((aminooxy)acetyl)tryptophanyl arginine methyl ester (aoWR) [7] was established to realize a highly sensitive and quantitative assay without any separation steps (Fig. 2).

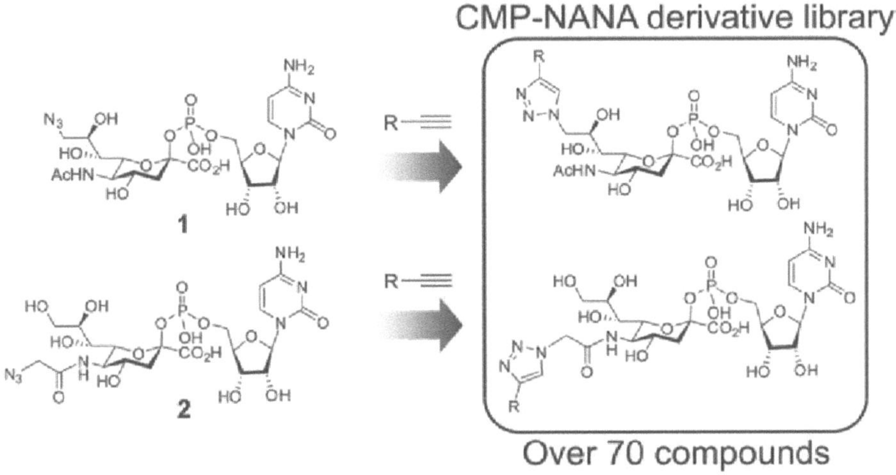

Fig. 1. Construction of the cytidine monophosphate *N*-acetylneuraminic acid (*CMP-NANA*) derivative library by the use of the "click reaction," which gives more than 70 compounds efficiently

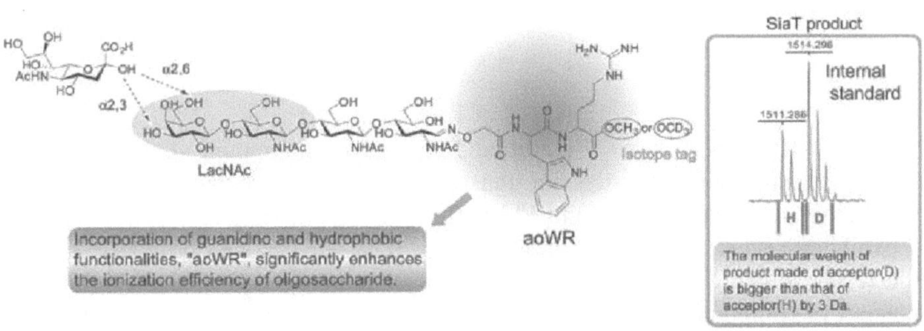

Fig. 2. The particular glycosyl acceptor of sialyltransferases (*SiaTs*) for the measurement of matrix assisted laser desorption adsorption ionization-time of flight mass spectrometry (MALDI-TOF) mass spectrometry (MS). The deuterium methyl ester moiety at the C-terminus makes a mass difference of 3 Da. *aoWR*, N^{α}-((aminooxy)acethyl) tryptophanyl arginine methyl ester

The tryptophanyl-arginine (WR) moiety of the aoWR is responsible for the sensitivity, by enhancing the ionization efficiency in positive-mode analysis. Moreover, the methyl ester moiety at the C-terminus of the aoWR allows for quantitative analysis, by comparing the ion intensity of the product obtained from the methyl ester substrate with that of the internal standard product obtained from the deuterated (d_3-methyl) analogue substrate. The simultaneous m/z information given by MALDI-TOF MS enables the quantitative and high-throughput analysis of the inhibition rate and also gives the transfer information of the inhibitor candidate itself as donor substrate.

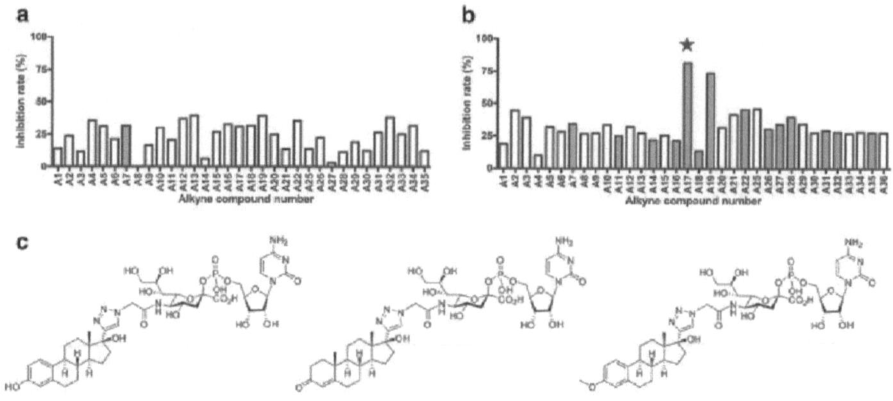

Compound 3 (2 + A17) Compound 4 (2 + A18) Compound 5 (2 + A19)

Fig. 3a–c. The results of high-throughput screening to determine the inhibitory activity of 5-Ac modified compounds and some of their chemical structures. In **a** and **b**, the *Y-axes* show inhibition rates, and the *X-axes* show the alkyne compound number combined with compound **2**. The *white bars* mean that the compound was recognized as the substrate, the *gray bars* mean the compound worked as an inhibitor. **a** The inhibitory effect against α2,6-SiaT. Almost all the compounds were transferred as substrates. **b** The inhibitory effect against α2,3-SiaT. Compounds **3** and **5** in particular exhibited strong inhibition. **c** The chemical structures of compounds bearing cholesterol moieties. *Asterisk in **b*** denotes the compound that showed the strongest inhibition rate, compound **3**

The CMP-NANA based 9-modified library from compound **1** did not show significant inhibition properties but transferred as a substrate; on the other hand, as shown in Fig. 3, compounds **3** and **5** from the 5-Ac modified library from compound **2** exhibited potent and specific inhibition for α2,3-SiaT. It can be inferred, from the result showing the weak inhibition rate of compound **4**, that the inhibitory potency of compounds **3** and **5** is due in great part to the terminal phenoxy structure in the cholesterol moiety. Interestingly, compound **3**, which showed the strongest activity for α2,3-SiaT, was transferred to the acceptor substrate in the presence of α2,6-SiaT. The different behaviors of these two enzymes clearly indicate that there are structural or mechanistic differences between them, especially around the conformation space of the phenoxy group in the active site of the enzymes.

The IC_{50} value for α2,3-SiaT and the *K*m value for α2,6-SiaT of compound **3** were determined to be 8.2 μM and 128 μM, respectively.

Conclusion

Two types of CMP-NANA derivative libraries were constructed effectively, and rapid quantitative assays of these libraries were performed for α2,3- and α2,6-SiaT, using a MALDI-TOF MS-based method. As a result, a new compound, compound

3, was obtained as a potent and specific inhibitor of α2,3-SiaT. The result suggests that this method is suitable not only for exploring the inhibitors of SiaTs but also for exploring the inhibitors of other glycosyltransferases.

References

1. Lee YC, Lee RT (1995) Carbohydrate-protein interactions: basis of glycobiology. Acc Chem Res 28: 321–327
2. Kannagi R, Izawa M et al (2004) Carbohydrate-mediated cell adhesion in cancer metastasis and angiogenesis. Cancer Sci 95: 377–384
3. Dall'Olio F, Chiricolo M (2001) Sialyltransferases in cancer. Glycoconj J 18: 841–850
4. Rostovtsev VV, Green LG, Fokin VV et al (2002) A stepwise Huisgen cycloaddition process: copper (1)-catalyzed regioselective "ligation" of azides and terminal alkynes. Angew Chem Int Ed Engl 41: 2596–2599
5. Kolb HC, Sharpless KB (2003) The growing impact of click chemistry on drug discovery. Drug Discov Today 8: 1128–1137
6. Palcic MM, Keiko S (2001) Assays for glycosyltransferases. Trends Glycosci Glycotechnol 13: 361–370
7. Uematsu R, Furukawa J-I, Nakagawa H et al (2005) High throughput quantitative glycomics and glycoform-focused proteomics of murine dermis and epidermis. Mol Cell Proteomics. 4: 1977–1989

Keyword Index

N. Tamaki, Y. Kuge (eds.)
Molecular Imaging for Integrated Medical Therapy and Drug Development.

MIX
Papier aus verantwortungsvollen Quellen
Paper from responsible sources
FSC® C105338

If you have any concerns about our products,
you can contact us on
ProductSafety@springernature.com

In case Publisher is established outside the EU,
the EU authorized representative is:
Springer Nature Customer Service Center GmbH
Europaplatz 3, 69115 Heidelberg, Germany

Printed by Libri Plureos GmbH
in Hamburg, Germany